PROCESS CONTROL AND AUTOMATION

PROCESS CONTROL AND AUTOMATION

Edited by:

Prof. Dr. Zoran Gacovski

ARCLER
PRESS

www.arclerpress.com

Process Control and Automation

Prof. Dr. Zoran Gacovski

Arcler Press
224 Shoreacres Road
Burlington, ON L7L 2H2
Canada
www.arclerpress.com
Email: orders@arclereducation.com

ISBN: 978-1-77469-916-4 (Hardcover)

DECLARATION

Some content or chapters in this book are open access copyright free published research work, which is published under Creative Commons License and are indicated with the citation. We are thankful to the publishers and authors of the content and chapters as without them this book wouldn't have been possible.

ABOUT THE EDITOR

Dr. Zoran Gacovski's current position is a full professor at the Faculty of Technical Sciences, "Mother Tereza" University, Skopje, Macedonia. His teaching subjects include Modeling and Simulation, Software engineering, Human-Computer interaction, and his areas of research are: information systems, intelligent control, machine learning, graphical models (Petri, Neural and Bayesian networks), and human-computer interaction. Prof. Gacovski has earned his PhD degree at Faculty of Electrical engineering, UKIM, Skopje. In his career he was awarded by Fulbright postdoctoral fellowship (2002) for research stay at Rutgers University, USA. He has also earned best-paper award at the Baltic Olympiad for Automation control (2002), US NSF grant for conducting a specific research in the field of human-computer interaction at Rutgers University, USA (2003), and DAAD grant for research stay at University of Bremen, Germany (2008 and 2012). The projects he took an active participation in, are: "A multimodal human-computer interaction and modelling of the user behaviour" (for Rutgers University, 2002-2003) - sponsored by US Army and Ford; "Development and implementation of algorithms for guidance, navigation and control of mobile objects" (for Military Academy – Skopje, 1999-2002); "Analytical and non-analytical intelligent systems for deciding and control of uncertain complex processes" (for Macedonian Ministry of Science, 1995-1998). He is the author of 3 books (including international edition "Mobile Robots"), 20 journal papers, over 40 Conference papers, and he is also a reviewer/ editor for IEEE journals and Conferences.

TABLE OF CONTENTS

Section 3 Intelligent Process Control

LIST OF CONTRIBUTORS

Hosny Abbas
Department of Electrical Engineering, Assiut University, Assiut, Egypt

Samir Shaheen
Department of Computer Engineering, Cairo University, Giza, Egypt

Mohammed Amin
Department of Electrical Engineering, Assiut University, Assiut, Egypt

ShuYu Ding
School of Network and Communication Engineering, Chengdu Technological University, Chengdu, 611730 Sichuan, China

JianLi Liu
School of Geophysics, Chengdu University of Technology, Chengdu, 610059 Sichuan, China

MingHong Yue
Technical Support Department, Heilongjiang Sub-Bureau of Northeast China Regional Air Traffic Management Bureau, CAAC, Harbin, 150000 Heilongjiang, China

Ziyad Kas
Department of Electrical and Computer Engineering, Oakland University, Rochester, USA

Manohar Das
Department of Electrical and Computer Engineering, Oakland University, Rochester, USA

Jun Li
Shenzhen Power Supply Bureau Co. Ltd., Shenzhen 518000, China

Bo Geng
Shenzhen Power Supply Bureau Co. Ltd., Shenzhen 518000, China

Zhixian Lin
Shenzhen Power Supply Bureau Co. Ltd., Shenzhen 518000, China

Min Chen
Shenzhen Power Supply Bureau Co. Ltd., Shenzhen 518000, China

Liangyou Shao
Shenzhen Power Supply Bureau Co. Ltd., Shenzhen 518000, China

Yanmin Zhou
School of Electrical Engineering, Southeast University, Nanjing 210096, China

Yuqing Bao
School of Electrical Engineering, Southeast University, Nanjing 210096, China

Xisong Chen
School of Automation, Southeast University, Nanjing 210096, China

Jiawei Yang
State Key Laboratory of Process Automation in Mining & Metallurgy, Beijing 102628, China
Beijing Key Laboratory of Process Automation in Mining & Metallurgy, Beijing 102628, China
Beijing General Research Institute of Mining and Metallurgy, Beijing 100160, China
Northeastern University, Shenyang 110819, China

Zhijie Zhong
School of Automation, Southeast University, Nanjing 210096, China

Junyong Zhai
School of Automation, Southeast University, Nanjing 210096, China

Tian Lyu
School of Mechatronics Engineering, Beijing Institute of Technology, Beijing, China.

C. Murugamani
HOD-IT, Bhoj Reddy Engineering College for Women, Hyderabad, India

Santosh Kumar Sahoo
Department of Electronics & Instrumentation Engineering, CVR College of Engineering, Hyderabad, India

Pravin R. Kshirsagar
Department of Artificial Intelligence, G.H. Raisoni College of Engineering, Nagpur, India

Boppuru Rudra Prathap
Department of Computer Science and Engineering, CHRIST University, Bangalore, India

Saiful Islam
Civil Engineering Department, College of Engineering, King Khalid University, Abha 61411, Asir, Saudi Arabia

Quadri Noorulhasan Naveed
College of Computer Science, King Khalid University, Abha 61413, Saudi Arabia

Mohammad Rashid Hussain
College of Computer Science, King Khalid University, Abha 61413, Saudi Arabia

Bui Thanh Hung
Data Analytics & Artificial Intelligence Laboratory-DAAI Lab, Director of Artificial Intelligence & Information System Programme, Institute of Engineering-Technology, Thu Dau Mot University, 06 Tran Van On Street, Phu Hoa District, Thu Dau Mot City, Binh Duong Province, Vietnam

Dawit Mamiru Teressa
Department of Chemical Engineering, College of Biological and Chemical Engineering, Addis Ababa Science and Technology University, Addis Ababa, Ethiopia

Svitlana Maksymova
Department of Computer-Integrated Technologies, Automation and Mechatronics, Kharkiv National University of Radio Electronics, Kharkiv, Ukraine.

Rami Matarneh
Department of Computer Science, Prince Sattam Bin Abdulaziz University, Al-Kharj, Saudi Arabia.

Vyacheslav V. Lyashenko
Department of Informatics, Kharkiv National University of Radio Electronics, Kharkiv, Ukraine.

Nataliya V. Belova
Department of Informatics, Kharkiv National University of Radio Electronics, Kharkiv, Ukraine.

Vladimir Ivancevic
Joint and Operations Analysis Division, Defence Science & Technology Group, Canberra, Australia.

Peyam Pourbeik
CEWD, Assured Communications, Survivable Networks, Defence Science & Technology Group, Adelaide, Australia.

Zishan Huang
School of Electrical and Electronic Engineering, Hubei University of Technology, Wuhan 430068, China

Waleed Dweik
Computer Engineering Department, University of Jordan, Amman, Jordan

Musa Abdalla
Mechanical Engineering Department, University of Jordan, Amman, Jordan

Yazan AlHroob
Computer Engineering Department, University of Jordan, Amman, Jordan

Anas AlMajali
Department of Computer Engineering, Faculty of Engineering, The Hashemite University, Zarqa, Jordan

Suad Alhaj Mustafa
Mechanical Engineering Department, University of Jordan, Amman, Jordan

Mohammad Abdel-Majeed
Computer Engineering Department, University of Jordan, Amman, Jordan

Irina Cojuhari
Department of Software Engineering and Automatics, Technical University of Moldova, Chisinau, Republic of Moldova.

Aaron K. Adik
Department of Electrical and Computer Engineering, Lakehead University, Thunder Bay, Canada.

Wilson Wang
Department of Mechanical Engineering, Lakehead University, Thunder Bay, Canada.

Ali S. Allahloh
Department of Electrical Engineering, Aligarh Muslim University, Aligarh, India

Mohammad Sarfraz
Department of Electrical Engineering, Aligarh Muslim University, Aligarh, India

Atef M. Ghaleb
Industrial Engineering Department, College of Engineering, Alfaisal University, 11533 Riyadh, Saudi Arabia

Sobhi Mejjaouli
Industrial Engineering Department, College of Engineering, Alfaisal University, 11533 Riyadh, Saudi Arabia

Liudmyla Yaroshchuk
Automation Hardware and Software Department, National Technical University of Ukraine "Igor Sikorsky Kyiv Polytechnic Institute", Kiev, Ukraine

Yevheniia Tiurina
Automation Hardware and Software Department, National Technical University of Ukraine "Igor Sikorsky Kyiv Polytechnic Institute", Kiev, Ukraine

Heping Chen
Ingram School of Engineering, Texas State University, San Marcos, USA

Samuel Stavinoha
Ingram School of Engineering, Texas State University, San Marcos, USA

Michael Walker
Ingram School of Engineering, Texas State University, San Marcos, USA

Biao Zhang
ABB Corporate Research Center, Windsor, USA

Thomas Fuhlbrigge
ABB Corporate Research Center, Windsor, USA

Alexandrine H. Karusine
School of Engineering, Information Engineering, Huzhou University, Huzhou, China.

Chenhuan Cui
School of Engineering, Information Engineering, Huzhou University, Huzhou, China.

Nathanael O. Larsey
School of Engineering, Information Engineering, Huzhou University, Huzhou, China.

Kanyarugano Tanguy-Nolain
School of Engineering, Information Engineering, Huzhou University, Huzhou, China.

Kai Wang and Wanqing Li
Department of Electrical Engineering and Automation, College of Electrical Engineering & New Energy, China Three Gorges University, Yichang, China.

LIST OF ABBREVIATIONS

AI	Artificial Intelligence
AMD	Advanced Machine-Metameric Dimension
AOSA	Adaptive-One-Step- Ahead
ARC	Adaptive Robust Control
ARM7	Advanced RISC Machines 7
ASR	Automatic Speech Recognition
AWOSA	Adaptive-Weighted One-Step-Ahead
BACSs	Building Automation and Control Systems
BAS	Building Automation System
BCMP	Block for Calculating Model Parameters
BEMS	Building Energy Management Systems
BOEMRE	Bureau of Ocean Energy Management, Regulation and En forcement
CAD	Computer-Aided Design
DAAM	Distributed Address Assignment Mechanism
DAI	Distributed Artificial Intelligence
DAQ	Data Acquisition
DARPA	Defence Advanced Research Projects Agency
DCS	Distributed Control System
DF	Directory Facilitator
DFT	Discrete Fourier Transform
DO	Disturbance Observer
DOFs	Degrees-of-Freedom
EE	End-Effector
EHS	European Home Systems
ETS	Engineering Tool Software
FFD	Full Functional Device
GCP	Grinding and Classification Process
GMM	Gaussian Mixture Model

GUI	Graphical User Interface
HDD	Hard Disk Drive
HetIoT	Heterogeneous Industrial Internet of Things
HOSA	Higher-Order Spectral Analysis
HSV	Hue Saturation Value
HTTP	Hypertext Transfer Protocol
HVAC	Heating, Ventilation, Air-Conditioning
IAE	Integral Absolute Error
IEEE	Institute of Electrical and Electronics Engineers
IIoT	Integrating the Internet of Things
IMAN	Immune Multi-agent Network
IoT	Internet of Things
ISE	Integral Square Error
JADE	Java Agent Development Environment
JRE	Java Run-time Environment
JSGF	Java Speech Grammar Format
JTAG	Joint Test Action Group
KAIST	Korea Advance Institute of Science and Technology
KNX	Konnex
LEDs	Light Emitting Diodes
L-M	Levenberg-Marquardt
LPCCs	Linear Predictive Cepstral Coefficients
LPG	Petroleum Gas
LQR	Linear Quadratic Regulator
LSE	Least squires estimator
MAC	Model Algorithmic Control
MAS	Multi-Agent Systems
MCU	Microcontroller Unit
MFCCs	Mel Frequency Cepstral Coefficients
MPC	Model Predictive Control
MPHC	Model Predictive Heuristic Control
MPU	Motion Processing Unit
NF	Neurofuzzy

NLP	Natural Language Processing
NN	Neural Networks
OFFSET	Offensive Swarm-Enabled Tactics
OG	Oils and Greases
OPC	Open Process Control
OSA	One-Step-Ahead
PD	Proportional-Derivative
PDF	Probability Density Function
PI	Proportional-Integral
PID	Proportional Integral Derivative
PLC	Programmable Logic Controller
PUA	Permanently Unmanned Area
ReLUs	Rectified Liner Units
ResNet	Residual Neural Network
RF	Radiofrequency
RFD	Reduce Function Device
RLS	Recursive Least Squares
RSW	Resistance Spot Welding
RTC	Real-Time Clock
SCADA	Supervisory Control and Data Acquisition
SD	Speaker Diarization
SI	Speaker Identification
SI-ASR	Speaker-Independent Automatic Speech Recognition
SLAM	Self-Localization and Mapping
SMA	Shape Memory Alloy
SR	Speaker Recognition
SSM	State-Space Model
SV	Speaker Verification
SVM	Support Vector Machine
SVPWM	Space Vector Pulse Width Modulation
TBM	Technical Building Management
TCO	Technological Control Object
TMA	Technical Means of Automation
TOF	Time-of-Flight

TSDO	Terminal Sliding-mode-based Nonlinear Disturbance Observer
UART	Universal Asynchronous Receiver Transmitter
UAVs	Unmanned aerial vehicles
UGVs	Unmanned Ground Vehicles
UV	Unmanned Vehicle
VAD	Voice Activity Detection
WOG	Waste Oil and Grease
WUW	Wake-Up Word
WPM	Weighted Product Method

PREFACE

The definition of a technical process is as: technical process is an event through which the state of matter, energy or information changes. This change of state can be understood as a transition from the initial state to the final state. This transition is determined by the time course of interactive events in the system, whereby matter, energy or information is transformed, transported or stored. In order to realize the continuous conversion of the inflow of matter, energy or information into the corresponding outflow (of matter, energy or information), the process control and regulation is required. Equally, for the purpose of monitoring the process, it is necessary to measure/observe the dimensions of the process.

In the so-called "conventional" techniques (non-programmable techniques), separate devices are used for measurement, control and regulation, which contribute to the emergence of independent professional disciplines: measurement techniques, control techniques and regulation techniques. In practice – we sometimes encounter a name MCR (Measuring, Control and Regulation devices). With the application of process computers and microelectronic components with a high degree of integration, the above mentioned 3 technical disciplines become an integral part of a new professional field called process automation, or process control, or process informatization. This new field of expertise, regardless of its name, involves the interaction of three "subjects":

- Personnel who monitor process events via appropriate display media (e.g. monitors) and who lead the process (control the process). If it is an automated process, the role of the staff is to intervene only in exceptional situations (e.g. in an emergency or during abnormal work regimes).
- Systems for collecting, processing and displaying information and for creating control actions on the process. In addition to this system, a corresponding communication system is often used.
- Technical process with appropriate measuring tools (sensors) and setting devices.

This edition covers different topics from industrial process control, robotic process control, intelligent process control, and automation applied in different sectors.

Section 1 focuses on industrial process control, describing simple, flexible, and interoperable SCADA system based on agent technology; use of ZigBee wireless communication technology in industrial automation control; an electrothermal model based adaptive control of resistance spot welding process; multiagent consensus control strategy considering whole-process thermodynamic characteristics of air conditioning process; and process control of ball mill based on MPC-DO.

Section 2 focuses on robotic process control, describing a control method for SMA robotic actuators; wireless communication for robotic process automation using machine learning technique; voice control for an industrial robot as a combination of various robotic assembly process models; and Hamiltonian servo: control and estimation of a large team of autonomous robotic vehicles.

Section 3 focuses on intelligent process control, describing automatic intelligent control system based on intelligent control algorithm; skeleton of implementing voice control for building automation systems; data-driven model identification and control of the inertial systems - intelligent control and automation; an intelligent system for real-time condition monitoring of tower cranes; and IIoT-based intelligent process control for crude oil separation: investigating the impact of model-based control and genetic algorithms.

Section 4 focuses on automation applied in different sectors, describing simulation of the industrial oil adsorption purification process for automation tasks; opportunities and challenges of robotics and automation in offshore oil & gas industry; the impact of automation in painting furniture parts; and application of electrical automation technology in power system.

SECTION 1
PROCESS CONTROL AND AUTOMATION

CHAPTER 1

Simple, Flexible, and Interoperable SCADA System Based on Agent Technology

Hosny Abbas[1], Samir Shaheen[2], and Mohammed Amin[1]

[1]Department of Electrical Engineering, Assiut University, Assiut, Egypt

[2]Department of Computer Engineering, Cairo University, Giza, Egypt

ABSTRACT

SCADA (Supervisory Control and Data Acquisition) is concerned with gathering process information from industrial control processes found in utilities such as power grids, water networks, transportation, manufacturing, etc., to provide the human operators with the required real-time access to industrial processes to be monitored and controlled either locally (on-site) or remotely (i.e., through Internet). Conventional solutions such as custom SCADA packages, custom communication protocols, and centralized

Citation: Abbas, H. , Shaheen, S. and Amin, M. (2015), "Simple, Flexible, and Interoperable SCADA System Based on Agent Technology". Intelligent Control and Automation, 6, 184-199. doi: 10.4236/ica.2015.63018.

architectures are no longer appropriate for engineering this type of systems because of their highly distribution and their uncertain continuously changing working environments. Multi-agent systems (MAS) appeared as a new architectural style for engineering complex and highly dynamic applications such as SCADA systems. In this paper, we propose an approach for simply developing flexible and interoperable SCADA systems based on the integration of MAS and OPC process protocol. The proposed SCADA system has the following advantages: 1) simple (easier to be implemented); 2) flexible (able to adapt to its environment dynamic changes); and 3) interoperable (relative to the underlying control systems, which belongs to diverse of vendors). The applicability of the proposed approach is demonstrated by a real case study example carried out in a paper mill.

Keywords: SCADA, Real-Time Monitoring, Process Control, Agent Technology, Multi-Agent Systems, Open Process Control (OPC)

INTRODUCTION

A SCADA system is responsible for gathering information and real-time data from variety of plants and providing this data to operators located at anywhere at any time. Furthermore, SCADA systems can be considered as critical information systems; their criticality comes from the fact that they are currently vital components of most nations' critical infrastructures. They control pipelines, water and transportation systems, utilities, refineries, chemical plants, and a wide variety of manufacturing operations. Failure of controlled systems can lead to direct loss of life due to equipment failure or indirect losses due to failure of critical infrastructure controlled by SCADA. SCADA systems have evolved in parallel with the growth and sophistication of modern computing technology. That means that SCADA is technology-dependent and it is not a standalone science but it is the result of integrating variety of applied sciences such as communication, computers, software engineering, networking, security, etc. Therefore and according to the evolution of technology, SCADA developers architect SCADA by selecting the appropriate technologies and mechanisms for handling SCADA challenges such as complexity resulted from the continuous increasing of the size of SCADA systems. SCADA systems should be scalable because their components and the amount of exchanged data increase with rapid rate. Further, a SCADA system should be flexible and able to adapt to internal or external changes. SCADA is no longer restricted to concern real-time monitoring and control inside a factory, but it currently has new concerns.

It has been moved from the local sense to the global one. For example, in advanced countries, a global SCADA system is used for supervising and real-time monitoring of the power grid (thousands of electric power generation and distribution stations).

Designing, monitoring and controlling of modern industrial systems is getting more challenging as a consequence of the steady growth of their size, complexity, level of uncertainty, unpredictable behavior, and interactions. Unfortunately, the conventional SCADA systems are not capable of providing information management and high-level intelligent approaches. That is because achieving those functionalities requires comprehensive information management support and coordination among system devices, and the control of many different types of task, such as data transportation, data display, data retrieval, information interpretation, control signals and commands, documentation sorting and database searching…,etc. These tasks operate at different timescales and are widely distributed over the global system and its subsystems. Without sophisticated software architectures and hardware structures, it is impossible to handle these tasks efficiently, safely and reliably, with the possibility of online reconfiguration and flexibly embedding applications [1] [2] .

It becomes obvious that computing systems, especially those related to modern industrial applications such as SCADA systems, are becoming increasingly interconnected and more difficult to maintain. Due to the increase in the size, complexity and the number of components, it is no longer practical to anticipate and model all possible interactions and conditions that the system may experience at design time. Similarly, the systems are becoming too large and too complex for system managers to maintain them at run-time [3] .

The agent-based approach seems to be the promising solution. The rapid development of the field of agent- based systems offers a new and exciting paradigm for the development of sophisticated programs in dynamic and open environments [4] . The agent-based approach is considered as a new software engineering architectural style for the development of complex, decentralized and open software applications. What distinguishes the agent-based approach from other traditional approaches is its unique ability to handle simultaneously many challenges of the current software applications specially those applications which are highly distributed and their working environments are highly dynamic and uncertain. MAS provide a suitable paradigm for decentralized systems in which autonomous individuals engage in flexible high-level interactions. By an agent-based system, we mean the one in which the key abstraction used is that of an agent. We therefore expect an agent-based system to be both designed and implemented in terms of agents. An agent-based system may contain any non-zero number of agents. The multi-agent case where a system is designed and implemented as several interacting agents, is both more general and significantly more complex than the single-agent case. Originally, MAS emerged as a scientific area, from the previous research efforts in Distributed Artificial Intelligence (DAI) started in the early eighties. MAS are now seen as a major trend in R & D, mainly related to artificial intelligence and distributed computing techniques. This research has attracted attention in many application domains where difficult and inherently distributed problems have to be tackled [5] . A MAS is defined as a set of interacting autonomous agents in a common environment in order to solve a common, coherent task. These agents try to achieve individual objectives which are sometimes conflicting. There are many definitions of the meaning of agent we found in their literature but the widely accepted definition of an agent is that which has been stated by Wooldridge and Jennings [6] who defined an agent as:

"... a hardware or (more usually) software-based computer system that enjoys the following properties: ?autonomy: agents operate without the direct intervention of humans or others, and have some kind of control over their actions and internal state; ?social ability: agents interact with other agents (and possibly humans) via some kind of agent-communication language; ?reactivity: agents perceive their environment, (which may be the physical world, a user via a graphical user interface, a collection of other agents, the Internet, or perhaps all of these combined), and respond in a timely fashion to changes that occur in it; ?pro-activeness: agents do not simply act in response to their environment, they are able to exhibit goal-directed behavior by taking the initiative".

A MAS is autonomous, means that there is no external entity which controls this system. This property is enforced because agents inside the system are autonomous. Inside a MAS, data (knowledge) are distributed inside all its agents. Moreover, the control is decentralized (there is no supervisor). MAS allow the design and implementation of software systems by using the same ideas and concepts that are the very founding of human societies and habits. These systems often rely on the delegation of goals and tasks among autonomous software agents, which can interact and collaborate with others to achieve common goals [7] . It provides an approach to solve a software problem by decomposing the system into a number of autonomous entities embedded in an environment in order to achieve the functional and quality requirements of the system [8] . The agent-based computing has been hailed as "the next significant breakthrough in software development" [9] , and "the new revolution in software" [10] .

Currently, agents are the focus of intense interest on the part of many sub-fields of computer science and artificial intelligence. Agents are being used in an increasingly wide variety of applications, ranging from comparatively small systems such as email filters to large, open, complex, mission critical systems such as air traffic control. At first sight, it may appear that such extremely different types of system can have little in common. And yet this is not the case: in both, the key abstraction used is that of an agent [11] . MAS are claimed to be especially suited to the development of software systems that are decentralized, can deal flexibly with dynamic conditions, and are open to system components that come and go. That is why they are used in domains such as manufacturing control, automated vehicles, and e-commerce markets.

There are many reviews and surveys tackled with the applications of agent and multi-agent systems in variety of application domains. For instance, in [11] the authors described the suitability of intelligent and autonomous agents to model and develop applications for certain types of software system which are inherently more difficult to correctly design and implement than others. And they subdivide these systems into three classes: open systems, complex systems, and ubiquitous computing systems. Moreover, they classified these systems or applications according to the application domain such as industrial applications (i.e. manufacturing, process control…, etc.), commercial applications (i.e. information management, e-commerce…, etc.), medical applications (i.e. patient monitoring, health care…, etc.), and entertainment (i.e. games, interactive cinema…, etc.). The authors tackled many types of agent-based applications and described generally their functionality. They tackled the design and development methods and challenges of MAS from a general viewpoint independent from a specific application domain.

In [12] the author presented a summary of the state-of-the-art of the Distributed Artificial Intelligence (DAI) applied to Intelligent Manufacturing and presented main applications along with different technologies applied in these areas. Also, he presented areas of agent technology applications and the agent development tools. Moreover, he described briefly and generally the agents and multi-agent architecture types and design approaches.

In [13] the authors reviewed multi-agent systems power engineering applications and stated that the flexibility offered by an open architecture of agents with good social ability easily led to the design of a fault-tolerant system. Also, they reviewed the application of MAS as a modeling and simulation approach and its application in grid computing and web services composition, etc. Moreover they presented a bibliographical analysis of agent research aiming to provide an indication of the active areas of agent research with respect to power systems and related applications. They concentrated on the functionalities and the different power systems aspects which can be designed and developed by MAS.

Shehory [14] stated that one aspect of multi-agent systems that had been only partially studied was their role in software engineering and especially their merit as a software architecture style. In his report he provided analysis guidelines which supported designers in their assessment of the suitability of MAS as a solution to computational problem they addressed. Moreover he discussed the architectural properties that should be considered when analyzing such systems and he supported his work with case studies of several MAS.

Still there is a debate about the identity of multi-agent systems: is it a radically new way for systems engineering? Is it a new software engineering modeling style? Weyns et al. [8] stated that the trend in agent-oriented software engineering was to consider multi-agent systems as a radically new way of engineering software, and this position isolated agent-oriented software engineering from mainstream software engineering and could be one important reason why MAS were not widely adopted in industry yet. What we understand from this is that the agent-based modeling should be considered similar to the object-oriented modeling both of them are software engineering modeling styles. In other words, MAS are considered now as a novel general-purpose paradigm for software development.

Moreover, Agents and multi-agent systems constitute one of the most prominent and attractive technologies in computer science at the beginning of this new century. Agents and multi-agent systems technologies, methods, and theories are currently contributing to many diverse domains. Jennings and Wooldridge [11] stated that intelligent agents were a new paradigm for developing software applications. Agent technology is seen as a fundamentally important new tool for building a wide array of systems (i.e. open systems, complex systems, and ubiquitous computing systems). A number of software tools exist that allow a user to implement software systems as agents, and as societies of cooperating agents, by tools we mean agent platforms. An agent platform provides a basis for the implementation of MAS, and the means to manage agent execution and message passing. For the sake of interoperability, it is intended that the agent platform architecture should be implemented by using the Foundation of Intelligent and Physical Agents [15] specifications of agent platforms abstract architecture. FIPA is an IEEE Computer Society standards organization that promotes agent-based technology and the interoperability of its standards with other technologies. FIPA, the standards organization for agents and MAS was officially accepted by the IEEE as its eleventh standards committee on June 8, 2005. The specifications define an abstract agent platform, a number of services that must or may be provided by such a platform and a standard communications language.

From the above discussion it is clear that the agent technology can offer a feasible approach for handling the challenges and requirements of modern SCADA systems [16] . This paper proposes a simple but flexible and interoperable SCADA system based on the agent-based approach. The proposed system-to-be can be used for providing not only a local access but also a remote access through the Internet to local control processes.

BACKGROUND

Traditionally, SCADA developers adopted the web-based approach to realize a remote real-time monitoring of production control processes, a survey of some web-based SCADA systems can be found in [17] . But the problem of the web-based applications which use the web browser as a remote client and a web server as the onsite server through which the remote operators have access to control processes, is an example of centralized control architecture which has disadvantages such as:

- Single point of failure problem;
- These systems are not flexible because they cannot adapt to changes such as increasing the number of system components;
- These systems are not scalable because increasing system components can result in web server overload and crash.

Figure 1 demonstrates why the web-based SCADA systems are not flexible, not scalable, and are not able to adapt to working environments dynamic changes. As shown in Figure 1(a) a remote operator uses the web browser to connect to an onsite web server to access a control process. In case the system changes by increasing the number control processes and the number of remote operators as shown in Figure 1(b) the web server which has limited resources such as the computation power, memory and the network bandwidth, is vulnerable to be overloaded and crash.

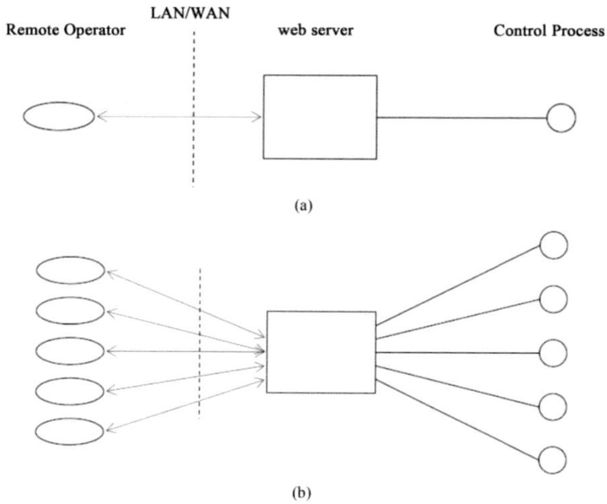

Figure 1. Traditional centralized web-based SCADA architecture.

Many of SCADA developers considered the web-based SCADA systems as a solution to some of modern SCADA challenges and shortcomings [17] [18] . Although they succeeded to develop web-based SCADA applications that had a number of important quality attributes such as efficiency, interoperability, and real-time monitoring, they failed to obtain other important attributes such as scalability and adaptivity. The reason is that the Hypertext Transfer Protocol (HTTP) uses a client-server architecture with which it is difficult to achieve higher scalability degree because of the dominant adopted centralized control regime. The author of [19] emphasized that the future infrastructures will contain a huge amount of data is generated by real world devices and needs to be integrated and processed within a specific context and communicated on demand and on time. As a result, traditional approaches aiming at the efficient data inclusion in enterprise services need to be changed. The main challenges facing current and future SCADA systems are generally related to quality attributes. The extent to which the system possesses a desired combination of quality attributes such as scalability, usability, performance, reliability, and security indicates the success of the design and the overall quality of the software system. New design and development approaches should be adopted for building future complex and critical SCADA systems. The agent-based approach considers a SCADA system as a multi-agent system. A multi-agent system provides a powerful computational technology, for which dynamic aspects are based on interactions between autonomous agents, rather than centralized control. To do their required functions, agents have the ability to communicate with each other, with their environment, and with human operators. Agents are particularly adapted to complex systems modeling where environments are unpredictable [20] . Figure 2 provides an abstract architecture of the agent-based real-time monitoring. As shown, the system is flexible and can easily adapt to changes such as increasing the number of control processes which can easily assigned to new agents, and also increasing the number of remote operators by assigning a remote agent to each operator. The agent-based approach adopts a decentralized control and behavior, which had proven to be the appropriate approach for handling system complexity and unpredictable work environments.

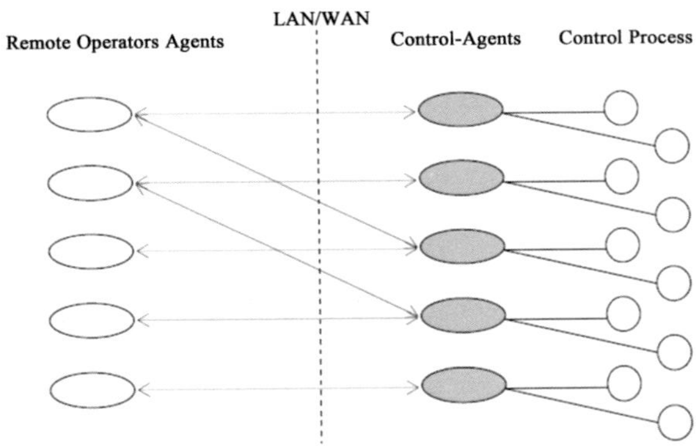

Figure 2. The decentralized agent-based SCADA.

The agent-based approach can be considered as the promising solution for the design and development of future SCADA because it has the ability to handle future SCADA challenges such as complexity, scalability, and flexibility, etc. Although these quality attributes are not unique to MAS, but combining them in a single system is unique to MAS. This combination results in the suitability of MAS for solving problems where information, location and control are highly distributed, heterogeneous, autonomous components comprise the system, the environment is open and dynamically changing, and uncertainty is present [14] [21] .

The agent-based approach reinforces decentralization which is the effective way for obtaining higher degree of scalability. Today, agents are being applied in a wide range of industrial applications, for example, process control, manufacturing, air traffic control, etc. These applications have been relatively successful, suggesting that the multi-agent approach is a promising method for the implementation of industrial automation systems and that encourages us to adopt MAS for the design and development of SCADA systems. Metzger et al. [22] surveyed the applications of the agent technology in the industrial process control and concluded that the agent technology is particularly popular in the manufacturing domain, while the applications in other domains of industrial control are scarce. They related their conclusions to the lack of the technology support on the part of control instrumentation vendors. In manufacturing automation, the process consists of discrete and countable components and actions. The natural approach is to assign the software agents to each of the components and each of the actions performed. On the other hand, the process automation deals with the continuous physical phenomena, such as chemical reactions. When a process automation system is designed, the phenomena are represented as mathematical models, for which control algorithms are chosen in order to keep the process parameters within a desired range. Therefore, in a single continuous control loop, there is not much place for any additional computational techniques, including the agent technology.

THE PROPOSED APPROACH

The proposed approach is based on the integration of MAS and OPC process protocol [23] , realizing this integration enables us to achieve two goals. First, it will be possible to transfer the process data from the process domain to the information domain (MAS), where executive management can use this data for decision making. Second, it will be possible to take the benefit of control devices interoperability provided by the OPC process protocol. Interoperability is assured through the creation and maintenance of non-proprietary open standards specifications. OPC initially meant Ole for Process Control, but after it becomes familiar for achieving control systems interoperability it was redirected to mean Open Process Control. OPC is open connectivity in industrial automation and the enterprise systems that support the industry. The first OPC standard specification resulted from the collaboration of a number of leading worldwide automation suppliers working in cooperation with Microsoft. Originally based on Microsoft's OLE COM/DCOM technologies, the specification defined a standard set of objects, interfaces and methods for use in process control and manufacturing applications to achieve interoperability. There are now hundreds of OPC Data Access (OPC DA) servers and clients.

Figure 3 shows the basic architecture of the OPC protocol usage. The OPC server is connected to the PLC (programmable Logic Controller) which is responsible of directly controlling a certain control process. By this way the OPC server can be considered as the driver of the PLC, in the same sense as the printer driver. The communication between the OPC server and the PLC is vendor-specific and depends on the custom technologies used by the vendors for manufacturing their control systems. On the other hand, the communication between the OPC server and software applications adopts the COM/DCOM technologies created by Microsoft. That is considered as a limitation because it means that the OPC protocol is bonded only to Microsoft Windows operating systems. In this research and as shown in figure a Siemens S7-400 PLC is used as a real control system but the same system and approach can be used with other control systems manufactured by other vendors such as ABB, Honeywell, etc. That is possible because of the vendors interoperability provided by the standard OPC protocol. This feature represents the second goal (interoperability) of the aimed two goals mention in the beginning of this section.

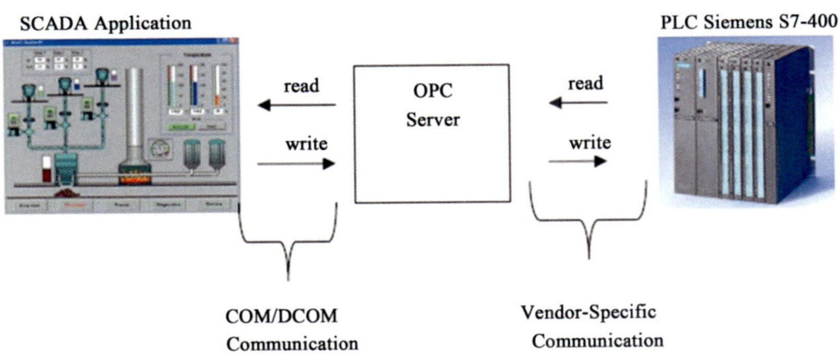

Figure 3. Basic OPC protocol architecture.

Most MAS platforms are implemented in Java programming language [24] , therefore to establish a connection between a Java agent and an OPC server, a Java-COM bridge or adapter is required as shown in Figure 4. For the purpose of this thesis the JEasyOPC Java-OPC adapter [25] [26] was selected to interface a Java MAS agent with OPC servers and of course the other bridges can be used in a similar way. JEasyOPC is a Java OPC client that is now greatly enhanced. It uses a JNI layer coded in Delphi. The current version supports both OPC DA 2.0 and OPC DA 3.0. JEasyOPC is free source and can be downloaded easily from the Internet. Appendix A provides some information about realizing the Java-OPC interactions through JEasyOPC.

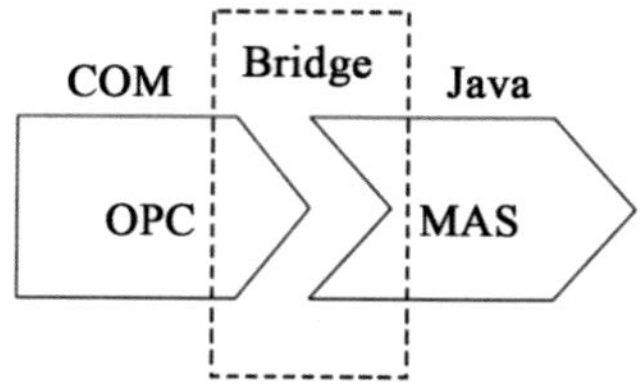

Figure 4. Integrating MAS with OPC protocol.

Figure 5 demonstrates the layered nature of the system-to-be. The system contains four layers (or worlds) with three interfaces in between. The layers are (from bottom to up): the physical process layer, control systems layer, OPC communication layer, and the agents' layer which represents the application world. The figure also demonstrates the appropriate interface between any two layers. For example, the interface between the physical process and its control system is the field interface, which is constructed using analog/digital signals transferred through wires, or it can be a digital field bus. Further, the interface between the control system and the OPC driver layer is vendor specific. Finally, the interface between the OPC layer and the agents' world layer a COM/ DCOM interface. In this paper, we are interested in the later (top) layer and interface (Shown in Dashed line in Figure 5).

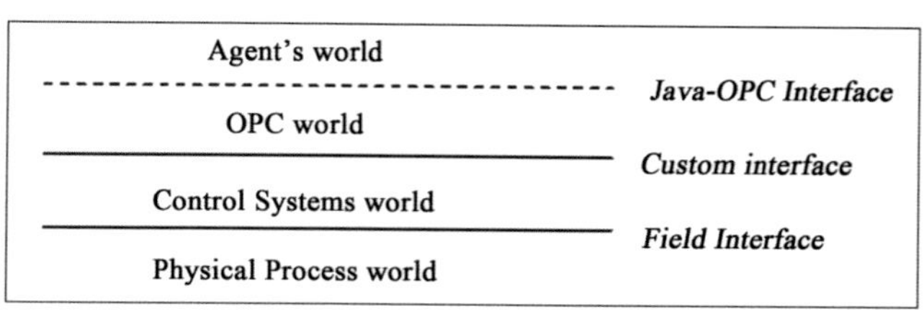

Figure 5. System layers and the interfaces in between.

To connect non-agent Java applications to OPC servers, it is required to download the JEasyOPC library, which includes a sample Eclipse [27] java project that contains basic sample examples programmed to connect to a default OPC server (Matrikon. OPC. Simulation). These examples can be modified as required, for instance the Makitron OPC server can be replaced with a Siemens one (i.e. OPC. Simatic Net). Furthermore, the examples don't provide a graphical user interface (GUI) and therefore the developer has to modify them to provide a suitable GUI. Before running Eclipse, a recent Java run-time environment (JRE) should be installed on the operating system. Using a Java development environment such as Eclipse frees the developer from caring about modifying related system variables such as CLASPATH and PATH as it does these issues automatically.

From the other hand, to connect a Java agent (under Eclipse) to an OPC server, it is required first to install a MAS platform such as JADE (Java Agent Development Environment) [28] . Jade is a software framework fully implemented in Java language. It simplifies the implementation of multi-agent systems through a middleware that claims to comply with the FIPA specifications and through a set of tools that supports the debugging and deployment phase. The agent platform can be distributed across machines with different operating systems and the configuration can be controlled via a remote GUI. The configuration can be even changed at run-time by creating new agents and moving agents from one machine to another one as and when required. The only system requirement is the Java Run-Time version 5 or later. JADE is distributed in open source. To run Jade under Eclipse, the developer should add Jade libraries to Eclipse Java build path (project à prosperities à Java Build path à Libraries à add external Jars), then through the Windows file system find Jade.jar file in the Jade home as shown in Figure 6. Now Eclipse is ready for creating a new java class that extends jade. core. Agent class and start programming the required agent let us call it OPC-Agent.

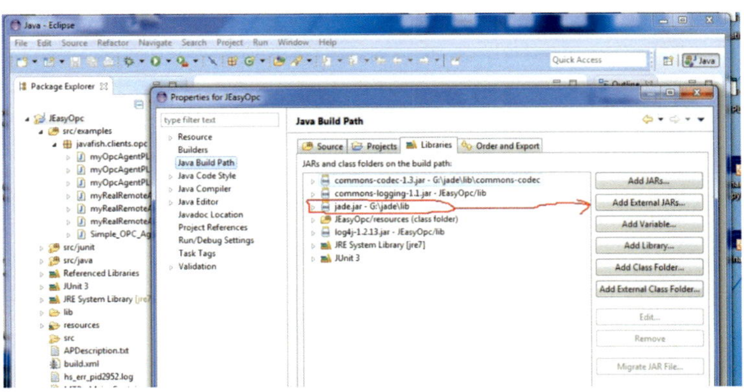

Figure 6. Running JADE from ECLIPSE.

REAL CASE STUDY EXAMPLE

The case study is concerned with the real adoption of the agent-based approach for connecting locally and remotely to real control processes. The case study was executed in a paper mill, specifically in the finishing area within a paper mill. The paper mill finishing area comprises three separated stations as follows:

1. The Winder station, which takes a 6 meters width paper spool as its input and gives a smaller width paper rolls (i.e., 6 × 1 meter) as output.
2. The wrapping station, which is used for paper rolls packaging and labeling.
3. The Salvage winder station for preparing sample rolls; it is a smaller Winder station.

Each control process is controlled by a Siemens S7-400 PLC; also the three stations are connecting a local LAN including the operator stations and the control systems. Further, each control process contains many process variables that should be monitored continuously, for instance, for the winder station the following variables required to be monitored in real-time: Machine Speed (m/min), Paper Tension (N/m), Drums Torques (N/m), and so on. Figure 7 shows the winder station while it is running. The architecture of the experimental practical project is shown in Figure 8.

As shown in the figure, all hosts are connected through a LAN with the possibility to be connected to the Internet to enable real-time data accessing from outside the paper mill. Furthermore, the normal remote Jade agents are considered as the process real-time data consumers while each one can be hosted on a different host in the LAN or in the Internet. In this case study project, it is assumed that:

1. Each PLC of a control process is assigned to an OPC-Agent, which in turn registers its services to the JADE DF (Directory Facilitator) agent, represents the yellow page service in the JADE platform. Therefore, in the case study example there are three OPC-Agents, one for each real PLC system.
2. The remote operator agents initially don't know which OPC-Agent is responsible of which PLC system. Accordingly, after start running they contact the JADE DF agent to search for the required OPC-Agents, as illustrated in Figure 9.

3. After identifying the required OPC-Agents, the operator agents send an ACL message with per-formative (Request) to the OPC-Agent asking her to continuously send back the real-time process data of its assigned PLC. This request is done only once then the OPC-Agent will continuously send the required real-time process data through ACL messages with per-formative (Inform), this mechanism is called subscription behavior.
4. Each operator agent can access many OPC-Agents simultaneously, in other words each remote agent can be used to supervise and monitor more than one control process.
5. The project JADE platform was named as SCADA, so an agent with local name "H1" has a full name "H1@SCADA".
6. The implemented agents were made simple with light GUI because it is just an illustrative experimental example. In real applications, the agents GUI may become complex and user friendly. Figure 10 shows a simple remote operator agent continuously receives changed data from an OPC-agent situated on a host inside the mill. The remote agent designed to provide real-time monitoring, alarm service, and trend service, as shown in the figure. This feature represents our first goal (remote real-time monitoring) as we mentioned in the beginning of this section. This feature is very important for executive management stuff, which might be located far from the factory to be able to take the suitable quick decision in the suitable time.

Figure 7. Qena paper winder station.

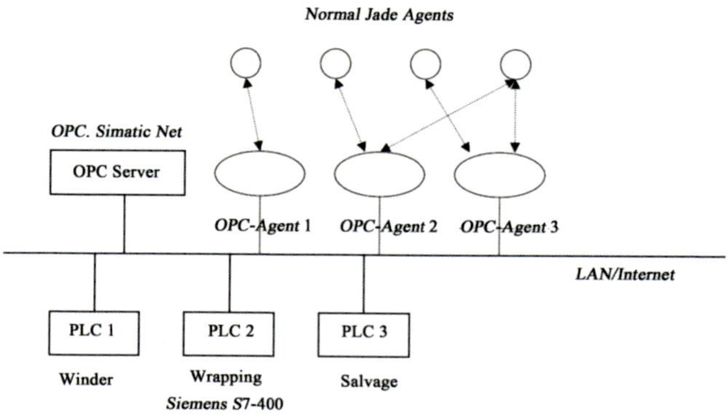

Figure 8. Architecture of the concrete example.

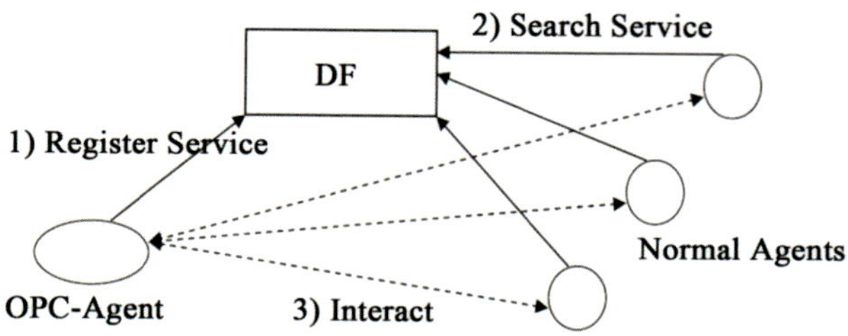

Figure 9. Exploiting JADE yellow page service.

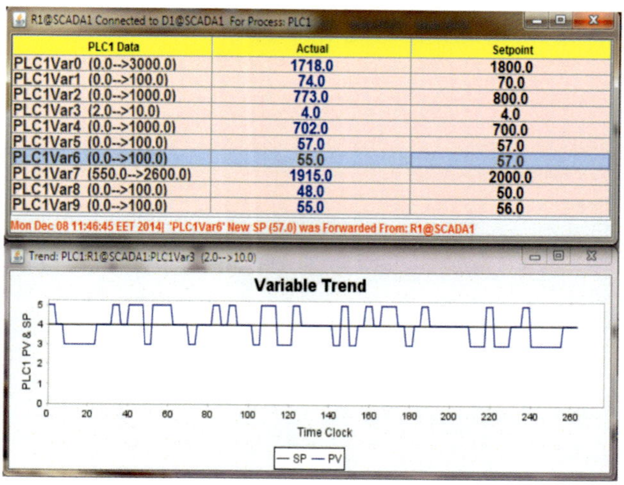

Figure 10. A remote operator agent connected to an onsite OPC-agent and provides real-time process monitoring, Alarm service, and Trend service.

It is possible to run an operator agent from any host in the LAN or even from the Internet. The point will be with the agents launching commands which defined in the Eclipse class arguments setting (run asàrun configuration à new configuration à arguments), as shown in Figure 11. For example (considering javafish. clients.opc as the package name):

- For initially booting and running an OPC-Agent called H1 in the main container of the JADE platform called SCADA: -gui-name SCADA -local -agents H1: javafish.clients.opc.myOpcAgent
- For running an operator agent called R1 on the same host as the OPC-Agent H1: -container-agents R1:javafish.clients.opc. myRealRemotAgent1
- For running an operator agent called R2 from another host in the LAN Where the host with IP (192.168.100.31) is the one on which the Jade platform main container is situated: -container-host 192.168.100.31 R2:javafish.clients.opc.myRemoteAgent

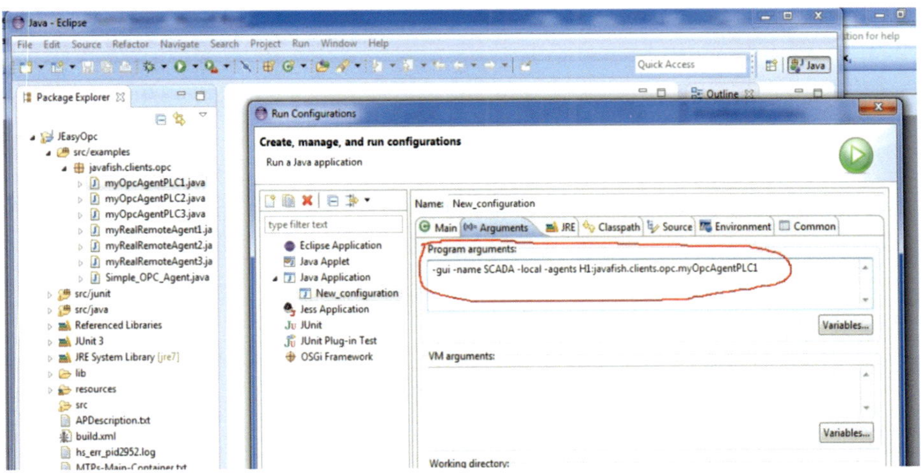

Figure 11. Configuring a JADE agent in Ellipse.

For debugging purposes, the JADE platform provides a built-in sniffer tool, which is used to visualize messages as a low-level UML sequence diagram, was used as shown in Figure 12, which demonstrates how JADE agents interacts together through exchanging asynchronous messages. The sniffer tool is very powerful for tracing the system-to-be behaviors in run-time.

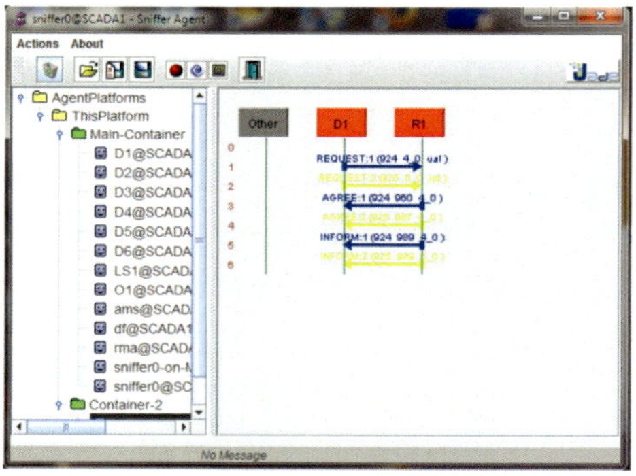

Figure 12. Jade sniffer tool shows message flow between two agents.

Returning to the concerned case study, Figure 13 presents the system sequence diagram. As shown in the figure the remote operator agent depends on the yellow page service provided by the JADE directory facilitator (DF) to find the control agent responsible of a required control process. The operator agent will wait for the DF until it provides it with the required control process. After getting a response from the DF the remote agent starts interacting with the control agent to get the process real-time data and presenting this data to the human operator in text or graphical format. As we have illustrated before the remote operator agent can be hosted locally onsite or remotely outside, i.e. using the Internet.

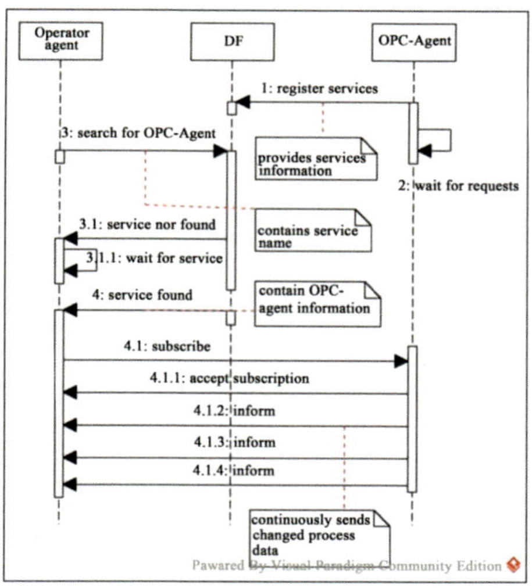

Figure 13. System sequence diagram demonstrated how a remote agent finds a control agent to interact with.

DEPLOYMENT AND TESTING

The proposed approach solves the problem of accessing the OPC server (which is a COM/DCOM server) from the Internet because using DCOM communication is not recommended through the Internet because of its delay time and configuration security problems. The agent-based approach depends on the HTTP protocol, which is a firewall-friendly protocol. Figure 14 shows the concrete project while it is running. The figure includes two running OPC-Agent in the left side:

- WinderOpcAgent 1;
- WrappingOpcAgent 1;

And three remote normal agents in the right side:

- WinderRemoteAgent 1;
- WinderRemoteAgent 2;
- WrappingRemoteAgent 2;

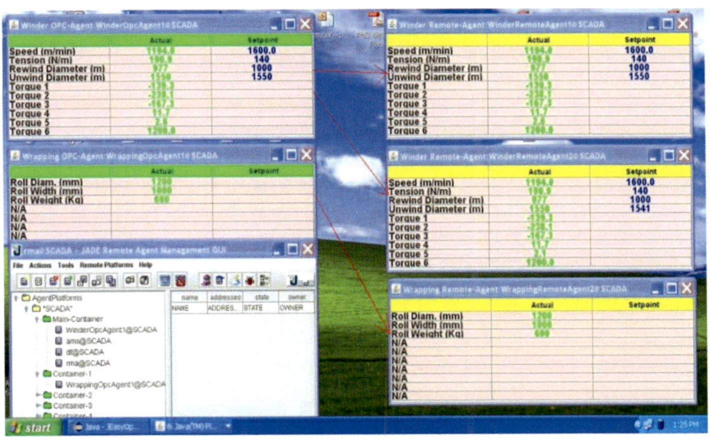

Figure 14. Running the concrete example: Two OPC-Agents (left side) and three normal Jade Agents (right side).

Figure 15 demonstrates how the OPC-Agents continuously send real-time data to the remote operator agents as a result of subscription requests from the operator agents. The figure shows two OPC-Agents and three remote agents. Two remote operator agents subscribed to one of the two OPC-Agents and one subscribed to the other one. The number of OPC-Agents and remote agents can be increased easily by instantiating these agent types as required and anywhere in the LAN or the Internet. Note that it's not necessary for the OPC-Agents to have a GUI but the operator agents should have a user friendly GUI. In this experiment, a simple GUI is designed for both agent types.

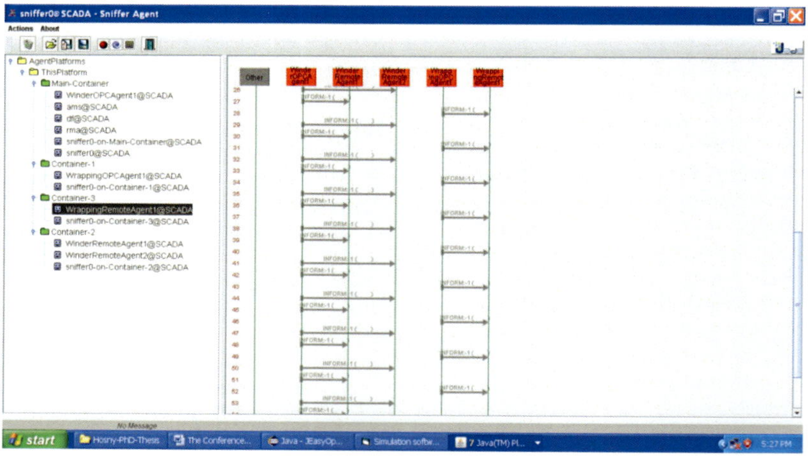

Figure 15. JADE's built-in sniffer tool, three normal jade agents, two of them subscribe to Winder OPCAgent and one subscribes to Wrapping OPCAgent.

CONCLUSION

Multi-agent systems propose solutions to highly distributed problems in dynamic and open computational domains. SCADA is one of these systems which are highly distributed, decentralized, open, and requires high degree of scalability. The agent-based approach should be adopted for developing flexible and scalable SCADA systems. In this research, the agent-based approach was adopted to develop a simple, flexible, and interoperable SCADA system. The achieved interoperability is realized by the adoption of the OPC technology for process communication, by this way it is possible to run the developed system with any type of control systems (i.e. PLC, DCS, CNC, etc.) independent of any vendor, and the same Java-OPC interface will be used with only some few modifications. By flexible, we mean that the system is able to adapt its dynamic working environment. The applicability of the agent-based approach is demonstrated by developing a practical project carried out in a paper mill. Furthermore, the proposed approach is cost-effective compared with custom SCADA packages. The future work will be the adoption of the agents' technology for building large-scale and highly distributed SCADA systems.

ACKNOWLEDGEMENTS

Hosny Abbas thanks the managers of Quena Paper Company (Egypt), Mr. Shazely Abd-El Azeem (General Manager) and Mr. Abd-El Hameid Omar (Manager of Automation Sector) for their encouragement and their acceptance to allow us to carry out this research in the company at their responsibilities and supervisory.

APPENDIX A: JAVA-OPC INTERACTION

The JEasyOPC project is established on LGPL (GNU Library or Lesser General Public License). The project can be downloaded from SourceForge.net as actual release (2.xx.xx) or night revision from SVN repository. The project is built on Eclipse 3.2.x Open Source IDE. The release is distributed as a zip-file for a quick download. In a zip-file, there are these important directories and files:

- jeasyopc.jar: the final library for usage in your application.
- src.jar: the source of library for a preview of library classes.
- eclipse-project\JEasyOpc.zip: zip-file with whole JEasyOPC project for Eclipse. There are all examples, JUnit tests, all sources!
- Doc: the directory includes documentation.
- Resources: the configuration files of JEasyOPC library. These resources have to be included in CLASSPATH of your project. There are all important information about usage of logging, internationalization and dll-library path (property library.path).

Interface with OPC servers:

```
Import javafish.clients.opc.JCustomOpc;
Import javafish.clients.opc.JEasyOpc;
Import javafish.clients.opc.JOpc;
Import javafish.clients.opc.asynch.AsynchEvent;
Import javafish.clients.opc.asynch.OpcAsynchGroupListener;
Import javafish.clients.opc.browser.JOpcBrowser;
Import javafish.clients.opc.component.OpcGroup;
Import javafish.clients.opc.component.OpcItem;
//Initialization
jopc_meas=new JEasyOpc(hostName, serverName, groupName+
"_meas");
```

```
//Reading
gotItem= jopc_meas.synchReadItem(statusGroup, (OpcItem) ItemToRead);
//Writing
jopc_command.synchWriteItem(commandGroup, item);
```

As an example, consider a JADE application contains two agents: one is connected through LAN to the OPC server and continuously read changed process variables from the server (call is OPC-Agent); the other is a remote agent interacts with the OPC-Agent by message passing to get the latest changed process variables. The following Java program presents a possible implementation of the OPC-Agent (to save paper size, exceptions handling is not included in the code):

```
package javafish.clients.opc;
// Jade Imports
import jade.core.*;
import jade.core.behaviours.TickerBehaviour;
import jade.lang.acl.ACLMessage;
// JEasyOPC imports
import javafish.clients.opc.component.*;
import javafish.clients.opc.exception.*;
public class Simple_OPC_Agent extends Agent
{
// OPC Declarations
private JEasyOpc jopc;
private OpcGroup group;
private OpcItem item1,item2;
private OpcItem responseItem1,responseItem2;
// Agent Setup function
protected void setup() {
// connecting to Siemens OPC.SimaticNet OPC server
jopc = new JEasyOpc("localhost", "OPC.SimaticNET", "JOPC1");
JOpc.coInitialize();
// OPC Group Creation
```

```
item1 = new OpcItem("s7:[@LOCALSERVER]db1,w0", true, "");
item2 = new OpcItem("s7:[@LOCALSERVER]db1,w2", true, "");
group = new OpcGroup("group1", true, 400, 0.0f);
group.addItem(item1);
group.addItem(item2);
jopc.addGroup(group);
// Starting the OPC Server
jopc.start();
addBehaviour(new TickerBehaviour(this, 1500) {
protected void onTick() {
responseItem1 = jopc.synchReadItem(group, item1);
responseItem2 = jopc.synchReadItem(group, item2);
System.out.println("item1="+ responseItem1.getValue());
System.out.println("item2="+ responseItem2.getValue());
ACLMessage msg1=new ACLMessage(ACLMessage.INFORM);
// Process the message
msg1.addReceiver(new AID("H2", AID.ISLOCALNAME));
msg1.setLanguage("English");
msg1.setContent(responseItem1.getValue()+responseItem2.
getValue());
send(msg1);
}});}
```

In this example the OPC server is supposed to be OPC. SimaticNET, which is used to connect to Siemens control systems. Also the OPC server is supposed to be installed on the same machine as the sample client application which means that the communication between the client and the server takes place through COM, if the client and server are hosted by different machines then DCOM will be used. The remote agent is a simple Jade agent required to interact with the OPC-Agent by massage passing. First it might send only one starting message to the OPC-Agent and continuously, through a cyclic behavior, it receives messages from the OPC- Agent and retrieves the OPC data from these messages and prints them to the console. For more information about the application of JEasyOPC, interested readers are invited to read the documents and manuals of JEasyOPC.

REFERENCES

1. Chakrabarti, S., Kyriakides, E., Bi, T., Cai, D. and Terzija, V. (2009) Measurements Get Together. IEEE Power and Energy Magazine, 7, 41-49. http://dx.doi.org/10.1109/mpe.2008.930657
2. Buse, D.P. and Wu, Q.H. (2007) IP Network-Based Multi-Agent Systems for Industrial Automation: Information Management, Condition Monitoring and Control of Power Systems. Springer, New York.
3. Kota, R., Gibbins, N. and Jennings, N.R. (2009) Decentralised Structural Adaptation in Agent Organisations. In: Vouros, G., Artikis, A., Stathis, K. and Pitt, J., Eds., Organized Adaption in Multi-Agent Systems, Springer, Berlin Heidelberg, 54-71. http://dx.doi.org/10.1007/978-3-642-02377-4_4
4. Luck, M. (1999) From Definition to Deployment: What Next for Agent-Based Systems? The Knowledge Engineering Review, 14, 119-124. http://dx.doi.org/10.1017/S0269888999142048
5. Oliveira, E., Fischer, K. and Stepankova, O. (1999). Multi-Agent Systems: Which Research for Which Applications. Robotics and Autonomous Systems, 27, 91-106. http://dx.doi.org/10.1016/S0921-8890(98)00085-2
6. Woolbridge, M. and Jennings, N.R. (1995) Agent Theories, Architectures, and Languages: A Survey. In: Woolbridge, M. and Jennings, N.R., Eds., Intelligent Agents, Springer-Verlag, Berlin, 1-39. http://dx.doi.org/10.1007/3-540-58855-8_1
7. Di Marzo Serugendo, G., Gleizes, M.-P. and Karageorgos, A. (2011) Self-Organizing Software, from Natural to Artificial Adaptation. Springer, New York.
8. Weyns, D., et al. (2006) Multi-Agent Systems and Software Architecture. AAMAS'06, Hakodate, May 2006, 8-12.
9. Sargent, P. (1992) Back to School for a Brand New ABC. The Guardian Newspaper, 12 March 1992, 28.
10. Guilfoyle, C. and Warner, E. (1994) Intelligent Agents: The New Revolution in Software. Ovum Report.
11. Jennings, N.R. and Wooldridge, M. (1998) Applications of Intelligent Agents. Springer, Berlin.

12. Madejski, J. (2007) Survey of the Agent-Based Approach to Intelligent Manufacturing. Journal of Achievements in Materials and Manufacturing Engineering, 21, 67-70.
13. McArthur, S.D.J., Catterson, V.M. and Hatziargyriou, N.D. (2007) Multi-Agent Systems for Power Engineering Applications—Part I: Concepts, Approaches, and Technical Challenges. IEEE Transactions on Power Systems, 22, 1743-1752.
14. Shehory, O. (1998) Architectural Properties of Multi-Agent Systems, Technical Report CMU-RI-TR-98-28. The Robotics Institute, Carnegie Mellon University, Pittsburgh.
15. Foundation for Intelligent Physical Agents (FIPA) (2000) FIPA Agent Management Specification. http://www.fipa.org/specs/fipa00023/
16. Abbas, H.A. (2014) Future SCADA Challenges and the Promising Solution: The Agent—Based SCADA. International Journal of Critical Infrastructures, 10, 307-333. http://dx.doi.org/10.1504/IJCIS.2014.066354
17. Abbas, H.A. and Mohamed, A.M. (2011) Review in the Design of Web Based SCADA Systems Based on OPC DA Protocol. International Journal of Computer Networks, 2, 266-277.
18. Mohamed, A.M. and Abbas, H.A. (2011) Efficient Web Based Monitoring and Control System. Proceedings of the 7th International Conference on Autonomic and Autonomous Systems, Venice, 22-27 May 2011, 18-23.
19. Karnouskos, S. and Colombo, A.W. (2011) Architecting the Next Generation of Service-Based SCADA/DCS System of Systems. Proceedings of the 37th Annual Conference on IEEE Industrial Electronics Society, Melbourne, 7-10 November 2011, 359-364. http://dx.doi.org/10.1109/iecon.2011.6119279
20. Vale, Z.A., Morais, H., Silva, M. and Ramos, C. (2009) Towards a Future SCADA. Proceedings of the Power & Energy Society General Meeting, Calgary, 26-30 July 2009, 1-7.
21. Jennings, N.R. (2001) An Agent-Based Approach for Building Complex Software Systems. Communications of the ACM, 44, 35-41. http://dx.doi.org/10.1145/367211.367250
22. Metzger, M. and Polakow, G. (2011) A Survey on Applications of Agent Technology in Industrial Process Control. IEEE Transactions on Industrial Informatics, 7, 570-581. http://dx.doi.org/10.1109/TII.2011.2166781

23. OPC Foundation (2010) OPC DA 3.0 Specification [DB/OL].
24. Russell, J.P. (2001) Java Programming for Absolute Beginner. Prima Publishing, Roseville.
25. (2015) http://sourceforge.net/projects/jeasyopc/
26. Diaconescu, E. and Spirleanu, C. (2012) Communication Solution for Industrial Control Applications with Multi-Agents Using OPC Servers. Proceedings of the 2012 International Conference on Applied and Theoretical Electricity (ICATE), Craiova, 25-27 October 2012, 1-6.
27. (2015) https://www.eclipse.org/
28. Bellifemine, F., Poggi, A. and Rimassi, G. (1999) JADE: A FIPA-Compliant Agent Framework. Proceedings of the Practical Applications of Intelligent Agents and Multi-Agents, April 1999, 97-108.

CHAPTER 2

The Use of ZigBee Wireless Communication Technology in Industrial Automation Control

ShuYu Ding[1], JianLi Liu[2], and MingHong Yue[3]

[1]School of Network and Communication Engineering, Chengdu Technological University, Chengdu, 611730 Sichuan, China

[2]School of Geophysics, Chengdu University of Technology, Chengdu, 610059 Sichuan, China

[3]Technical Support Department, Heilongjiang Sub-Bureau of Northeast China Regional Air Traffic Management Bureau, CAAC, Harbin, 150000 Heilongjiang, China

ABSTRACT

With the accelerating process of world industrialization, countries pay more attention to industrial modernization. In order to solve the problems

Citation: S. Ding, J. Liu, M. Yue, "The Use of ZigBee Wireless Communication Technology in Industrial Automation Control", Wireless Communications and Mobile Computing, vol. 2021, Article ID 8317862, 11 pages, 2021. https://doi.org/10.1155/2021/8317862.

of traditional control technology and system with low accuracy, high energy consumption, and high delay in industrial automation control and the difficulty in adapting to the new industrial production process, the new industrial automation control system is discussed and studied. The industrial automation control is studied, and a new monitoring system is designed for automation control by using ZigBee wireless communication technology. The design of the system data communication scheme is completed by the in-depth study and summary of ZigBee wireless communication technology. The hardware structure and software structure of ZigBee wireless communication module are designed, and the wireless monitoring system using ZigBee is established. Taking the production workshop of a pharmaceutical factory as the monitoring object, the working stability and data transmission efficiency of the model are tested. The experiment shows that the ZigBee wireless transmission model has lower temperature and better stability than the wired transmission system in the working process. Compared with the traditional wired control system, the temperature of ZigBee wireless control system is lower when it works. The difference between the highest temperature and the lowest temperature is 0.1°C. In the overall work flow, the system temperature is lower than that of the wired control. ZigBee wireless control has good stability. Subsequently, the method of manually setting obstacles is used to test the effectiveness of data transmission. The experiment found that in the absence of barrier, the maximum effective transmission distance was 40 meters, and the packet loss rate was very low, which could ensure the timeliness of data. Therefore, this system can be applied to industrial automation control. The research results can provide some reference for the design of new control systems in other industrial control fields. The improved system can also be applied to other industries. ZigBee is a short-range technology for wireless sensor networks, and it is a two-way communication technology with low power consumption, low data rate, and low cost. It can be applied to automatic control and remote control. This provides a platform for wireless networking and control of small and cheap devices. ZigBee will not compete with connection technologies such as Wi-Fi. Meanwhile, it will greatly stimulate the progress of the industrial market.

INTRODUCTION

As a big manufacturing country, China's industry occupies a large proportion in the economic system and has gradually become an important pillar industry of China's economic development. From the current development process of developed countries, industrialization is an indispensable process [1]. Therefore, China has always attached great importance to industry. With the emergence of various high-tech in the world, the times require China to transform from an industrial power to an industrial power. But there is still a big gap between China and developed countries in terms of industrial capacity and technology, so industrial automation has become an essential goal [2].

At present, there are two types of industrial automation technology: One is programmable logic controller (PLC). Zhao used PLC control technology in electronic automatic control. The PLC power automation control system was designed, and the system was simulated and tested [3]. The other is the distributed control system (DCS). Aparna (2020) used DCS control technology to design a nonlinear interactive multi-input, multioutput system battery [4]. With the advancement of technology and the improvement of computer performance, the automation control of industrial automation and wireless and wired communication gradually merge and achieve data synchronization. Fernandez et al. used time-sensitive networks to design the industrial application technology of media access control for wired communication [5]. Although the technical security of wired transmission is strong and the delay is low, the data has high timeliness and reliability [6]. However, in actual situations, the wiring method of wired transmission is easily restricted by the scene. Once the line is damaged, it may cause damage to the system and data [7]. Therefore, wireless communication technology has been gradually applied in industrial automatic control systems, and many kinds of sensing technologies have been derived. ZigBee is a new type of short-distance wireless technology [8]. Abdulkarem et al. [9] introduced a new technology for structural health monitoring by wireless sensor networks that offers advantages over traditional wired systems. The advantage of the system is to reduce the installation and maintenance costs of the structural health monitoring system. Open research issues in wireless sensor networks for structural health monitoring are discussed. Li and Liu [10] established a reasonable mathematical model. Additionally, aiming at the problem of poor coverage of existing algorithms, they proposed a monitoring area coverage optimization algorithm by the sensing probability around nodes. The experiments show that the algorithm has good performance, which can reduce the moving distance of nodes, improve the coverage range of the network, and extend the service time of the whole network. Because of this, the application of ZigBee technology in industrial automatic control is being studied.

Firstly, relevant documents are retrieved, and the theoretical knowledge of ZigBee technology is summarized. By copying the concept learning through the structure frame diagram of ZigBee technology, the three structures of ZigBee network are analyzed, and the address allocation rules of ZigBee network are expounded. The optional network structure of ZigBee technology is studied. Subsequently, using ZigBee, communication modules and wireless sensor models were designed. Combined with the requirements of industrial control technology, a model of the production workshop of the pharmaceutical factory was established. The data transmission efficiency and working state stability of ZigBee wireless transmission technology are analyzed. The stability of ZigBee wireless transmission technology is expressed by the temperature of the system. In this way, the performance stability of ZigBee industrial automation control system can be judged. The innovation of the method proposed lies in exploring the efficiency of wireless transmission by setting obstacles during wireless transmission and using temperature to demonstrate the stability of ZigBee wireless technology.

MATERIALS AND METHODS

ZigBee Technology

ZigBee protocol communication equipment is divided into three types: coordinator, router, and terminal equipment. Generally, each complete ZigBee network consists of a coordinator and numerous routers and terminals. [11] The coordinator is generally by a full functional device (FFD) [12]. The function of coordinator is to initialize the network and form the ZigBee network. First, the coordinator scans the surrounding channel and selects the high energy value as the alternative channel. Then, the coordinator filters the channels with fewer nodes to establish the network and sets the corresponding network parameters. It can also be used as a router, mainly controlling the entire network, so stable energy supply is needed. Routers are generally set in the middle area of ZigBee network, responsible for the discovery and selection of data when passing, and the most suitable path for data transmission. The function of terminal equipment is relatively simple [13], which is generally at the end of the network and is responsible for data collection and transmission at the edge of the network. Terminal equipment can save energy by sleeping. The network structure of ZigBee is shown in Figure 1.

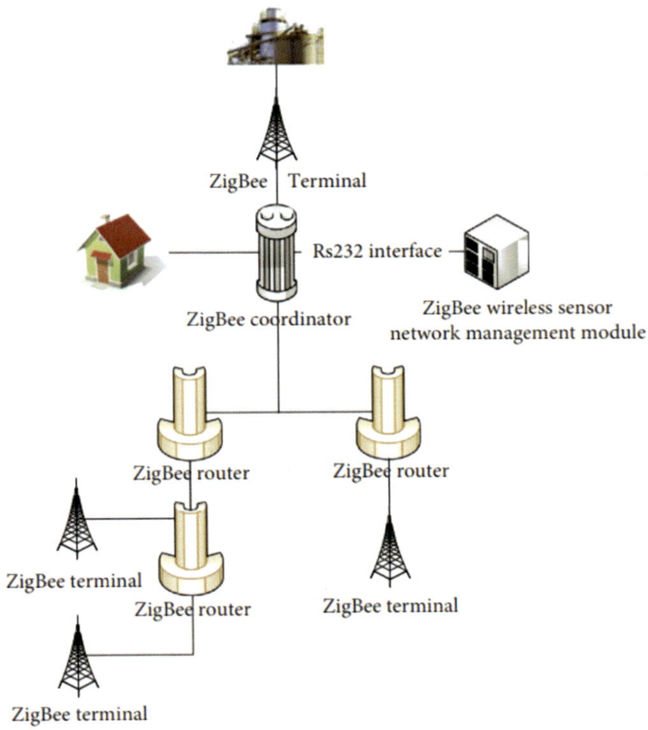

Figure 1. Basic structure diagram of ZigBee network.

In Figure 2, there are generally three ZigBee network structures, a star structure, a mesh structure, and a tree structure.

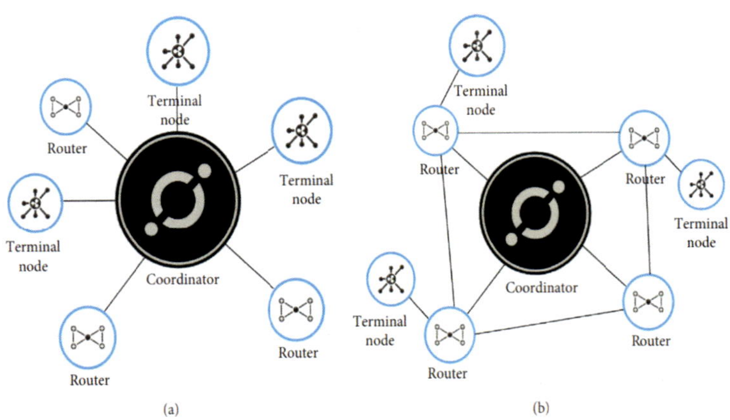

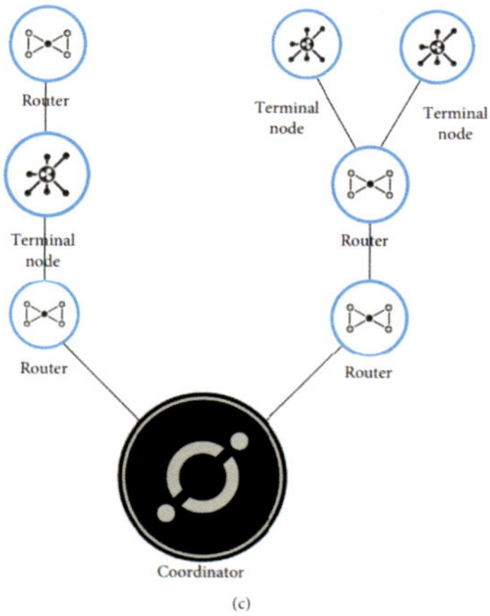

Figure 2. Three technologies of ZigBee wireless communication network ((a) is a star structure, (b) is a mesh structure, and (c) is a tree structure).

The star structure [14] is to connect several terminal devices and nodes to the same coordinator. This device can be another complete system or a simple networked device. Since the star structure is multiple devices connected to the same coordinator, it is generally used in short-distance communication scenarios, such as various intelligent electrical and general control systems in the home. The star structure system management is very simple. When one device fails, it will not affect the normal operation of other devices. The disadvantage of the star structure system is that it is easy to cause too much burden on the coordinator and cause network congestion.

The mesh structure [15] is a directional path diagram that connects a coordinator and many nodes or devices, and the internal devices and nodes can communicate autonomously. Therefore, the mesh structure has the characteristics of high flexibility, is a structure that can support a variety of hop nodes, and is suitable for large-scale network structures. The disadvantage of the mesh structure system is that it requires a large amount of data and information to support operation, which increases energy consumption.

In the tree structure [16], the position of the coordinator is at the root, playing the same role as the root of the tree. We connect the remaining nodes as child nodes to form a multilayered one-to-many parent-child hierarchical relationship and transmit data according to the parent-child relationship. The tree structure is usually used for long-distance data transmission. Because the tree structure does not have the routing function, the data can only be transmitted layer by layer in the order of the assigned nodes, so the tree structure consumes less energy, but the transmission delay will be slightly higher.

There are generally two ZigBee address allocation rules. One is the 64-bit Institute of Electrical and Electronics Engineers (IEEE) extended address [17], and its coordinator network address is 0. The other is the address of the parent node when the node is connected to the network through the distributed address assignment mechanism (DAAM) [18]. $C_{skip}(d)$ is a parameter in the DAAM allocation rule. It is the address offset of the child node, which means the interval between the coordinator and the node at depth d, which is left for the node address at depth d + 1 space. The address of the first child node that joins the parent node is parent node + 1, and the calculation of $C_{skip}(d)$ satisfies

$$C_{\text{skip}}(d) = \begin{cases} 1 + C_m * (L_m - d - 1), R_m = 1, \\ \dfrac{1 + C_m - R_m - C_m * R_m{}^{L_m - d - 1}}{1 - R_m}, \text{Other}, \end{cases} \quad (1)$$

where C_m is the maximum number of child nodes that the parent node can load. R_m is the maximum number of nodes that the child node can load. L_m is maximum depth of ZigBee network. Refer to the DAAM allocation rules, assuming that the address of the parent node is A_p, then, the address of the nth access parent node is

$$A_n = A_p + 1 + C_{\text{skip}}(d) * (n - 1). \quad (2)$$

If the nth node is the end point, the address is

$$A_n = A_p + 1 + C_{\text{skip}}(d) * R_m + n. \quad (3)$$

ZigBee routing protocol [19] is the core technology of ZigBee. During data transmission, a node can transmit to its parent node or child node along the tree structure according to the connection between the destination address and the current address. Assuming that the current node address is A, the network depth is d, and the end point address is D, then, the method for judging whether the end point is a child node is as shown in

$$A < D < A + C_{\text{skip}}(d). \quad (4)$$

If the end node address satisfies Equation (4), it means that the end node is the child node of the current node, and the next URL N can be calculated, as in Equation (5):

$$N = A + 1 + \left[\frac{D - (A + 1)}{C_{\text{skip}}(d)} \right] \cdot C_{\text{skip}}(d), \quad (5)$$

where $C_{skip}(d)$ is a parameter in DAAM allocation rules, which is the address offset of the child node. It means the interval between coordinator and node in depth d. It is the space left for the node address with depth d + 1, and the first child node to join the parent node is the parent node + 1. The address of the node is A, the network depth is d, and the endpoint address is D. Whenever the source node of the router transmits data to the destination node, the router of the source node does not store the information of the destination node and enters the route discovery stage. The source node writes own information and routing requirements into the system and transmits it to the entire network. When the intermediate node accepts it for the first time, it will retain the information in the packet, combine it with the address information of the previous node to form a reverse routing path from this node to the source node, and then transmit it to the broadcast. When the end node receives the information for the first time, it returns the packet to reply. If the node is not receiving it for the first time, it will be discarded as spam. Figure 3 shows the three transmission modes of route discovery, reverse routing, and forward routing.

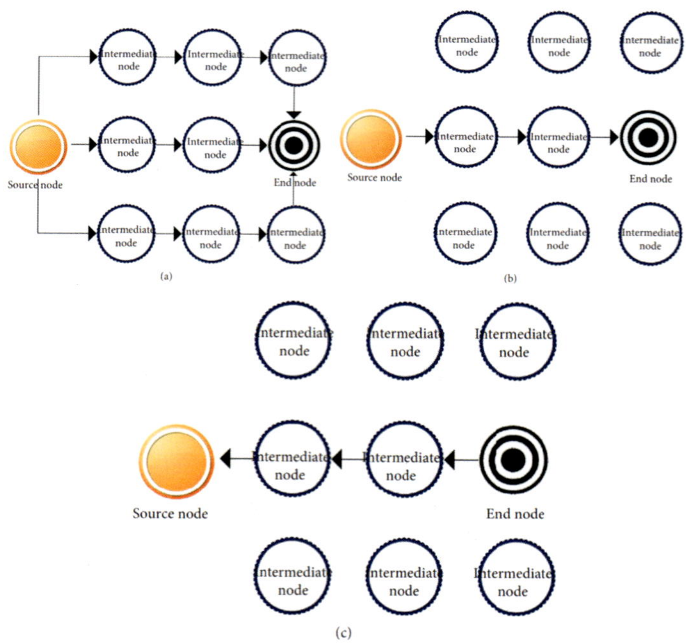

Figure 3. Three router discovery processes ((a) is the usual router discovery process, (b) is the router reverse data transmission process, and (c) is the router forward data transmission process).

Design and Establishment of ZigBee Wireless Sensor System

In the design of the ZigBee system structure, the current wired communication network of a pharmaceutical factory is improved, and the use of a wireless system can effectively solve the problem of complex wiring. A pharmaceutical factory production workshop was used as a model, and a wireless sensor was designed to monitor the status of the workshop. Figure 4 is an overall structural diagram. First, the sensor collects the data transmitted by the field networked devices and processes the information transmitted by the router. The router is responsible for network data transmission to ensure stable and correct data transmission between various devices. The coordinator is responsible for collecting data and sending it to the host computer through the serial port. The whole system can record and query the situation and data on site.

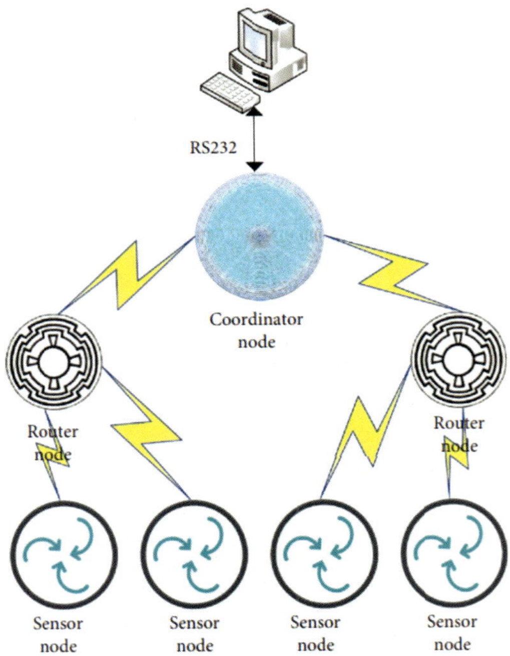

Figure 4. Overall structure of the ZigBee wireless sensor.

The terminal node of the console has rich functions, can monitor the parameters of the current environment and the state of the equipment in real time, and can also set the parameters of the control system. Figure 5 is the hardware of the console. Among them, Advanced RISC Machines

7 (ARM7) processor is a high-speed flash memory [20]. Real-Time Clock (RTC) is a program that can set time [21]. It can run 32-bit system at the maximum clock rate. Joint Test Action Group (JTAG) [22] is an interface for testing chips and internal nodes.

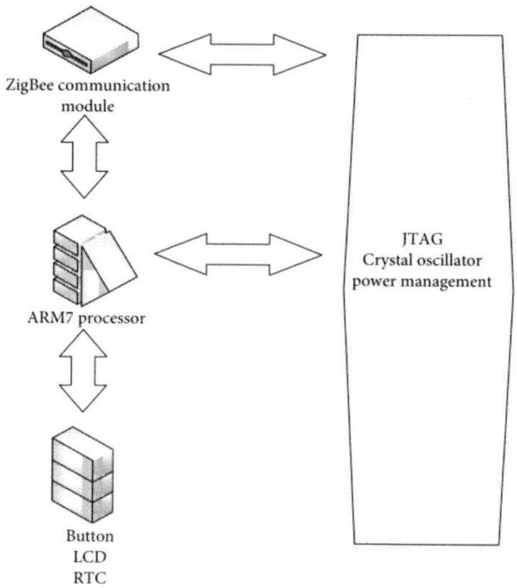

Figure 5. ZigBee console hardware structure diagram.

In the establishment of communication network, in the wireless sensor, the responsibility of each node is to collect, calculate, and communicate the data in the system. Judging from the category of sensing equipment, sensors can be divided into two categories. One is a full function device (FFD). FFD has strong data processing capabilities and huge storage space for routing information. The other is reduce function device (RFD) [23]. The energy consumption of RFD is low, and the storage space is correspondingly small. Using a full function device as a system gateway is the most important and common role of RFD. Simplified functional devices are used as sensor nodes to transmit data in mobile phones and transmit them to nodes in the same network layer. The two devices, RFD and FFD, work together. When the network coverage needs to be expanded, incorporating RFD equipment into the FFD can significantly reduce the energy consumption of the entire system.

The wireless sensor network routing monitoring centre (computer) and many terminal devices constitute [24], and it is necessary to use a computer to automatically control multiple devices at the same time. Therefore, the star structure network is used, and the structure diagram is shown in Figure 6. The monitoring centre includes a computer and a master node. The master node is a full-featured service device with ZigBee, whose main function is to aggregate, transmit, and communicate data from multiple child nodes. The connected device node can use RFD or FFD. The most basic ZigBee node is the RFD node. The energy consumption of device nodes using ZigBee wireless communication technology is very low, and one AA size battery can supply energy for a period. Independent ZigBee communication equipment generally has an effective transmission distance of less than 75 meters. To expand the communication coverage, multiple nodes can be used together.

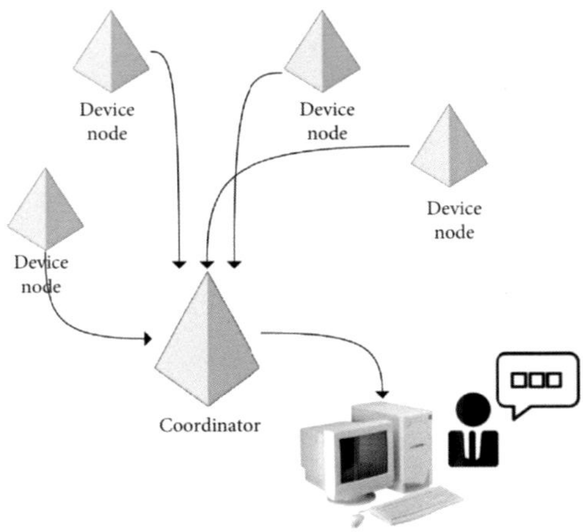

Figure 6. Structure diagram of communication network built with star structure.

In the design of equipment nodes, the support platform for wireless sensors is built by device nodes. This kind of device node is essentially a miniature embedded system, which includes a microcontroller unit (MCU) [25] and MCU modules of external devices, and data collection components. The structure of the power supply and communication components is shown in Figure 7. The responsibility of the data collection module [26] is to collect data within the specified monitoring range and can collect different data according to actual application areas, such as temperature, light intensity, and precipitation. The MCU is responsible for the routing protocol of the control node. Its main job is to reach the requirements of the agreement and ensure the stability of the network. The power module is powered by two 1.5 V alkaline batteries, which can be replaced with a smaller silver cell battery according to the actual situation. The reserved communication interface is used, and the data collection unit is set on the end node, and it is set independently. The scalability of the equipment can be improved, and the difficulty of debugging the equipment can be reduced.

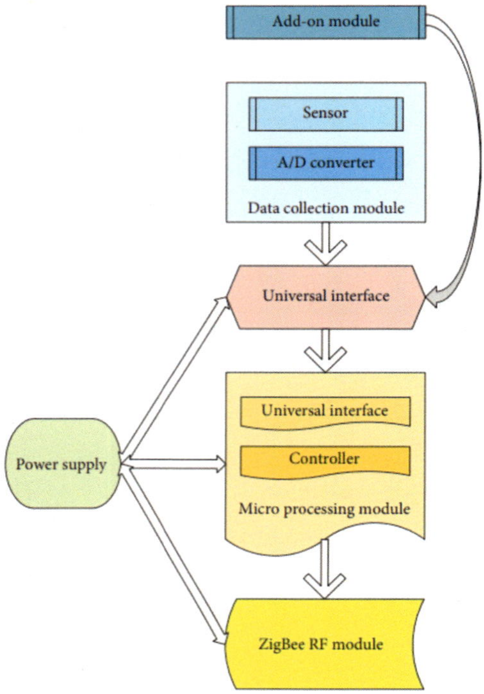

Figure 7. ZigBee sensor node composition diagram.

In the software structure of sensor nodes, the software structure can be divided into three layers according to the use, operating system, system service, and application management. As shown in Figure 8, the access entry of the components and hardware implementing the task is controlled by the operating system. Through the control of the operating system, the functions of storage management, hardware start-up, stop processing, and task adjustment can be realized. System services typically include service components and network protocol components that provide services to corresponding applications. The nature of the system service section can be seen as an intermediate layer. From the overall analysis, node software architecture design must include communication protocols between nodes, data collection and processing of data content.

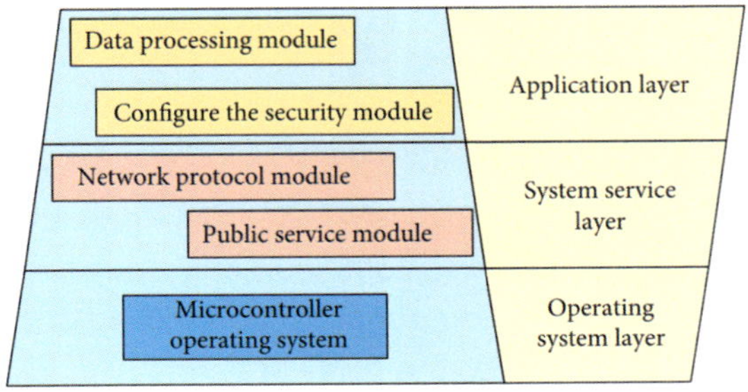

Figure 8. ZigBee wireless sensor node software composition diagram.

In the structure of communication protocol, from a specific direction, the network communication protocol can be seen as a two-dimensional architecture. In Figure 9, the communication protocol structure includes the application layer, transmission layer, network layer, data connection layer, and physical layer [27].

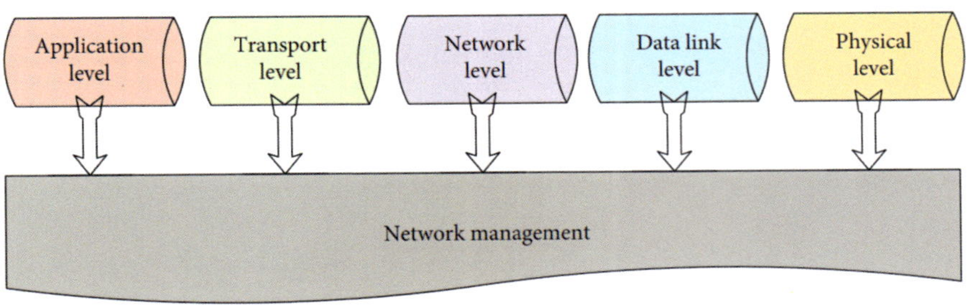

Figure 9. Architecture of ZigBee wireless sensor communication protocol.

The physical layer is responsible for providing physical channels for data transmission. In addition, the physical layer can also assist other systems to complete the channel frequency setting and energy monitoring. Usually, the physical layer's job is to turn on or off the wireless receiver. The data link is responsible for data frame monitoring, media contact and error control, multipath multiplexing of data flow, and building one-to-one or one-to-many stable link paths. Among them, the media access control is the most important function of the connection layer. The radiofrequency (RF) module of the node is controlled by the media access protocol, which is responsible for solving the resource allocation problem of wireless communication between sensors. Since the nodes will move, conflicts may occur between multiple nodes, and the interference of the environment will improve the error rate of the network channel of the wireless sensor. The main monitoring data of the sensor are related to energy. Therefore, the data connection layer must design a media access control method to improve the effectiveness of energy monitoring.

In the network layer control routing generation, ideally, wireless sensor networks should have on attribute address recognition function. Moreover, the network layer should provide a way to connect with external networks (such as the Internet), control system networks, etc.

In data transmission and input, Figure 10 is the data transmission process. Firstly, the transmitted data should be stored in the first in and first out cache. When processing packets with uncertain scale, the length bytes appear in the first byte. If the data size is fixed, the first byte of the first in and out cache should be the endpoint node. When valid data is stored in the first in and first out cache, synchronous bytes are sent, followed by valid loads.

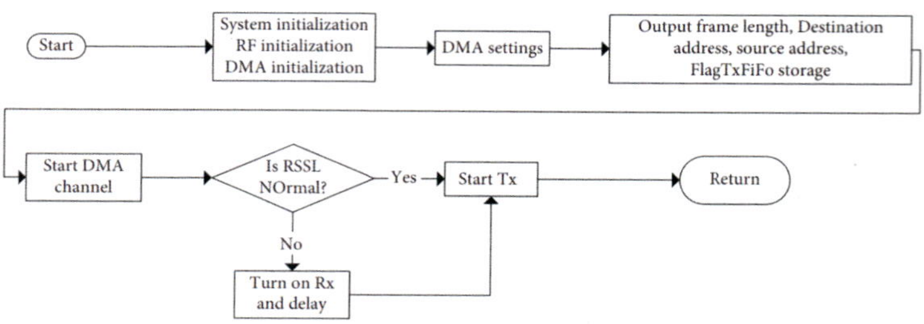

Figure 10. ZigBee wireless sensor data transmission process.

Figure 11 shows the data output process. Inverse to the data transmission process, the wireless signal receives and transmits data from the line and then is processed by the hardware to achieve the exchange of data and signals. Then, data is stored in the data memory area.

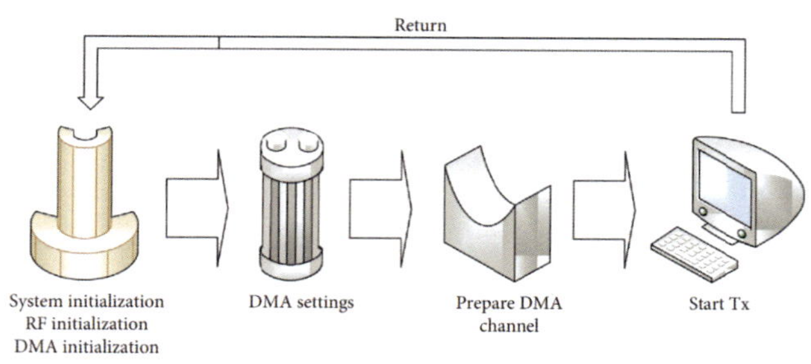

Figure 11. Data output process of ZigBee wireless sensor.

When the device is in the data receiving state, the preamble and synchronization bytes in the data will be detected by the mediator and the receiving device. After matching the data, the mediator will receive the payload and performs data analysis. If the length of the data is fixed, the receiver will receive the data with reference to the default length byte. If the length of this part of the data is variable, the receiver will first save the length of the first byte record and then receive the data according to the same length.

After the match is successful, the first payload is accepted by the mediator. If the data length at this time is fixed, only the byte data of the default length will be accepted. If the data is in variable length mode at this time, the data that matches the length scale of the data will be saved.

Wireless Sensor Performance Research and Testing

The performance of the designed and built wireless transmitter needs to be verified. Two aspects can best reflect the performance of transmission technology, one is stability, and the other is transmission efficiency. The stability of the ZigBee wireless transmitter will be reflected by the real-time temperature during operation. If the temperature is too high, it means that the system is prone to be in an unstable state, and there is a possibility of failure at any time. We check the temperature through the node temperature monitoring data collected by the ZigBee wireless sensor network and compare it with the temperature monitoring data of the wired network sensor to analyze the stable performance of the ZigBee wireless sensor. Under the working conditions of high load, each component is in the state of full power operation. The machine needs high voltage to ensure operation. Long-time high-voltage operation will lead to system temperature rise. At high temperature, the mechanical properties of metal decrease rapidly. This will shorten the service life of each component, such as central processing chip, image processor, and computer graphics card. Persistent high temperatures can also lead to increased chemical performance in some specific systems, resulting in machine paralysis. In the circuit, too high temperature will also trigger off protection, resulting in the entire system to stop working, resulting in loss of production facilities.

Interfere with the transmission path by inserting obstacles in the middle of the data output device and the data receiving device. Then, the effective distance of data transmission is calculated. In the process of data transmission, whether the data is correct and whether the real-time performance can be guaranteed will determine whether the system can be used for actual automatic control. Whether the equipment parameters of the experimental site can be collected and transported correctly is a key point that must be pay attention to. Data accuracy is generally reflected in two aspects. One is whether the data collection of the sensor node is the same as the actual data or the error is small, and whether the accuracy can meet the accuracy requirements of industrial production; the other is whether a packet loss phenomenon occurs during wireless transmission of data.

RESULTS

ZigBee Wireless Sensor and Wired Sensor Temperature Comparison Test

Comparing the temperature monitored by the two sensors, the data statistics of the two groups of sensors are shown in Figure 12.

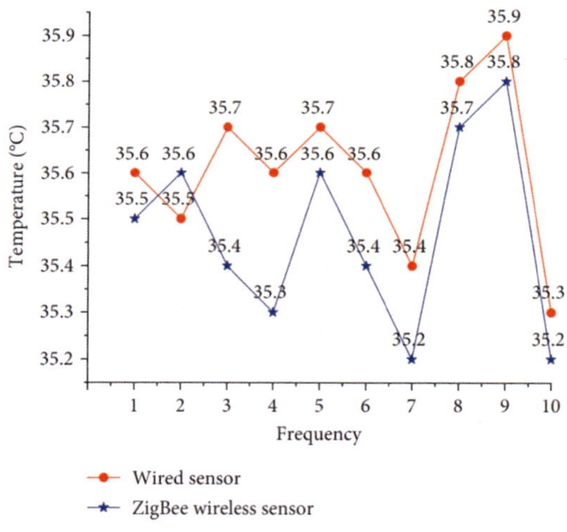

Figure 12. Comparison results of ZigBee wireless sensor and wired sensor operating temperature.

By analyzing the curve data, the temperature of the traditional wired control system is higher than that of the ZigBee control system when it works. Only in the second experiment, the ZigBee temperature is slightly higher than the wired control system. The maximum temperature of the wired control system is 35.9°C, and the minimum temperature is 35.3°C. The maximum working temperature of ZigBee control system is 35.8°C, and the minimum temperature is 35.2°C. The low working temperature shows that the system has high stability and will not be powered off due to overheating. Therefore, it indicates that ZigBee control system has better stability than traditional wired control system. In the whole working process, ZigBee can maintain the temperature range of the components that is not too hot and cannot be moved due to too low temperature. So, it can better operate the whole system. ZigBee allows the system to operate at full power without breaking due to overheating.

ZigBee Wireless Sensor Data Transmission Efficiency and Distance Test

After obtaining the result that the stability of the ZigBee wireless sensor is higher than that of the wired sensor, it is necessary to test the data transmission efficiency of the ZigBee wireless sensor. In the actual application process, check whether there will be data packet loss. The research conducted experiments by increasing the distance of data transmission by setting obstacles in the data receiving, transmitting equipment, and appropriately extending the distance between the equipment. The results are shown in Figure 13.

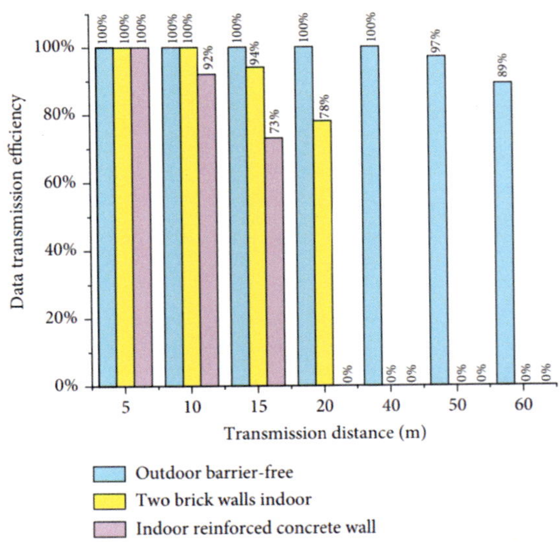

Figure 13. ZigBee wireless sensor data transmission efficiency test results.

Figure 13 shows that when there is no obstacle between the outdoor data receiving device and the transmitting device, the effective transmission efficiency of the ZigBee wireless sensor starts to drop from 40 meters. Therefore, the effective transmission distance of the ZigBee wireless sensor is about 40 meters to 50 meters when there are no obstacles.

When there are obstacles in the room that block the straight-line transmission of data, for example, when there are two brick walls blocking, the effective transmission distance of the ZigBee wireless sensor should be less than 15 meters as much as possible. When the transmission distance is farther than 20 meters, the transmission efficiency of the equipment drops greatly. When the transmission distance exceeds 40 meters, the transmission efficiency cannot be detected, so it is no longer suitable for data transmission.

Finally, when replacing the brick wall with a reinforced concrete wall, the effective transmission distance of the ZigBee wireless sensor does not exceed 10 meters. When the transmission distance exceeds 10 meters, the transmission efficiency drops significantly. When the transmission distance exceeds 20 meters, the transmission efficiency can no longer be detected.

Experiments show that compared with wired control, the ZigBee wireless control system is limited by transmission distance and will be blocked by obstacles between the receiver and the transmitter. But ZigBee has the advantage of fast deployment. The wired connection will occupy a large area of land because of the connection line, resulting in the disorder of the engineering site and the damage of the connection line. Wireless connection can effectively avoid these problems.

Thus, ZigBee wireless control cannot be applied in remote automatic control. However, wireless control has better controllability than wired control. Compared with wired control, wireless control is less constrained by terrain. When operating in some areas with complex environments, cable-controlled line deployment is a problem that must be overcome, and wireless transmission can well overcome this difficulty.

CONCLUSIONS

China is a country dominated by manufacturing. In the process of gradual industrial modernization, if China wants to quickly shorten the gap in industrial strength with developed countries, it needs to improve industrial manufacturing technology and automation control technology. ZigBee wireless communication technology is used as technical support and in the establishment of wireless sensors. Necessary components such as nodes and equipment terminals in the sensor network are designed and analyzed. Subsequently, the designed ZigBee wireless sensor is used to test the automated control of a pharmaceutical factory's production workshop. The real-time temperature monitoring during the working process of the equipment and the effective data transmission distance of the equipment are studied further. In the temperature comparison between the wireless sensor and the wired sensor, it is concluded that the stability of the ZigBee wireless sensor is better than that of the wired sensor. In the transmission efficiency test, the study is conducted under the conditions of outdoor barrier-free, indoor two brick walls, and indoor reinforced concrete walls, and the maximum effective data transmission distance of three wireless sensors is obtained. Combining the two test results, wireless sensors using ZigBee wireless communication technology can be used in industrial automation control systems. At present, whether wireless sensors can be applied to industrial automation control is regarded as a research focus and measures to improve the transmission efficiency of wireless sensors have not been studied. For some large-scale industrial tasks, it is difficult to use wired transmission when the transmission distance is long. Transmission lines not only waste a lot of resources but also cause safety hazards. In this case, long-distance wireless data transmission is required. Therefore, how to improve the effective transmission distance of ZigBee wireless sensors will be the focus of subsequent research. Moreover, in the second part of the results, the environmental factors affecting wireless transmission are studied. How to solve the impact of wireless transmission environment has not been discussed. Therefore, in the future, improving the timeliness of ZigbBee wireless control system will be discussed.

REFERENCES

1. H. Haraguchi, B. Martorano, and M. Sanfilippo, "What factors drive successful industrialization? Evidence and implications for developing countries," *Structural Change and Economic Dynamics*, vol. 49, pp. 266–276, 2019.
2. U. Hacioglu and G. Sevgilioglu, "The evolving role of automated systems and its cyber-security issue for global business operations in industry 4.0," *International Journal of Business Ecosystem & Strategy (2687-2293)*, vol. 1, no. 1, pp. 1–11, 2019.
3. B. Zhao, "The design and research of electric automation control system based on PLC," *Journal of Frontiers of Society, Science and Technology*, vol. 1, no. 6, pp. 169–176, 2021.
4. V. Aparna, "Application of DCS for level control in nonlinear system using optimization and robust algorithms," *ADCAIJ: Advances in Distributed Computing and Artificial Intelligence Journal*, vol. 9, no. 1, pp. 29–50, 2020.
5. Z. Fernandez, Ó. Seijo, M. Mendicute, and I. Val, "Analysis and evaluation of a wired/wireless hybrid architecture for distributed control systems with mobility requirements," *IEEE Access*, vol. 7, pp. 95915–95931, 2019.
6. S. Fu, J. Wu, H. Wen, Y. Cai, and B. Wu, "Software defined wireline-wireless cross-networks: framework, challenges, and prospects," *IEEE Communications Magazine*, vol. 56, no. 8, pp. 145–151, 2018.
7. G. Debita, P. Falkowski-Gilski, M. Habrych et al., "Subjective and objective quality evaluation study of BPL-PLC wired medium," *Elektronika ir Elektrotechnika*, vol. 26, no. 3, pp. 13–19, 2020.
8. S. Shrestha and S. Shakya, "Technical analysis of ZigBee wireless communication," *Journal of trends in Computer Science and Smart technology (TCSST)*, vol. 2, no. 4, pp. 197–203, 2020.
9. M. Abdulkarem, K. Samsudin, F. Z. Rokhani, and M. F. A Rasid, "Wireless sensor network for structural health monitoring: a contemporary review of technologies, challenges, and future direction," *Structural Health Monitoring*, vol. 19, no. 3, pp. 693–735, 2020.
10. Q. Li and N. Liu, "Monitoring area coverage optimization algorithm based on nodes perceptual mathematical model in wireless sensor networks," *Computer Communications*, vol. 155, pp. 227–234, 2020.

11. D. Marrero, A. Suárez, E. Macías, and V. Mena, "Extending the battery life of the ZigBee routers and coordinator by modifying their mode of operation," *Sensors*, vol. 20, no. 1, p. 30, 2020.
12. S. G. Varghese, C. P. Kurian, V. I. George, A. John, V. Nayak, and A. Upadhyay, "Comparative study of ZigBee topologies for IoT-based lighting automation," *IET Wireless Sensor Systems*, vol. 9, no. 4, pp. 201–207, 2019.
13. J. Xiao and J. T. Li, "Design and implementation of intelligent temperature and humidity monitoring system based on ZigBee and WiFi," *Procedia Computer Science*, vol. 166, pp. 419–422, 2020.
14. V. M. Mishra and A. Kumar, "Zigbee internode communication and FPGA synthesis using mesh, star and cluster tree topological chip," in *Wireless Personal Communications*, pp. 1–19, Springer, 2021.
15. Y. R. Hamdy and A. I. Alghannam, "Evaluation of ZigBee topology effect on throughput and end to end delay due to different transmission bands for IoT applications," *Journal of Communications Software and Systems*, vol. 16, no. 3, pp. 254–259, 2020.
16. Y. Yu, B. Xue, Z. Chen, and Q. Zv, "Cluster tree topology construction method based on PSO algorithm to prolong the lifetime of ZigBee wireless sensor networks," *EURASIP Journal on Wireless Communications and Networking*, vol. 2019, no. 1, 2019.
17. C. E. Acosta, D. Cali, and C. Espinosa, "Autoconfiguration with global addresses using IEEE 802.15.4 standard in multi-hop networks," *Enfoque UTE*, vol. 12, no. 1, pp. 44–58, 2021.
18. H. Y. Chang, "A connectivity-increasing mechanism of ZigBee-based IoT devices for wireless multimedia sensor networks," *Multimedia Tools and Applications*, vol. 78, no. 5, pp. 5137–5154, 2019.
19. S. Meka and B. Fonseca, "Improving route selections in ZigBee wireless sensor networks," *Sensors*, vol. 20, no. 1, p. 164, 2020.
20. L. Yan and S. Jianwei, "Monitoring and fault diagnosis system of wind–solar hybrid power station based on ZigBee and BP neural network," *Australian Journal of Mechanical Engineering*, vol. 16, supplement 1, pp. 54–60, 2018.
21. K. Hussein, H. J. Hadi, M. R. Abdul-Mutaleb, and Y. S. Mezaal, "Low cost smart weather station using Arduino and ZigBee," *Telkomnika*, vol. 18, no. 1, pp. 282–288, 2020.

22. J. M. Gómez, J. C. Mondéjar, J. R. Gómez, and J. M. Martínez, "Developing an IoT forensic methodology. A concept proposal," *Forensic Science International: Digital Investigation*, vol. 36, article 301114, 2021.
23. C. H. Ke, S. Y. Hsieh, T. C. Lin, and T. H. Ho, "Efficiency network construction of advanced metering infrastructure using ZigBee," *IEEE Transactions on Mobile Computing*, vol. 18, no. 4, pp. 801–813, 2019.
24. K. Xia, J. Ni, Y. Ye, P. Xu, and Y. Wang, "A real-time monitoring system using ZigBee and 4G communications for photovoltaic generation," *CSEE Journal of Power and Energy Systems*, vol. 6, no. 1, pp. 52–63, 2020.
25. A. R. Kasar and S. Tiwari, "ZigBee on wireless sensor network," *International Journal of Scientific Research and Engineering Trends*, vol. 4, pp. 1327–1332, 2018.
26. V. Saranya, S. Shankar, and G. R. Kanagachidambaresan, "Energy efficient data collection algorithm for mobile wireless sensor network," *Wireless Personal Communications*, vol. 105, no. 1, pp. 219–232, 2019.
27. H. Xie, B. Yang, Z. Ren, K. Mu, X. Zhao, and B. Li, "The information security transmission method for intelligent examination based on ZigBee communication," *International Journal of Information and Communication Technology*, vol. 19, no. 3, pp. 258–274, 2021.

CHAPTER 3

An Electrothermal Model Based Adaptive Control of Resistance Spot Welding Process

Ziyad Kas and Manohar Das

Department of Electrical and Computer Engineering, Oakland University, Rochester, USA

ABSTRACT

Resistance Spot Welding (RSW) is a process commonly used for joining a stack of two or three metal sheets at desired spots. The weld is accomplished by holding the metallic workpieces together by applying pressure through the tips of a pair of electrodes and then passing a strong electric current for a short duration. Inconsistent weld and insufficient nugget size are some of the common problems associated with RSW. To overcome these problems, a new adaptive control scheme is proposed in this paper. It is

Citation: Kas, Z. and Das, M. (2015), "An Electrothermal Model Based Adaptive Control of Resistance Spot Welding Process". Intelligent Control and Automation, 6, 134-146. doi: 10.4236/ica.2015.62014.

based on an electrothermal dynamical model of the RSW process, and utilizes the principle of adaptive one-step-ahead control. It is basically a tracking controller that adjusts the weld current continuously to make sure that the temperature of the workpieces or the weld nugget tracks a desired reference temperature profile. The proposed control scheme is expected to reduce energy consumption by 5% or more per weld, which can result in significant energy savings for any application requiring a high volume of spot welds. The design steps are discussed in details. Also, results of some simulation studies are presented.

Keywords: Resistance Spot Welding, Adaptive Control, Nugget Formation, Energy Saving

INTRODUCTION

In resistance spot welding, the welding process begins by applying pressure on a stack of metal sheets, held together between a pair of electrodes. A weld current is then passed through the electrodes, causing resistive heating of the metal workpieces and the formation of a welded joint or nugget, as shown in Figure 1. The formation of a weld nugget strongly depends on the electrical and thermal properties of the sheet and coating materials [1] . Since the contact resistance near the faying surface is much higher than the resistance of the sheets and electrodes, most of the heating is concentrated near the faying surface, causing melting and formation of a nugget there. Depending on the thickness and type of material, welding current ranges from 1,000 to 20,000 amperes or more, while the voltage typically is between 1 and 30 volts [2] .

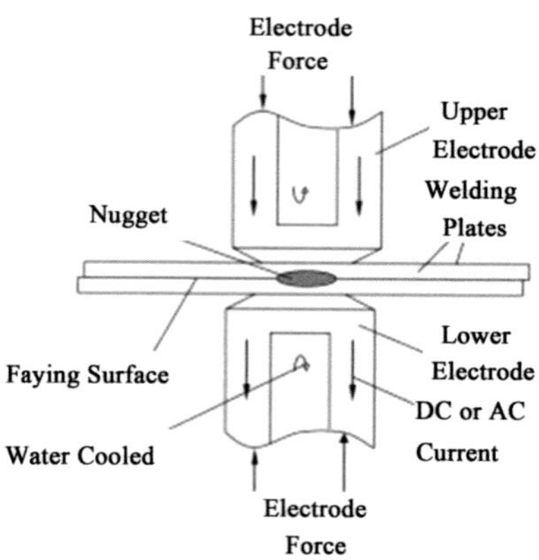

Figure 1. Resistance spot welding system.

A Resistance Spot Welding cycle consists of three main stages as follows:

Stage 1: Squeeze time, which is the time when electrodes press the welded workpieces together.

Stage 2: Weld time, which is the time when welding current is applied producing heat at the faying surface of the workpieces and thus creating a weld nugget.

Stage 3: Hold time, which is the time when electrode force still presses the workpieces together and cools the weld down after the welding current is switched off.

One of the most common applications of resistance spot welding is in the automobile manufacturing industry, where it is used almost universally to weld the sheet metals to form the car body and parts. A typical automotive vehicle today requires about 4000 - 6000 spot welds per vehicle. Considering a worldwide annual production volume of 80 million automotive vehicles, an energy saving RSW controller can result in significant energy savings and reduce carbon footprint accordingly.

During the past two decades, a number of studies have been carried out to improve the RSW process, which focuses on monitoring and control of weld parameters to improve weld quality. The RSW control techniques proposed to date include Proportional-Integral (PI) [3] , Proportional-Derivative (PD) [4] , Proportional-Integral- Derivative (PID) [5] , Fuzzy [6] - [8] , Neural Networks (NN) [9] [10] , or a combination of Fuzzy and NN [11] . The main drawback of these techniques is that they do not take into account the thermal dynamics of the RSW process, i.e. they do not utilize dynamical models that govern the heat transfer and nugget formation in the RSW process. Also, these systems don't take into account any welding process variations, such as variations in coating materials, electrode degradation, and weld force variations.

In this paper, a novel approach to RSW control is presented. This approach has not been explored by other researchers. We start with a simplified heat balance model of a RSW process proposed in [12] and [13] , and then use it to design a controller. This thermal model of the heat balance is a function of nugget growth and it determines the temperature variation during welding time. This model is used later to design an adaptive-one-step- ahead (AOSA) controller and an adaptive-weighted one-step-ahead (AWOSA) controller that compensate for unknown process variations and track a desired reference temperature profile. Finally, some simulation results that show the performance of the proposed controllers are presented and compared to the performance of a PID controller. Simulation results show that AOSA and AWOSA controllers are capable of tracking a reference temperature profile when the weld parameters are unknown, as well as reduce the energy needed to make a weld by 6%.

ELECTROTHERMAL DYNAMICAL MODEL OF A RSW NUGGET FORMATION PROCESS

To start with, we consider a simplified heat balance model of a RSW process, presented in [13] . The simplified dynamical model of a RSW process determines the heat balance in the system as a function of nugget temperature. For a simplified nugget model, shown in Figure 2, the heat balance can be described by the following equations:

The total heat generation rate, $\dot{Q}_g(t)$ is given by

$$\dot{Q}_g(t) = I^2(t)R(t) \tag{1a}$$

$$R(t) = R_w + R_c + R_e \tag{1b}$$

where $I(t)$ denotes the welding current, and $R(t)$ denotes the total resistance consisting of the resistance of work pieces, R_w, contact resistance, R_c, and electrode resistance, R_e. Since R_w and R_e are very small compared to the total contact resistance R_c, R_w and R_e can be neglected in (1b).

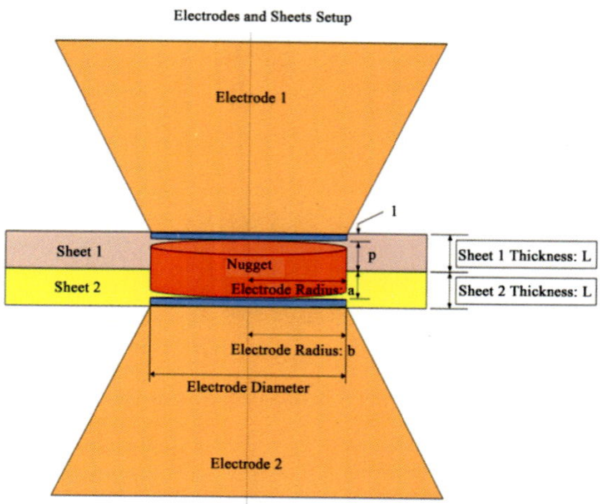

Figure 2. A simplified model of a weld nugget.

The total contact resistance can then be described as,

$$R_c = R(t)_{\text{electrode-sheet}} + R(t)_{\text{sheet-sheet (faying surface)}} \tag{1c}$$

A linear relationship between the resistance and temperature is assumed to model the heat generated as a function of temperature. Thus,

$$R(t)_{\text{electrode-sheet}} = \rho \frac{2l_1}{A} \tag{1d}$$

$$R(t)_{\text{sheet-sheet(faying surface)}} = \rho \frac{2p}{A} \tag{1e}$$

$$\rho = \rho(T) = \rho_\circ \left[1 + \alpha_r (\theta - \theta_\circ)\right] \tag{1f}$$

where ρ denotes the resistivity of the material, l_1 denotes the distance from the melting interface to electrode contact surface, p denotes the penetration, A is the cross sectional area, ρ_o denotes the resistivity at reference temperature θ_o, θ and α_r are the temperature to be controlled and the temperature coefficient respectively.

Substituting (1f) in (1d) and (1e) we get

$$R(t)_{\text{electrode-sheet}} = c_1\theta(t) + c_2 \tag{1g}$$

$$R(t)_{\text{sheet-sheet(faying surface)}} = c_3\theta(t) + c_4 \tag{1h}$$

where

$$c_1 = \frac{2l_1\rho_\circ\alpha}{A} \tag{1i}$$

$$c_2 = \frac{2l_1\rho_\circ}{A}(1 - \alpha_r\theta_\circ) \tag{1j}$$

$$c_3 = \frac{2p\rho_\circ\alpha}{A} \tag{1k}$$

$$c_4 = \frac{2p\rho_\circ}{A}(1 - \alpha_r\theta_\circ) \tag{1l}$$

Substituting (1g) and (1h) in (1a) we get

$$\dot{Q}_g(t) = I^2(t)\left(c_1\theta(t) + c_2 + c_3\theta(t) + c_4\right) \tag{1m}$$

$$= c_5 I^2(t)\theta(t) + c_6 I^2(t) \tag{1n}$$

where

$$c_5 = c_1 + c_3 \tag{1o}$$

$$c_6 = c_2 + c_4 \tag{1p}$$

The heat of fusion required for nugget formation is given by:

$$H_f = H\Delta V_n \tag{2a}$$

$$\Delta V_n = \pi a^2 p \tag{2b}$$

where H denotes the heat of fusion per unit volume, ΔV_n denotes the nugget volume, and p, a denote the penetration and nugget radius respectively. Substituting (2b) in (2a) and normalizing over the weld duration, Δt, we get the heat of fusion per unit time:

$$\frac{H_f}{\Delta t} = H\pi a^2 p = c_7 \tag{2c}$$

Neglecting the heat loss in the surroundings and the electrodes, the heat required to raise temperature by dθ (t) is given by

$$\mathrm{d}Q_T(t) = \rho C_p \mathrm{d}\theta(t)\Delta V \tag{3a}$$

where ρ denotes the density, C_p denotes the specific heat, V is the volume, and dθ (t) is the temperature rise. We rewrite (3a) as:

$$\mathrm{d}Q_T(t) = c_8 \mathrm{d}\theta(t) \tag{3b}$$

where

$$c_8 = \rho C_p \pi a^2 p \tag{3c}$$

The total heat loss rate is given by

$$\dot{Q}_L(t) = \dot{Q}_a(t) + \dot{Q}_r(t) \tag{4a}$$

$$= k_1 \pi a^2 \left[\frac{\theta(t) - \theta_1}{l_1} + \frac{10\theta(t)\beta L}{b\sqrt{\alpha}} \right]$$

$$= \left(\frac{k_1 \pi a^2}{l_1} + \frac{10 k_1 \pi a^2 \beta L}{b\sqrt{\alpha}} \right) \theta(t) - \frac{k_1 \pi a^2 \theta_1}{l_1}$$

$$= c_9 \theta(t) - c_{10} \tag{4b}$$

where

$$c_9 = \left(\frac{k_1 \pi a^2}{l_1} + \frac{10 k_1 \pi a^2 \beta L}{b\sqrt{\alpha}} \right) \tag{4c}$$

$$c_{10} = \frac{k_1 \pi a^2 \theta_1}{l_1} \tag{4d}$$

In the above equations, $\dot{Q}_a(t)$ and $\dot{Q}_r(t)$ denote the axial and radial loss rates, respectively; k_1 represents thermal conductivity, a is the nugget radius; $\theta(t)$, θ_1, represent the melting temperature and the interface temperature at the work piece respectively; l_1 is the distance from the melting interface to the electrodes contact area; β represents the final penetration to work piece thickness ratio; L is the sheet thickness; b,α represent the electrode radius and thermal diffusivity of work piece respectively.

The heat balance equation over time (t, t + dt) is given by

$$\dot{Q}_g(t) = \frac{H_f}{\Delta t} \mathrm{d}t + \mathrm{d}Q_T(t) + \dot{Q}_L(t)\mathrm{d}t \tag{5}$$

Substituting (1n), (2c), (3b), and (4b) in (5) and rearranging it, we get

$$c_8 \frac{\mathrm{d}\theta(t)}{\mathrm{d}t} = c_5 I^2(t)\theta(t) + c_6 I^2(t) - c_9 \theta(t) + c_{10} - c_7 \tag{6a}$$

or, equivalently,

$$\frac{\mathrm{d}\theta(t)}{\mathrm{d}t} = c_{11} I^2(t)\theta(t) + c_{12} I^2(t) - c_{13}\theta(t) + c_{14} \tag{6b}$$

where

$$c_{11} = c_5 / c_8 \tag{6c}$$

$$c_{12} = c_6 / c_8 \tag{6b}$$

$$c_{13} = c_9 / c_8 \tag{6c}$$

$$c_{14} = \left(c_{10} - c_7\right) / c_8 \tag{6d}$$

For the sake of notational convenience, let $y(t) = \theta(t)$ and $u(t) = I^2(t)$. Then (6b) can rewritten as

$$\frac{\mathrm{d}y(t)}{\mathrm{d}t} = c_{11}u(t)y(t) + c_{12}u(t) - c_{13}y(t) + c_{14} \tag{7}$$

Equation (7) represents a bilinear electrothermal dynamical model of a RSW process. Note that this simplified model neglects the heat required to raise the temperature of the electrodes and the nugget surroundings. Also, it assumes that most of the heating occurs near the faying surface due to its high contact resistance. The size of the workpieces is assumed to be infinite in the radial direction and the nugget shape is assumed to be a disk growing radially and axially in the same proportions. The nominal nugget diameter is assumed to be $4.5\sqrt{L}$, where L is the sheet thickness.

Using a first order Euler approximation for $\frac{\mathrm{d}y}{\mathrm{d}t}$ with a sampling period T_s, the following discrete time equation is derived from the system Equation (7):

$$\frac{y(k+1) - y(k)}{T_s} = c_{11}u(k)y(k) + c_{12}u(k) - c_{13}y(k) + c_{14} \tag{8a}$$

or

$$y(k+1) = Ay(k) + Bu(k) + Cu(k)y(k) + D \tag{8b}$$

where

$$A = 1 - c_{13}T_s \tag{8c}$$

$$B = c_{12}T_s \tag{8d}$$

$$C = c_{11}T_s \tag{8e}$$

$$D = c_{14} T_s \tag{8f}$$

Also, k denotes the discrete time index $(k = 0, 1, 2, \cdots)$ and kT_s denote the sampling instances. The above electrothermal model is characterized by four unknown parameters, namely, A, B, C, and D.

DESIGN OF A RSW CONTROLLER

To develop a control scheme for controlling the nugget temperature of the RSW model presented by Equation (8a), we realize that it presents a bilinear system characterized by some unknown parameters. These parameters can vary from weld to weld, and in most cases we have no prior knowledge of the parameter values. In view of this, we propose to use an adaptive OSA and WOSA controllers.

The proposed adaptive control scheme involves measurement of the inputs and outputs of the system, estimation of unknown system parameters using a recursive least squares (RLS) parameter estimation algorithm, and computation of a control signal based on the estimated parameter values. Also, the temperature of the weld nugget is monitored indirectly by assuming it to be proportional to the contact resistance.

Adaptive OSA and WOSA Controllers

In an adaptive controller, the sampled measurements, u(k) and y(k), are used to estimate the model parameters, A,B,C and D in Equation (8b), using a recursive parameter estimation method, such as recursive least square (RLS). The estimated values of these parameters are then used to compute the OSA/WOSA control signals.

Parameter Estimation

First we write model Equation (7) in the following form:

$$y(k+1) = \varphi(k)^{\mathrm{T}} X^* \tag{9a}$$

where

$$\varphi(k)^{\mathrm{T}} = [\, y(k-1) \quad u(k-1) \quad y(k-1) * u(k-1) \tag{9b}$$

$$X^* = [A \quad B \quad C \quad D]^{\mathrm{T}} \tag{9c}$$

Next, the estimated value of $\theta°$ is computed recursively using the following RLS algorithm:

$$\hat{\theta}(k)=\hat{\theta}(k-1)+\frac{P(k-2)\varphi(k-1)}{1+\varphi(k-1)^{T}P(k-2)\kappa(k-1)}\left[y(k)-\varphi(k-1)^{T}\hat{\theta}(k-1)\right];\quad k\geq 1 \tag{10a}$$

$$P(k-1)=P(k-2)-\frac{P(k-2)\varphi(k-1)\varphi(k-1)^{T}P(k-2)}{1+\varphi(k-1)^{T}P(k-2)\varphi(k-1)} \tag{10b}$$

$$\hat{X}(0)=[\gamma\quad 0\quad 0\quad 0]^{T} \tag{10c}$$

$$P(-1)=\sigma I \tag{10d}$$

where $\gamma > 0$ is a small number and $\sigma > 0$ is chosen to be large. Also, $\hat{C}(k)$ is always constrained to be non-negative, i.e.,

$$\hat{C}(k)>\varepsilon>0 \quad \text{for all } k \tag{10e}$$

Given an estimate $\hat{X}(k)$ of X^{*}, we define the predicted output at time k +1 as:

$$\hat{y}(k+1)=\varphi(k)^{T}\hat{X}(k) \tag{11}$$

Adaptive-One-Step-Ahead Tracking Controller

One-step-ahead (OSA) control scheme for linear systems has been well investigated in [14] . An OSA controller attempts to bring the predicted output, y(k +1) at time k +1, to the desired value, $y^{*}(k+1)$ in one step. Thus, it minimizes the following cost function:

$$J_{1}(k+1)=\frac{1}{2}\left[y(k+1)-y^{*}(k+1)\right]^{2} \tag{12}$$

The corresponding OSA control law is given by [14] :

$$\bar{u}(k)=\frac{y^{*}(k+1)-Ay(k)-D}{B+Cy(k)} \tag{13}$$

The above control signal needs to be constrained by the maximum current delivery capacity of the controller, u_{max}, as follows:

$$u(k)=\begin{cases}\bar{u}(k), & \text{if } 0<\bar{u}(k)<u_{\max}\\ 0, & \text{if } \bar{u}(k)\le 0\\ u_{\max}, & \text{if } \bar{u}(k))\ge u_{\max}\end{cases} \tag{14}$$

The adaptive OSA controller uses the estimate, $\hat{X}(k)$ in Equation (11) to compute the control signal, u(k), from the following adaptive version of Equation (13) above:

$$\bar{u}(k)=\frac{y^*(k+1)-\hat{A}(k)y(k)-\hat{D}(k)}{\hat{B}(k)+\hat{C}(k)y(k)} \tag{15}$$

where $\hat{A}(k),\hat{B}(k),\hat{C}(k),$ and $\hat{D}(k)$ denote the estimated values of A, B, C, and D, respectively, at time k.

One of the potential drawbacks of OSA controllers is excessive control efforts that often result from attempting to bring y(k +1) to $y^*(k+1)$ in one step. To address this potential problem, an AWOSA controller is discussed below.

Adaptive Weighted One-Step-Ahead Controller

The excessive effort to bring the output y(k +1) to the desired value $y^*(k+1)$ in one step using AOSA may

result in an unfavorable saturation of the input. The adaptive weighted one-step-ahead controller attempts to seek a tradeoff between tracking accuracy and control effort by considering a slight generalization of the cost function (12) to the form (16) given below. Thus, it minimizes the following cost function:

$$J_2(k+1)=\frac{1}{2}\left[y(k+1)-y^*(k+1)\right]^2+\frac{\lambda}{2}u(k)^2 \tag{16}$$

where, $0<\lambda<1$ is chosen to provide a desired tradeoff.

The minimization of the cost function in (16) leads to the weighted one-step-ahead control law [14] :

$$\bar{u}(k)=\frac{\left(B+Cy(k)\right)\left(y(k+1)-Ay(k)-D\right)}{\left(B+Cy(k)\right)^{2}+\lambda} \tag{17}$$

The above control law is also constrained by the maximum current delivery capacity, u_{max}, as shown in Equation (14) above. The choice of λ provides a desired tradeoff between tracking accuracy and control effort. A small λ results in good tracking but requires high level of control effort. A large λ, on the other hand, reduces control efforts at the cost of tracking accuracy.

The adaptive WOSA controller uses the estimate, $\hat{X}(k)$, in Equation (11) to compute the control signal, u(k) from the following adaptive version of Equation (17) above:

$$\bar{u}(k)=\frac{\left(\hat{B}(k)+\hat{C}(k)y(k)\right)\left(y(k+1)-\hat{A}(k)y(k)-\hat{D}(k)\right)}{\left(\hat{B}(k)+\hat{C}(k)y(k)\right)^{2}+\lambda} \tag{18}$$

where $\hat{A}(k),\hat{B}(k),\hat{C}(k),$ and $\hat{D}(k)$ denote the estimated values of A, B, C, and D, respectively, at time k.

SIMULATION RESULTS AND DISCUSSION

This section presents the results of a simulation study showing the performance of the system with the proposed AOSA and AWOSA controllers and also compare them with a PID controller. Each controller is designed for tracking a reference temperature profile.

The reference temperature profile is a good indicator of the weld quality. Therefore, it is desirable to keep the temperature variation close to a desired variation curve, which may be experimentally predetermined for the good welds. A typical reference temperature profile for good weld is shown in Figure 3 below [1] . Basically, such a curve is characterized by a fast rise of temperature to melting point, melting of the workpieces at the faying surface area which causes a slight drop in temperature, followed by a cooling zone that results from removal of weld current. The actual nugget temperature is measured during the weld cycle using the relationship described by Equation (1f). Depending on the tracking error signal, the welding current is adjusted so as to reduce the temperature error. For these simulations, we have selected two sheets of mild steel with the same thickness as the materials to be welded. The force variation and electrode wear are considered as unknown process variables that impact the nugget size (diameter and penetration). The Figures below show the performance of the AOSA, AWOSA, and PID controllers due to 20% increase in nugget diameter and 50% increase in indentation from their desired values.

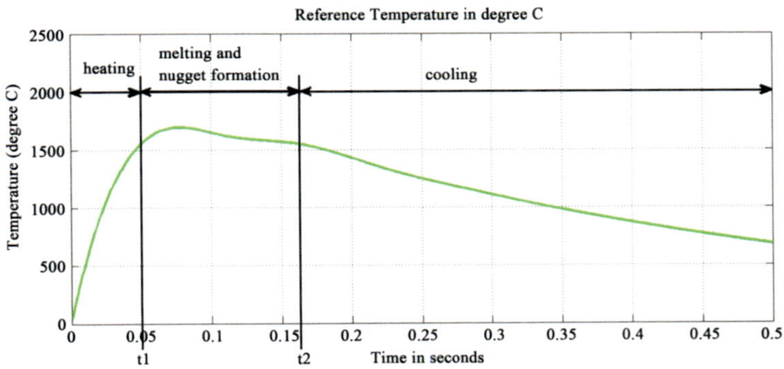

Figure 3. Desired reference temperature profile.

Figure 4 shows the performance of the AOSA controller using I_{max} = 12 KA, where I_{max} denotes the maximum current delivery capacity of the weld controller. We can see that the AOSA controller adapts to the parameter change and force the output temperature profile to follow the desired temperature profile. Also, we can see that the energy required for the weld is lower than that of the PID controller.

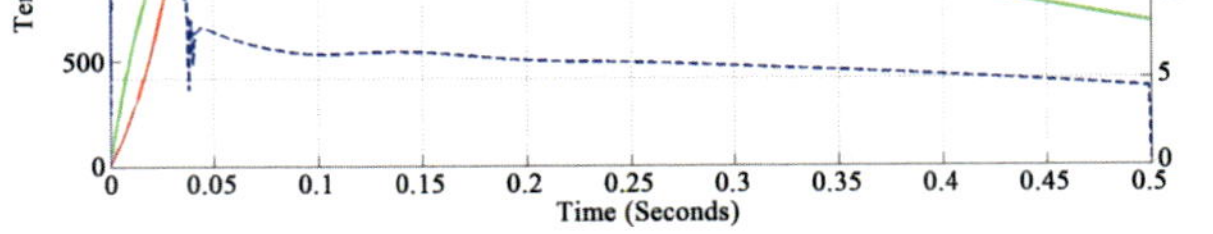

Figure 4. Performance of AOSA Controller with 20% increase in nugget diameter and 50% increase in indentation; I_{max} = 12 KA, Energy = 2583 W.

Figure 5 and Figure 6 show the performance of AWOSA controller using I_{max} = 12 KA with $\lambda = 0.1$ and

1, respectively. Here we notice that when λ is high, the output temperature profile does not follow the desired output temperature profile well. However, increasing λ results in decreasing the total energy required for the weld.

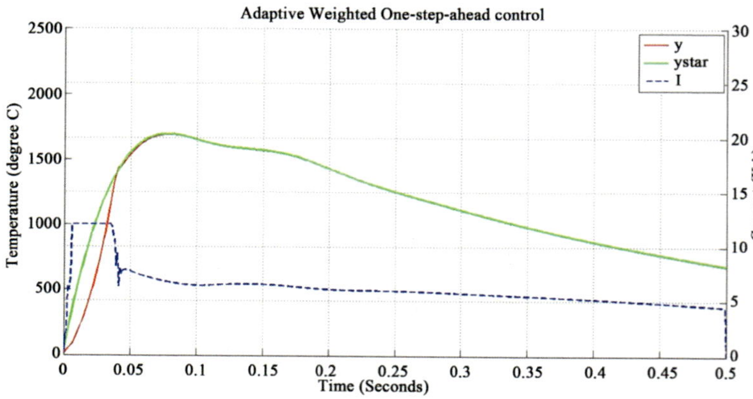

Figure 5. Performance of AWOSA Controller with 20% increase in nugget diameter and 50% increase in indentation; $\lambda = 0.1$, $I_{max} = 12$ KA, Energy = 2558 W.

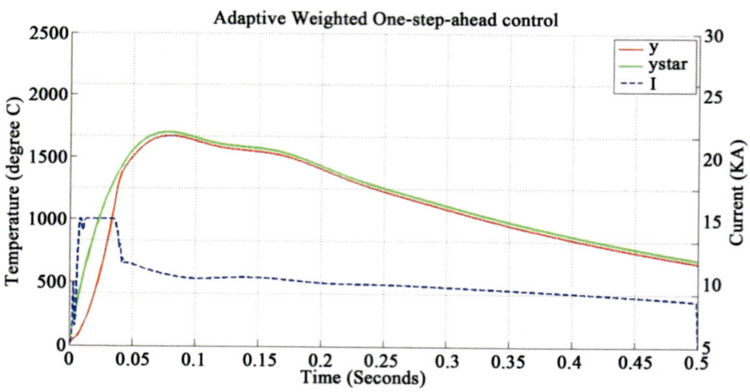

Figure 6. Performance of AWOSA Controller with 20% increase in nugget diameter and 50% increase in indentation; $\lambda = 1$, $I_{max} = 12$ KA, Energy = 2470 W.

Figure 7 shows the performance of the PID controller prior to any parameter change using $I_{max} = 12$ KA. After multiple trial and error attempts to get satisfactory results, the parameters of the PID controllers are: Proportional (P) = 0.5, Integral (I) = 26.56, Derivative (D) = 0.

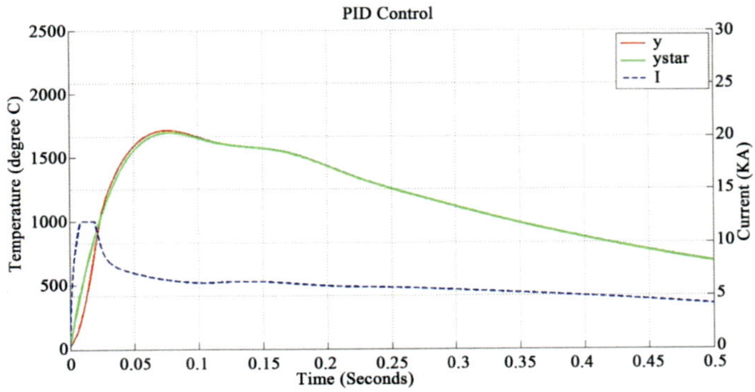

Figure 7. Performance of PID Controller prior to unknown parameter variations; I_{max} = 12 KA, Energy = 2393 W.

In Figure 8 we see that the PID controller looses track of the reference temperature profile due to weld parameters change. Also, we can see that PID controller requires more energy for the weld comparing to AOSA and AWOSA.

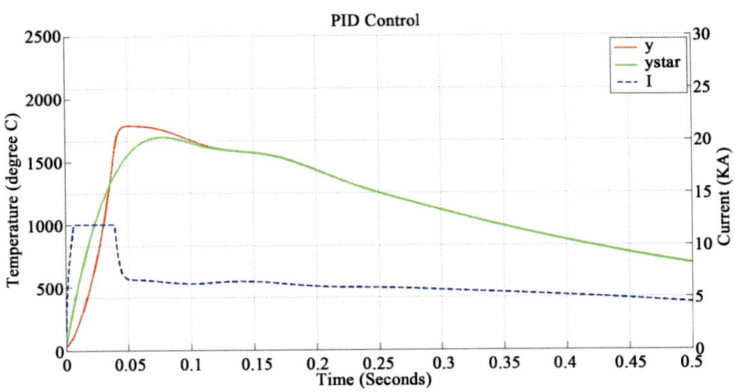

Figure 8. Performance of PID Controller with 20% increase in nugget diameter and 50% increase in indentation; I_{max} = 12 KA, Energy = 2632 W.

Comparing the simulation results for the three controllers, we can see that AOSA and AWOSA controllers compensate for the parameter variations and track the reference temperature profile quite well. Simulation results in Figure 5 for the AWOSA controller show satisfactory performance and a good tradeoff between tracking error and total energy required for the weld regardless of change in weld parameters. The output temperature profile follows the desired temperature profile reasonably well during the heating stage prior to the melting point. Also, we can see that the total energy required to make a weld using AWOSA is reduced by 6% comparing to the PID controller when I_{max} = 12 KA. This can result in significant energy savings for applications requiring a high volume of spot welds, such as manufacturing of automotive vehicles.

CONCLUSION

Ziyad Kas, Manohar Das This paper presents a new approach for designing adaptive OSA and WOSA controllers for resistance spot welding processes by utilizing a simplified electrothermal dynamical model of the process. Simulation results of AOSA and AWOSA performance are compared with those of a PID controller. These results indicate that using the proposed AOSA and AWOSA controllers, the nugget temperature profile is forced to track a desired reference temperature profile in presence of unknown parameter variations. Also, these controllers reduce the energy consumed to perform a spot weld, which can result in significant energy savings for applications requiring a high volume of spot welds, such as manufacturing of automotive vehicles.

APPENDIX

Boundedness of Nugget Temperature

Since a RSW is a time limited process ($\Delta t < 0.5$ sec usually), establishing a proof of asymptotic tracking would be meaningless. However, it is important to make sure that the nugget temperature remains bounded during time (0, Δt). A theoretical upper bound of the nugget temperature rise, $\Delta\theta$ (t) during time, (0, Δt), can be established as follows.

Notice the amount of heat absorbed = the amount of heat supplied ? the amount of heat loss

Suppose

$\Delta\theta$ = rise in temperature during time, (0, Δt)

Thus,

$$\text{Amount of heat obsorbed} = C_A \Delta\theta(t) \tag{19}$$

where C_A is a constant.

$$\text{Amount of heat supplied} = \int_0^{\Delta t} I^2(t) R(t)\,\mathrm{d}t \le \Delta t I_{\max}^2 R_{\max} \tag{20a}$$

where

$$R_{\max} = R_{\circ} + \alpha_r \Delta\theta \tag{20b}$$

and $I_{\max}$ denotes the maximum weld current.

$$\text{Amount of heat lost} = C_L \Delta\theta \tag{21}$$

where, C_L is a constant.

Thus,

$$C_A \Delta\theta \le \Delta t I_{\max}^2 \left(R_{\circ} + \alpha_r \Delta\theta\right) - C_L \Delta\theta \tag{22a}$$

or,

$$\Delta\theta \left(C_A - \Delta t I_{\max}^2 \alpha_r + C_L\right) \le \Delta t I_{\max}^2 R_{\circ} \tag{22b}$$

or,

$$\Delta\theta \le \frac{\Delta t I_{\max}^2 R_{\circ}}{C_A - \Delta t I_{\max}^2 \alpha_r + C_L} \tag{22c}$$

which proves the boundedness of the nugget temperature rise during weld time, $(0, \Delta t)$.

REFERENCES

1. Zhang, H. and Senkara, J. (2012) Resistance Welding Fundamentals and Applications. Taylor & Francis Group, Boca Raton.
2. Govik, A. (2009) Modeling of the Resistance Spot Welding Process. M.S. Thesis, Institute ofTechnology, Linkopings University, Linkoping.
3. Won, Y.J., Cho, H.S. and Lee, C.W. (1983) A Microprocessor-Based Control System for Resistance Spot Welding Process. Proceedings of ACC, San Francisco, 22-24 June 1983, 734-738.
4. Zhou, K. and Cai, L. (2014) A Nonlinear Current Control Method for Resistance Spot Welding. Proceedings of ASME Transactions on Mechatronics, 19, 559-569. http://dx.doi.org/10.1109/TMECH.2013.2251351
5. Salem, M. and Brown, L.J. (2011) Improved Consistency of Resistance Spot Welding with Tip Voltage Control. Proceedings of CCECE, Niagara Falls, 8-11 May 2011, 548-551. http://dx.doi.org/10.1109/ccece.2011.6030511
6. Chen, X., Araki, K. and Mizuno, T. (1997) Modeling and Fuzzy Control of the Resistance Spot Welding Process. Proceedings of SICE, Tokushima, 29-31 July 1997, 898-994.
7. El-Banna, M., Filev, D. and Chinnam, R.B. (2006) Intelligent Constant Current Control for Resistance Spot Welding. Proceedings of IEEE Conference on Fuzzy Systems, Vancouver, 16-21 July 2006, 1570-1577. http://dx.doi.org/10.1109/fuzzy.2006.1681917
8. Chen, X. and Araki, K. (1997) Fuzzy Adaptive Process Control of Resistance Spot Welding with a Current Reference Model. Proceedings of IEEE Conference on Intelligent Processing Systems, Beijing, 28-31 October 1997, 190-194.
9. Shriver, J., Peng, H. and Hu, S.J. (1999) Control of Resistance Spot Welding. Proceedings of ACC, San Diego, 2-4 June 1999, 187-191. http://dx.doi.org/10.1109/acc.1999.782766
10. Ivezic, N., Allen Jr, J.D. and Zacharia, T. (1999) Neural Network-Based Resistance Spot Welding Control and Quality Prediction. Proceedings of IPMM, Honolulu, 10-15 July 1999, 989-994. http://dx.doi.org/10.1109/ipmm.1999.791516
11. Messler Jr, R.W., Jou, M. and Li, C.J. (1995) An Intelligent Control System for Resistance Spot Welding Using a Neural Network and

Fuzzy Logic. Proceeding of IAC, Orlando, October 1995, 1757-1763. http://dx.doi.org/10.1109/ias.1995.530518

12. Kim, E.W. and Eagar, T.W. (1988) Parametric Analysis of Resistance Spot Welding Lobe Curve. SAE Technical Paper Series, Warrendale.
13. Kas, Z. and Das, M. (2014) A Thermal Dynamical Model Based Control of Resistance Spot Welding. Proceedings of IEEE EIT 2014, Milwaukee, 5-7 June 2014, 264-269. http://dx.doi.org/10.1109/eit.2014.6871774
14. Goodwin, G.C. and Sin, K.S. (1983) Adaptive Filtering Prediction and Control. Prentice-Hall, Englewood Cliffs.

CHAPTER 4

Multiagent Consensus Control Strategy considering Whole-Process Thermodynamic Characteristics of Air Conditioning Process

Jun Li[1], Bo Geng[1], Zhixian Lin[1], Min Chen[1], Liangyou Shao[1], Yanmin Zhou[2], and Yuqing Bao[2]

[1]Shenzhen Power Supply Bureau Co. Ltd., Shenzhen 518000, China

[2]School of Electrical Engineering, Southeast University, Nanjing 210096, China

ABSTRACT

Due to the distributed and decentralized characteristics of air conditioning load, the distributed control strategy has advantages for the air conditioning load to participate in the demand response. However, existing approaches focus on the dynamic control performance with very few considerations on

Citation: J. Li, B. Geng, Z. Lin, M. Chen, L. Shao, Y. Zhou, Y. Bao, "Multiagent Consensus Control Strategy considering Whole-Process Thermodynamic Characteristics of Air Conditioning Process", Journal of Electrical and Computer Engineering, vol. 2021, Article ID 5543298, 10 pages, 2021. https://doi.org/10.1155/2021/5543298.

the cost. To fill this gap, this paper proposes a multiagent consensus control method considering the whole-process response cost of air conditioning. Based on the thermodynamic characteristics of air conditioning load in the load reduction process and recovery process, the cost function curve of air conditioning load is established. Then, the multiagent consensus control strategy is adopted to send the power adjustment information to each air conditioner to realize the optimal control of the air conditioning load. The simulation results verify that the proposed method can take into account the whole-process response cost of air conditioning loads and result in smaller control cost than existing methods.

INTRODUCTION

With the increasing penetration of renewable energy, it is increasingly difficult for the power system to keep the active power balance by using generation side resources. More and more attention has been paid to the utilization of demand side resources to accommodate renewable energy.

As an important demand side resource, air conditioning load occupies a relatively high proportion in the power load and has the capacity of cold and hot storage. Focusing on the control strategy of air conditioning load, a lot of research works have been reported. Bao et al. [1] proposed an air conditioning load control strategy based on the state-queue model, through the broadcast control signal to achieve load tracking control of large number of air conditioning loads. In [2], a centralized control strategy based on supply-demand interaction is proposed, in which the household energy management device is used to send load shedding signals to the air conditioning load, and the large-scale air conditioning loads are controlled. In [3], a direct load control strategy of inverter air conditioners is proposed, which can control the air conditioning group by changing the set-point of air conditioning temperature. In [4], a direct load control method based on fuzzy logic is proposed, and a fuzzy controller with triangular membership function is designed to control the air conditioning load. In [5], a direct load control strategy based on Fourier transform is proposed to track the load demand curve of central air conditioning.

Due to the distributed characteristics of the air conditioning loads, the traditional centralized control method has problems in sending command to large amount of nodes. Conversely, the distributed control framework has advantages in sending control signals and improving the control system reliability. In recent years, many distributed control methods have been proposed, for example, the nonnegative edge consensus control method [6], the positivity-preserving consensus control method [7], and the event-triggered guaranteed cost consensus control method [8]. The distributed consensus method is applied to energy storage systems [9], the microgrids [10], the distributed generators [11], and so on.

Focusing on the distributed control strategy of air conditioning loads, Wang et al. [12] proposed a consensus control strategy of inverter air conditioner based on the distributed control framework. The adjustable temperature interval was selected as consensus variable and the same degree of control was realized for all the users. In [13], a distributed control strategy of air conditioning load for the large-scale photovoltaic-penetrated system is proposed. By designing a hierarchical distributed control strategy, the dual regulation of power and voltage is realized. Although the distributed consensus control strategy for air-conditioning load is designed in [12, 13], the control cost is not considered for the thermodynamic characteristics of air-conditioning load.

Table 1 gives a comparison of the abovementioned methods. It can be seen that none of them consider the whole-process cost of air conditioners. Since the indoor temperature is affected by the load control of air conditioning loads, it is necessary to investigate the whole-process cost according to the thermodynamic characteristics of air conditioning load. By this way, system operators can selectively regulate the air conditioning load with less negative impact, so as to minimize the impact of load control on user's comfort during the control process.

Table 1. Summary of the abovementioned methods.

Methods	Consensus algorithm?	For air conditioners?	Considering the whole-process cost of air conditioners?
[1–5]	×	√	×
[6–11]	√	×	×
[12, 13]	√	√	×

Aiming at the distributed control problem of air conditioning load, this paper proposes a multiagent consensus control method that considers the thermodynamic characteristics of the whole process of air-conditioning load which is proposed to minimize the cost function while achieving the control target, so as to reduce the impact on user's comfort. The contributions of this paper can be summarized as follows:

1. Establishing a whole-process cost function considering the thermodynamics of air conditioners
2. Applying the multiagent consensus control method to air conditioners in order to minimize the overall cost of the control

CONTROL STRATEGY OF THE AIR-CONDITIONING LOADS

Overall Framework of the Proposed Method

The framework of the proposed method is shown in Figure 1. Firstly, the equivalent thermal parameter (ETP) model of the air conditioners is established.

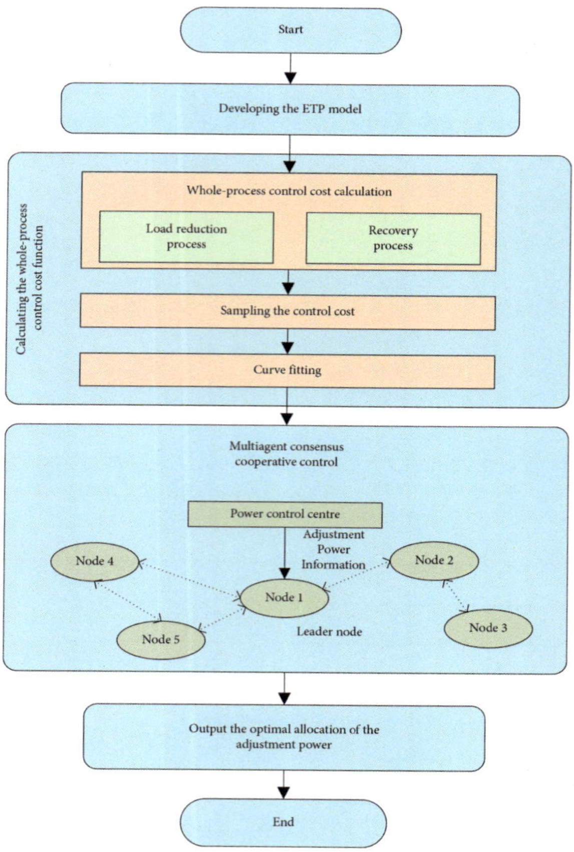

Figure 1. The overall flowchart of the proposed method.

Based on the ETP model, a whole-process control cost function of the air conditioner is defined according to the temperature deviation during the 2 processes, including the load reduction process and the recovery process. To simplify the cost function, the scatter plot of the control cost is sampled and the polynomial function is used to fit the scatter plot. Then, the multiagent consensus cooperative control algorithm is applied to the control problem, in order to get the optimal allocation of the adjustment power.

Thermodynamic Model of Air Conditioning Load

The thermodynamic model of a single air conditioner can be represented by an ETP model, as shown in Figure 2. It can be expressed as follows:

$$\dot{T}_a = -\frac{1}{RC}T_a + \frac{1}{RC}T_o \pm \frac{\eta}{C}P, \tag{1}$$

where T_a is the air temperature in the room, T_o is the outdoor temperature, R is the thermal resistance of the room, C is the heat capacity of the room, P is the electrical power of air conditioning load, and η is the thermal efficiency of air conditioning load. The plus and minus sign in front of η depends on whether the air conditioner is in the cooling or heating mode. The negative sign is taken when the air conditioner is in the cooling mode, and the positive sign is taken when the air conditioner is in the heating mode.

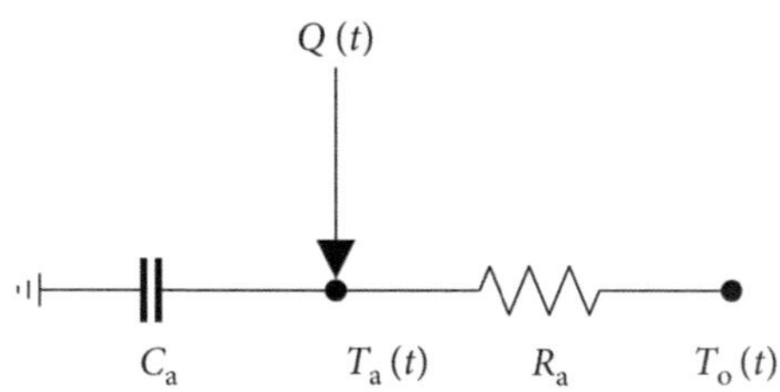

Figure 2. equivalent thermal parameter model of single air conditioner.

For fixed-frequency air conditioners, the power cannot be continuously adjusted. The power P of the air conditioner is a discrete variable, taking 0 or P_{max}. The power control is realized by duty cycle control, and its dynamic process is relatively complex. For inverter air conditioners, the power can be continuously adjusted within $[0, P_{max}]$. In this paper, a distributed control strategy is designed for the inverter air conditioner whose power can be adjusted continuously.

Establishment of Whole-Process Cost Function of Air Conditioning Load

Based on the equivalent thermal parameter model of air conditioning load, a cost function model considering the whole-process thermodynamic characteristics of air conditioning load is established through sampling and fitting. On the basis of the air conditioning parameters R, C, and η, the parameters indoor temperature T_a, outdoor temperature T_o, and initial air conditioning power P are initialized according to the actual scene. Under initial conditions, the indoor temperature T_a stabilizes around the initial temperature set point T_{set0}. When the load-curtailment event occurs, the air conditioning power P and the room temperature T_a will go through 2 processes, as shown in Figure 3:

i. Load reduction process: in order to respond to the control signal, the air conditioning load reduces the electrical load and the power has a step change of P_{dec}. As a result, the indoor temperature continues to rise.

ii. Recovery process: after the load reduction, the electric power of air conditioning load is increased to $P=P_{max}$ in order to restore the room temperature as soon as possible, until the temperature reduces to the initial temperature set point T_{set0}.

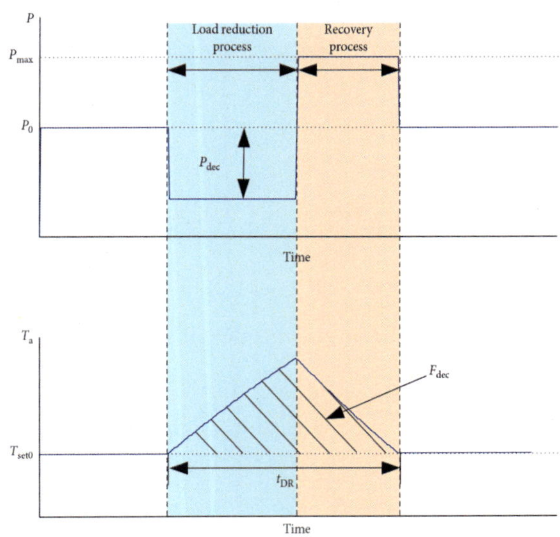

Figure 3. Example of air conditioning load control and reduction process and recovery process.

In the whole process, the indoor temperature rises first and then decreases, which has an effect on the comfort of users. The degree of influence is related to the amplitude and duration of temperature deviation. Therefore, the control cost F_{dec} of the air conditioning load can be defined as the square integral of the indoor temperature deviation $T_a - T_{\mathrm{set0}}$, which can be expressed as

$$F_{\mathrm{dec}} = \int_{t_0}^{t_0+t_{DR}} (T_a - T_{set0})^2 \mathrm{d}t, \tag{2}$$

where t_0 and t_{DR} are the start time of load reduction event and the duration of temperature deviation, respectively.

In order to establish the cost function of the air-conditioning load, the P_{dec}-F_{dec} scatter plot is formed by calculating F_{dec} under different P_{dec} conditions. A polynomial function is used to fit the P_{dec}-F_{dec} scatter plot, and the following fitting curve is obtained:

$$F_{\mathrm{dec}}(P_{\mathrm{dec}}) = a_m P_{\mathrm{dec}}^m + a_{m-1} P_{\mathrm{dec}}^{m-1} + \cdots + a_1 P_{\mathrm{dec}} + a_0, \tag{3}$$

where F_{dec} function is used as the basis of multiagent consensus control.

Multiagent Consensus Cooperative Control

Based on the cost function curve of each air conditioner, the multiagent consensus control strategy is adopted to realize the cooperative control of a large number of air conditioners, so as to achieve the goal of optimal total cost.

Suppose the cost function of the *i*th air conditioner can be expressed as

$$F_{\mathrm{dec}i}(P_{\mathrm{dec}i}) = a_m P_{\mathrm{dec}i}^m + a_{m-1} P_{\mathrm{dec}i}^{m-1} + \cdots + a_1 P_{\mathrm{dec}i} + a_0, \tag{4}$$

where $F_{\mathrm{dec}i}$ $(P_{\mathrm{dec}i})$ represents the cost of the *i*th air conditioner, $P_{\mathrm{dec}i}$ represents the power reduction of the *i*th air conditioner, and a_m–a_0 is the corresponding coefficient.

The goal of multiagent control is to minimize the total cost $F_{\mathrm{dec\_all}}$ which is expressed as follows:

$$\mathrm{Min}\, F_{\mathrm{dec\_all}} = \sum_i F_{\mathrm{dec}i}. \tag{5}$$

In the consensus algorithm, incremental cost is taken as the consensus variable, and the partial derivative of air conditioning cost to its load power is defined as the incremental cost of the load aggregator. For any load aggregator i, its incremental cost is

$$\mathrm{IC}_i = \frac{\partial F_{\mathrm{deci}}}{\partial P_{\mathrm{deci}}} = \lambda_i, \quad i = 1, 2, 3, \ldots, n. \tag{6}$$

For the nonleader node, the information update of air conditioning i is realized by the following formula:

$$\lambda_i(k+1) = \sum_{j=1}^{n} \delta_{ij}\lambda_j(k), \quad i = 1, 2, 3, \ldots, n, \tag{7}$$

where λ_i represents the state information (incremental cost) of air conditioner i, λ_j represents the state information (incremental cost) of the air conditioner adjacent to air conditioner i, n is the number of air conditioners, and δ_{ij} represents the element in the sparse iteration matrix and is defined by the following formula:

$$\delta ij = \frac{|l_{ij}|}{\sum_{j=1}^{n} l_{ij}}, \quad i = 1, 2, 3, \ldots, n, \tag{8}$$

where l_{ij} is the corresponding element of the Laplacian matrix.

For the leader node, the air conditioning load i needs to determine whether the reduced load power reaches a given value, and the power adjustment signal needs to be sent to other nodes through the leader node. Therefore, the information update of the leader node air-conditioning load i is realized by the following formula:

$$\lambda_i(k+1) = \sum_{j=1}^{n} \delta_{ij}\lambda_j(k) + \varepsilon\Delta P, \tag{9}$$

where ε is a positive coefficient called a convergence coefficient, which controls the convergence speed of the leader node, and ΔP is the total amount of power deviation, which can be expressed as the form of the deviation between the load reduction command and the actual load reduction power of the air conditioning loads:

$$\Delta P = P_{\mathrm{dec\_req}} - \sum_{i=1}^{n} P_{\mathrm{dec}i}, \tag{10}$$

where $P_{\mathrm{dec\_req}}$ is the power reduction command of all the air conditioners. Based on equations (4) and (6), the response power $P_{\mathrm{dec}i}$ of the air conditioner i can be obtained. The expression of P_{deci} depends on the order of the cost function. When the second-order cost function is used, P_{deci} is calculated by the following formula:

$$P_{deci} = \frac{\lambda_i - a_1}{2a_2}. \tag{11}$$

When the third-order cost function is used, P_{deci} can be calculated as

$$P_{deci} = \frac{-2a_2 + \sqrt{4a_2^2 - 12a_3(a_1 - \lambda_i)}}{6a_3}. \tag{12}$$

It is difficult to get the expression of P_{deci} directly when the cost function is a quartic function or a multiple function of more than four degrees.

Under the iteration of the above algorithm, the incremental cost λ_i and the power deviation ΔP tend to converge gradually. Finally, the incremental cost λ_i of all air conditioners reaches to consensus, and ΔP tends to zero.

TESTING RESULTS

The example considers a distributed air-conditioning control system of 6×4 in a building, as shown in Figure 4, with six floors and four occupants (air conditioning) on each floor and assumes that each air-conditioning controller can communicate with the air-conditioning controller of adjacent nodes (upstairs or next door) via a narrow-band Internet, which can form a 6×4 grid-shaped communication topology. Suppose that the node at the edge of the 1st floor is the leader node and can receive the regulation commands directly. The mean values of the considered air conditioners' parameters are listed in Table 2.

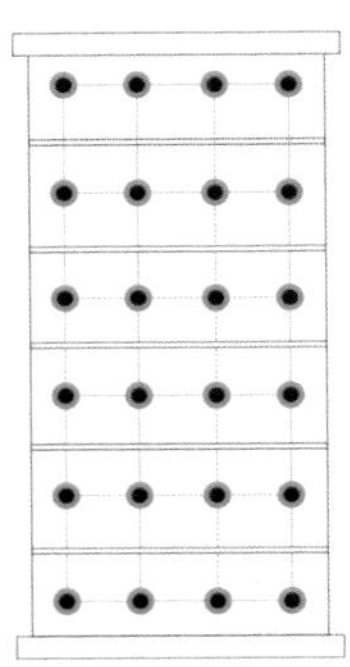

Figure 4. Communication topology of the air-conditioning controller.

Table 2. Selection of air-conditioning parameters.

Parameters	Mean value
R	2°C/kW
C	2 kWh/°C
η	2.5
T_o	32°C
T_{set0}	26°C

In order to implement distributed control, the cost function of single air conditioning load is established first. In order to determine the order of the polynomial fitting function, the P_{dec}–F_{dec} scatter plot and its fitting curves are shown in Figures 5(a) and 5(b). The integral square error (ISE) between the fitting curve and the P_{dec}–F_{dec} scatter are shown in Figure 5(c). It can be seen from Figure 5 that, with the increase of the order of polynomial fitting functions, the fitting error shows a downward trend. However, the increase of the order will lead to the complexity of the fitting function, which will increase the cost of parameter identification and distributed calculation.

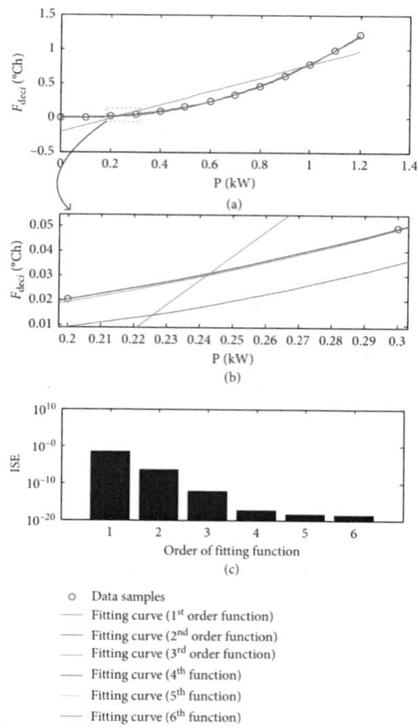

Figure 5. Fitting results of the cost F_{dec} with different orders. (a) Adjustment cost fitting curve with different orders of fitting function. (b) Adjustment cost

fitting curve with different orders of fitting function (zoomed in). (c) ISE with different orders of fitting function.

Using the third-order polynomial function to fit the cost of each air conditioner in the air-conditioning control system, the fitting function results of randomly selected 5 air conditioners are shown in Figure 6. It can be seen that the third-order polynomial function is able to describe the control cost of air conditioning load well.

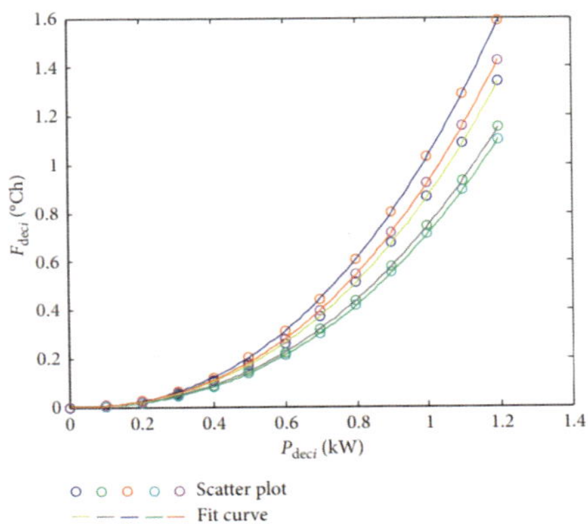

Figure 6. Scatter plot and fit curve of P_{dec}-F_{dec} of randomly selected 5 ACs.

Figure 7 shows the P_{dec}−F_{dec} scatter plot and its fitting curve when the outdoor temperature T_o is different. From Figure 7, it can be seen that, with the change of outdoor temperature T_o, the P_{dec}-F_{dec} scatter plot and its fitting curve are also different. The higher the outdoor temperature is, the larger the response capacity of air conditioning load can be provided, but the response cost is also increased. In addition, when T_o is different, the third-order polynomial function can fit P_{dec}−F_{dec} well, indicating that the fitting function has good robustness.

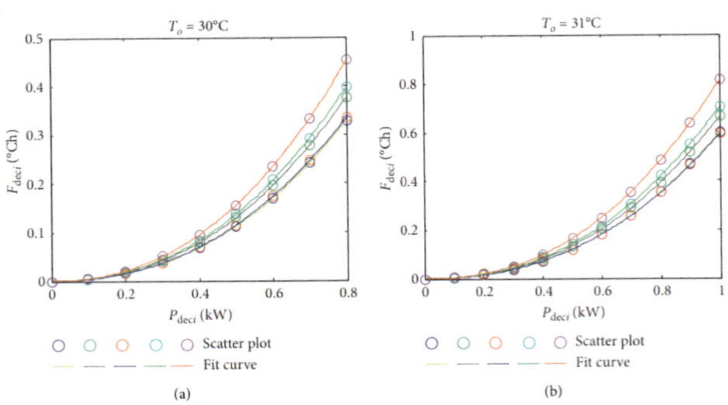

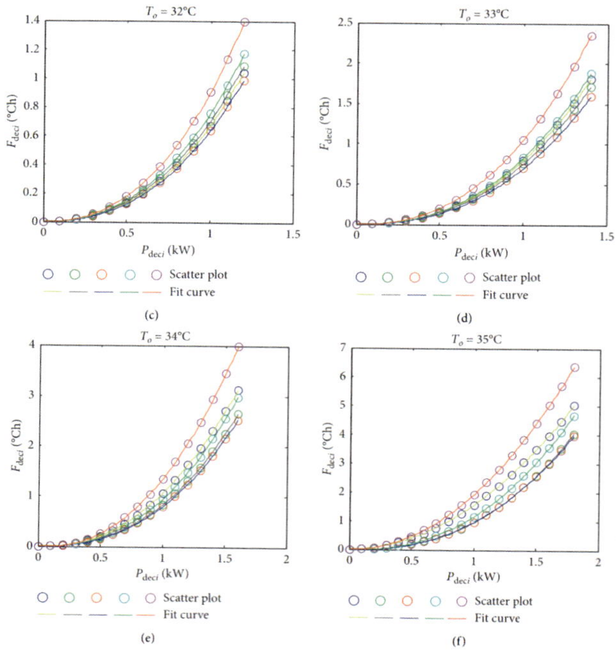

Figure 7. Scatter plot and fit curve of P_{dec}-F_{dec} with different outer temperatures.

Then, the effectiveness of the consensus control strategy in this paper is verified. In order to carry out comparative analysis, three distributed control methods are considered:

Method 1 (distributed consensus control strategy): load reduction power is selected as the consensus variable, and the average distribution of the adjustment power can be realized.

Method 2 (distributed consensus control strategy): the required proportion of load reduction is selected as the consensus variable, and the load reduction is in proportion to the original power consumption of the air conditioning load. This strategy follows the idea of the existing methods [13, 14].

Method 3 (the proposed method) (distributed consensus control strategy): the incremental cost of air conditioning load is selected as the consensus variable, and the optimal allocation of the adjustment power can be realized.

Consider the load-curtailment event with 28 kW power reduction command on a duration of one hour. The simulation results of methods 1~3 are shown in Figures 8–10, respectively.

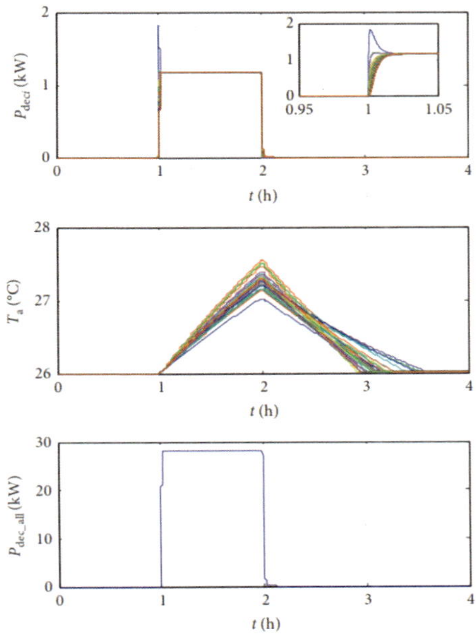

Figure 8. Simulation results of method 1.

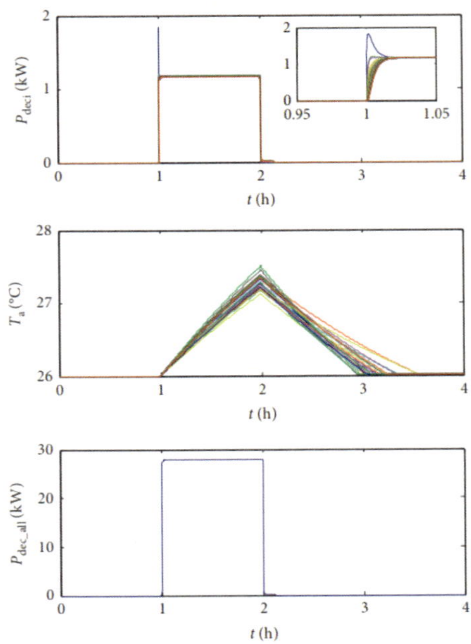

Figure 9. Simulation results of method 2.

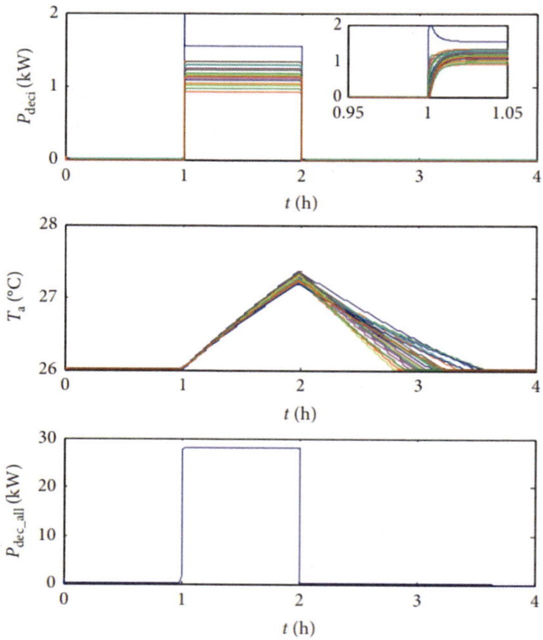

Figure 10. Simulation results of method 3.

Table 3 lists the total control cost F_{dec_all} of the three methods and integral absolute error (IAE) of power reduction error ΔP, which is expressed by the following formula:

$$\mathrm{IAE} = \int_{t_0}^{t_0+t_{DR}} \left(P_{\mathrm{dec\_req}} - \sum_{i=1}^{n} P_{\mathrm{deci}} \right) \mathrm{d}t. \tag{13}$$

Table 3. Comparison of performance from methods 1–3.

	F_{dec_all}	IAE
Method 1	29.2038	0.1120
Method 2	29.5383	0.1115
Method 3	28.5434	0.1047

IAE reflects the performance of control. The smaller the IAE is, the better control performance is.

It can be seen from the simulation results that the three control methods have achieved good results in tracking the load reduction signal and achieved the goal of reducing 28 kW power between 1 h and 2 h. The IAEs of the three methods are 0.1120, 0.1115, and 0.1047, respectively. However, from the aspect of control cost, method 1 and method 2 does not optimize the cost in the process of control (the cost is 29.2038 and 29.5383), while method 3 (the method in this paper) can achieve the optimal cost through the consensus of the incremental cost (the cost is 28.5434) which is significantly improved compared with method 1 and method 2, thus verifying the effectiveness of the proposed method.

Results and Discussion

The above testing results verify the effectiveness of the proposed method, including the cost function deriving and the cost function curve fitting, as well as the consensus control strategy. From the analysis of the cost function curve fitting, it can be seen that the third-order polynomial function can fit the cost function very well. By comparing the performance of the proposed method with existing methods, the proposed method exhibits better performance in minimizing the cost.

CONCLUSION

Considering the distributed characteristics and the inherent thermodynamic characteristics of air conditioning loads, this paper proposes a multiagent consensus control strategy considering the whole-process thermodynamic characteristics of air-conditioning loads.

The main findings of this paper can be summarized as follows:

(1) The whole thermodynamic characteristics of air conditioning loads includes two processes: the load reduction process and the recovery process, based on which the power-cost relationship can be established

(2) Third-order polynomial fitting can better reflect the power-cost relationship of air-conditioning load, and the incremental cost consensus control strategy can reflect the cost in the control process and therefore minimize the adjustment cost

Compared with the traditional method, the proposed method has the following advantages:

(1) The cost function can reflect the thermodynamic characteristics of the whole process (reduction process-recovery process) of air-conditioning loads

(2) The multiagent consensus control strategy based on incremental cost can realize the optimal power allocation of air-conditioning loads

Future work may develope a more complex online monitoring and control system for the air conditioners considering the time-varying parameters.

NOMENCLATURE

Ta	Air temperature in the room
T_o	Outdoor temperature
R	Thermal resistance of the room
C	Heat capacity of the room
P	Electrical power of air conditioning load
η	Thermal efficiency of air conditioning load
F_{deci}	Cost of the ith air conditioner
P_{deci}	Power reduction of the ith air conditioner
a_m-a_0	Coefficients of the cost function
λ_i	State information (incremental cost) of air conditioner i
δ_{ij}	Element in the sparse iteration matrix
l_{ij}	Element of the Laplacian matrix
ε	Convergence coefficient
P_{dec_req}	Power reduction command of all the air conditioners.

ACKNOWLEDGMENTS

This study was supported by Science and Technology Project of China Southern Power Grid Corporation (Research on key technologies of flexible load participation in regional power grid dispatching in megalopolis 090000KK52190230).

REFERENCES

1. Y. Bao, P. Chen, and M. Hu, "Control parameter optimization of thermostatically controlled loads using a modified state-queuing model," *CSEE Journal of Power and Energy Systems*, vol. 6, no. 2, pp. 394–401, 2020.
2. D. Yang, Z. Lin, and W. Lai, "Air conditioning load control strategy under the background of virtual power plant," *Power Demand Side Management*, vol. 22, no. 1, pp. 48–85, 2020.
3. B. Li, Z. Li, and B. Chen, "Air conditioning group control strategy of distribution network in power market," *Power System Automation*, vol. 43, no. 15, pp. 124–134, 2019.
4. Z. A. Shah, H. F. Sindi, A. Ul-Haq, and M. A. Ali, "Fuzzy logic-based direct load control scheme for air conditioning load to reduce energy consumption," *IEEE Access*, vol. 8, pp. 117413–117427, 2020.
5. B. Li, Z. Li, and B. Chen, "Load control strategy of central air conditioning in Intelligent Park Based on DFT [J]," *Power Grid Technology*, vol. 44, no. 7, pp. 2549–2559, 2020.
6. J. J. R. Liu, J. Lam, and K.-W. Kwok, "Further improvements on non-negative edge consensus of networked systems," *IEEE Transactions on Cybernetics*, vol. 9, no. 1, pp. 1–9, 2021.
7. J. J. R. Liu, J. Lam, and Z. Shu, "Positivity-preserving consensus of homogeneous multiagent systems," *IEEE Transactions on Automatic Control*, vol. 65, no. 6, pp. 2724–2729, 2020.
8. D. Yao, H. Li, R. Lu, and Y. Shi, "Event-triggered guaranteed cost leader-following consensus control of second-order nonlinear multiagent systems," *IEEE Transactions on Systems, Man, and Cybernetics: Systems*, vol. 10, no. 1, pp. 1–10, 2021.
9. J. Khazaei and Z. Miao, "Consensus control for energy storage systems," *IEEE Transactions on Smart Grid*, vol. 9, no. 4, pp. 3009–3017, 2018.
10. B. N. Alhasnawi, B. H. Jasim, and B. E. Sedhom, "Distributed secondary consensus fault tolerant control method for voltage and frequency restoration and power sharing control in multi-agent microgrid," *International Journal of Electrical Power & Energy Systems*, vol. 133, no. 2, Article ID 107251, 2021.

11. B. N. Alhasnawi, B. H. Jasim, B. E. Sedhom, E. Hossain, and J. M. Guerrero, "A new decentralized control strategy of microgrids in the internet of energy paradigm," *Energies*, vol. 14, no. 8, p. 2183, 2021.
12. B. Wang, T. Zhang, X. Hu, Y. Bao, and H. Su, "Consensus control strategy of an inverter air conditioning group for renewable energy integration based on the demand response," *IET Renewable Power Generation*, vol. 12, no. 14, pp. 1633–1639, 2018.
13. K. Meng, D. Wang, D. Wang et al., "Distributed control of thermostatically controlled loads in distribution network with high penetration of solar PV," *CSEE Journal of Power and Energy Systems*, vol. 3, no. 1, pp. 53–62, 2017.
14. M. Franceschelli, A. Pilloni, and A. Gasparri, "Multi-agent coordination of thermostatically controlled loads by smart power sockets for electric demand side management," *IEEE Transactions on Control Systems Technology*, vol. 29, no. 2, pp. 731–743, 2021.

CHAPTER 5

Process Control of Ball Mill Based on MPC-DO

Xisong Chen[1], Jiawei Yang[2,3,4,5], Zhijie Zhong[1], and Junyong Zhai[1]

[1]School of Automation, Southeast University, Nanjing 210096, China

[2]State Key Laboratory of Process Automation in Mining & Metallurgy, Beijing 102628, China

[3]Beijing Key Laboratory of Process Automation in Mining & Metallurgy, Beijing 102628, China

[4]Beijing General Research Institute of Mining and Metallurgy, Beijing 100160, China

[5]Northeastern University, Shenyang 110819, China

ABSTRACT

The grinding process of the ball mill is an essential operation in metallurgical concentration plants. Generally, the model of the process is established as

---

Citation: Xisong Chen, Jiawei Yang, Zhijie Zhong, Junyong Zhai, "Process Control of Ball Mill Based on MPC-DO", Mathematical Problems in Engineering, vol. 2021, Article ID 9994666, 14 pages, 2021. https://doi.org/10.1155/2021/9994666.

a multivariable system characterized with strong coupling and time delay. In previous research, a two-input-two-output model was applied to describe the system, in which some key indicators of the process were ignored. To this end, a three-input-three-output system is proposed to improve the model accuracy. Moreover, some practical and effective control strategies have been studied. The common control methods, including model predictive control (MPC), disturbance observer (DO), and so on, show poor performance when strong external and internal disturbances exist. In this paper, a composite control strategy based on MPC-DO is put forward to realize the control of the three-input-three-output ball mill system. The disturbances of the system consist of external disturbances including fluctuation of ore hardness and internal disturbances including model mismatches and strong couplings. The proposed MPC-DO controller includes a feedback control component based on MPC and a feed-forward compensation component based on DO. The simulation results indicate that the composite control scheme based on MPC-DO has good performance of tracking and anti-interference in process control of the ball mill.

INTRODUCTION

Grinding and classification process (GCP) is a key process in the beneficiation production process of the metallurgical industry. In general, the beneficiation process includes crushing (medium crushing and fine crushing), gravity separation, magnetic separation, GCP, flotation separation, concentration, and filtration.

Beneficiation refers to the removal of harmful elements contained in the raw ore to enrich the useful minerals and separate the various useful minerals from each other. The grade (percentage of useful mineral content) from underground mine is generally low and unable to be smelted directly. Therefore, ore beneficiation is required to remove most of the gangue and harmful components to obtain a high-grade concentrate. For example, if the grade of iron is increased by 1%, the yield of iron from blast furnace can be increased by 2.5%, and the coke ratio can be reduced by 1.5%. In addition, if the iron concentrate contains high content of sulfur and phosphorus, it will be harmful to the smelting process. So, the content of sulfur and phosphorus should be reduced during the beneficiation process. Obviously, beneficiation is of great significance to the development of mining industry and the full utilization of mineral symbiosis resources. GCP is the continuation of the crushing and separation process. The purpose of GCP is to make most of the useful components in the ore reach the monomer separation (micron level) through grinding and to avoid the phenomenon of overcrushing. After GCP, the particle size can meet the requirements of sorting operations (e.g., magnetic separation and magnetic flotation), creating conditions for the effective recovery of useful components in the ore [1–4].

Ball mill is a typical grinding equipment in the metallurgical concentration plants. Among all the operation elements in GCP, the grinding process of the ball mill is the most significant one, which consumes the most electricity of concentration plants. Product particle size of the ball mill affects the recovery rate of useful components in ore and tail disposal afterwards. Generally, the purpose of GCP is to achieve efficient control of product particle size. However, undesirable features (i.e., strong couplings between variables, disturbances, complicated dynamic characteristics, and time delays) always bring challenges to researchers [5]. In recent years, in addition to the traditional PID controller [6], a lot of advanced control strategies, such as MPC [7–12], adaptive control [11, 13, 14], neural network control [15], robust control [16, 17], optimal control [18], supervisory control [3], and expert control [19], have been put forward. MPC is one of the most popular approaches in the field of GCP. Since the model predictive heuristic control algorithm (MPHC) [20] was proposed in 1978, MPC has achieved great development and has derived a series of algorithms such as model algorithmic control (MAC) [21]. MPC is not only a purely theoretical product but also a practical control method.

The advantages of MPC are listed as follows [11]: (1) perfect capability of dealing with MIMO systems, (2) performing decoupling and handling time delays, and (3) simplicity of modeling.

In GCP, disturbance has a great impact on the performance of ball mill grinding circuits. Product particle size, circulating load, and mill solid concentration are always sensitive to different types of disturbances. The disturbances are separated into external disturbances and internal disturbances [22]. External disturbances contain variations of the properties of raw ore (e.g., hardness). Strong external disturbances may lead to continuous fluctuation of product particle size, circulating load, and mill solid concentration. Moreover, internal disturbances include model mismatches, unmodeled dynamics, and couplings. Internal disturbances are likely to contribute to bad dynamic performances and even unstable state of the whole system. It must be noted that the control strategies mentioned above show some limited abilities in GCP of the ball mill in the presence of strong disturbances as they are unable to handle disturbances directly through the design of controller. Although the controllers can resist disturbances slowly by feedback controller, the control schemes become inefficient when coming across strong disturbances. For the sake of improving the anti-interference performance of the ball mill, a feed-forward compensation component is introduced into controller design as a supplement to the traditional feedback controller. However, it is difficult to measure the disturbances directly through instruments. One possible solution is to research the technology of disturbance estimation. DO is an efficient way to estimate disturbances. DO was firstly raised by Ohishi et al. [23] in motion control. Control schemes based on DO have been promptly applied to other fields, such as robots [24, 25], hard disk drive (HDD) [26, 27], and so on. At present, DO has a wide range of applications. Pan [28] proposes a combination of adaptive robust control (ARC) and a novel terminal sliding-mode-based nonlinear disturbance observer (TSDO) to achieve robust output tracking for the saturated uncertain nonlinear systems, where the modeling inaccuracy and disturbance are integrated as a lumped disturbance. The reasons of successful application of DO lie in the following three advantages [11]: (1) the design of DO does not need to depend on accurate modeling of disturbance, (2) DO is able to be applied to deal with unmeasurable disturbance, and (3) DO does not have to distinguish between external and internal disturbances. As DO achieves complete suppression of disturbances, it has been used extensively in the design of feed-forward compensation.

For a long time, GCP control has been the research focus of mineral processing at home and abroad. PID control based on two-input-two-output model [22] is widely used in GCP. However, it has a problem that the accuracy of the model is not high enough. The purpose of this article is to design a controller to achieve effective control of GCP. In this paper, different from the traditional two-input-two-output model of the ball mill grinding process, a three-input-three-output system considering the influence of three factors (i.e., fresh ore feed, dilution water, and mill feed water) is put forward to improve the accuracy of modeling. Furthermore, a compound control strategy for process control of the ball mill is proposed to describe the grinding process precisely where disturbances exist. It consists of two types of controllers—a feedback regulation component based on MPC and a feed-forward compensation component based on DO. In comparison to other schemes, the proposed strategy considers both types of disturbances. As the MPC-DO controller inherits advantages of both MPC and DO, it shows great performance in the following three aspects: (1) good decoupling property, (2) ability to deal with processes with time delay, and (3) capability in dealing with strong disturbances.

GRINDING PROCESS OF BALL MILL

In this section, the characteristics of the ball mill grinding process are demonstrated. Then, the controlled variables and the manipulated variables are imposed, respectively. Finally, the relationships among all the variables and disturbances are introduced briefly.

GCP is a widespread process applied in metallurgical concentration plants for size reduction. Its role is releasing valuable components from raw ore [22, 29]. Ball mill is the typical equipment widely employed in GCP. The flow diagram of the ball mill system is illustrated in Figure 1, including the ball mill, pump sump, hydro-cyclones, and vibratory conveyor.

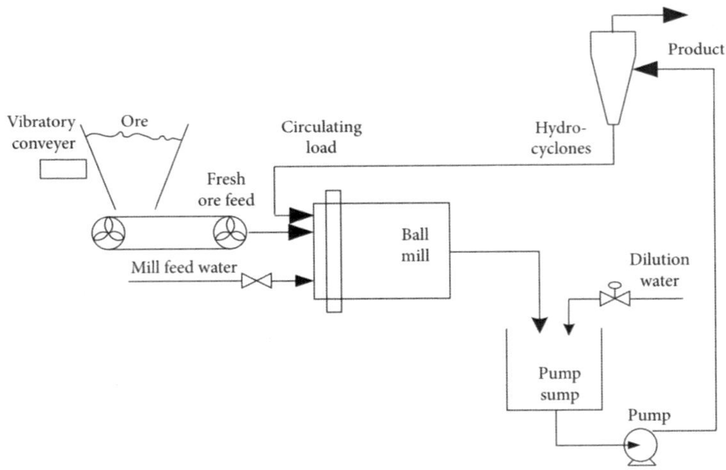

Figure 1. Flow diagram of the ball mill grinding process.

The system dynamics are described as follows. During the grinding process, the raw ore is firstly sent into the ball mill mixed with water. After that, the fresh raw ore is crushed to finer sizes by tumbling. The slurry with fine particles is then transferred to the pump sump. The dilution action adds water to the pump sump. Then, the slurry is pumped into the hydro-cyclones, which are used for further classification after dilution. After all the operations, the slurry is divided into an overflow stream comprising the substandard particles and an underflow stream comprising the substandard particles for recycling [22]. The fine particles, which meet the standard requirement of particle size, are the target product. The substandard particles are usually recycled back to the production line for regrinding and are selected as circulating load [3].

Product particle size y_1 (%) is regarded as the most significant controlled variable in the ball mill system. Raw ore has to be crushed to the specific size so that the fine particles are exposed in preparation for effective recovery during the subsequent technological processes. Besides, circulating load y_2 (t/h) is one of the controlled variables as it describes the amount of ore which is recycled to the production line. Moreover, mill solid concentration y_3 (%) is another controlled variable as the solid concentration must be controlled within a reasonable range. Generally, the scope of product particle size is 65%–75%, the scope of circulating load is 120 t/h–180 t/h, and the scope of mill solid concentration is 40% solids~60% solids. The upper limit of product particle size is set to prevent the excessively high particle size from causing the mill to be overloaded or the pipeline to be blocked. The upper limit of circulating load is set to prevent blockage of the slurry pipeline. The upper limit of mill solid concentration is set to ensure the relatively stable production volume. The lower limits of the controlled variables are set for economic considerations. During the grinding process, strong external disturbances may contribute to persistent fluctuations of product particle size in some cases. In this paper, we supplement the existing research results by rebuilding the model. The control objective is to realize the stability of controlled variables y_1, y_2, and y_3. The controlled variables can be adjusted by operating the manipulated variables, which are fresh ore feed rate u_1, dilution water feed rate u_2, and mill feed water feed rate u_3, respectively. The relationships among all the variables and disturbances in the ball mill system are introduced briefly in Figure 2.

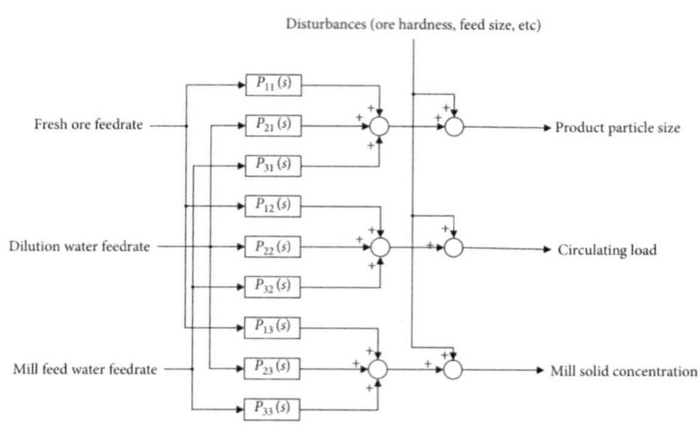

Figure 2. The relationships among the variables and disturbances.

These three controlled variables and three manipulated variables are completely determined by process analysis. The establishment of the model is based on process testing. Specifically, a step disturbance is applied to one of the manipulated variables near the equilibrium point in turn to observe the dynamic changes of the three controlled variables. By observing the dynamic changes, the change curves can be obtained, and the transfer function models can be obtained by curve fitting. The transfer function consists of a minimum-phase part and a time delay part.

DESIGN OF CONTROLLER

In this section, a compound strategy is put forward to deal with problems mentioned above in the ball mill grinding process. Principles of MPC and DO are firstly illustrated. After that, the MPC-DO controller is designed to realize the control objective.

Basic Introduction of MPC and DO

MPC algorithm has been applied extensively since it was proposed in 1980. In a system with m manipulated variables and p controlled variables, the quadratic performance objective function of the unconstrained MPC algorithm is listed as follows [11, 30]:

$$\min_{\Delta u_M} J(k) = [y_r(k) - \tilde{y}_P(k)]^T \mathbf{Q} [y_r(k) - \tilde{y}_P(k)] + [\Delta u_M(k)]^T \mathbf{R} [\Delta u_M(k)], \quad (1)$$

where $y_r(k)$ represents the setpoint of p controlled variables within P sampling intervals in the future, $\Delta u_M(k)$ stands for variations of m manipulated variables within M sampling intervals in the future, P and M denote the prediction horizon and control horizon, respectively, and $Q(P_p \times P_p)$ and $R(M_m \times M_m)$ are the error weighting matrix and input weighting matrix, respectively. *e model prediction values of outputs are represented as

where represents the setpoint of controlled variables within sampling intervals in the future, stands for variations of manipulated variables within sampling intervals in the future, and denote the prediction horizon and control horizon, respectively, and and are the error weighting matrix and input weighting matrix, respectively. The model prediction values of outputs are represented as

$$\tilde{\mathbf{y}}_N(k) = \tilde{\mathbf{y}}_0(k) + \mathbf{A}\Delta u_M(k), \quad (2)$$

where $\tilde{\mathbf{y}}_N(k)$ represents the model prediction vector and $\tilde{\mathbf{y}}_0(k)$ represents the initial predicted output vector without controller. A is a multivariable MPC dynamic matrix from the coefficients of unit step response. Because of the influence of measurement noise, the predicted outputs are usually imprecise and ought to be corrected by the real outputs. Construct e(k + 1) as the vector of predicted errors for the p controlled variables:

$$\begin{aligned}\mathbf{e}(k+1) &= \left[\mathbf{e}_1(k+1), \ldots, \mathbf{e}_p(k+1)\right]^T \\ &= \left[y_1(k+1) - \tilde{y}_{1,1}(k+1|k), \ldots, y_p(k+1) - \tilde{y}_{p,1}(k+1|k)\right]^T,\end{aligned} \tag{3}$$

where $\tilde{y}_{i,1}(k+1|k)\ (i = 1, \ldots, p)$ is the model prediction of p controlled variables at the (k + 1)th sampling interval. Therefore, the predicted outputs after correction can be listed as

$$\tilde{\mathbf{y}}_p(k) = \tilde{\mathbf{y}}_N(k) + \mathbf{He}(k+1), \tag{4}$$

where $\tilde{\mathbf{y}}_p(k)$ represents the predictive output after correction and H represents the corrected coefficient matrix. It can be solved from (1) that

$$\Delta u_M(k) = \left(\mathbf{A}^T\mathbf{QA} + \mathbf{R}\right)^{-1}\mathbf{AQ}\left[y_r(k) - \tilde{y}_p(k)\right]. \tag{5}$$

Generally, dynamic decoupling requires accurate modeling of process and brings challenges to industrial production. In comparison, static decoupling has more applications. However, static decoupling degrades the performance of control as it does not take the dynamic characteristics of process into consideration. MPC shows superb performance of decoupling. But it shows some limitations when strong disturbances exist. Therefore, DO is needed to compensate for disturbances.

In the field of motion control, DO has achieved many successful applications. DO can achieve better tracking and estimation of disturbance factors (external disturbance, model mismatching, and so on) that affect the system. This type of observer does not rely on precise disturbance analysis modeling and can organically combine observation results with controller design. Therefore, the state coupling between the channels that affects the performance of the system is regarded as the disturbance of each channel, and the corresponding disturbance observer for each channel is designed to suppress the internal disturbance and the external disturbance of the channel. In this paper, the technology of DO is introduced into the process control field. Combined with the design of feedback controller based on MPC, a new compound control scheme is put forward.

Controller Based on MPC-DO

MPC, as the feedback regulation part of the controller, has been well integrated by the Model Predictive Toolbox of Simulink. As a result, the focus of the design is DO.

*e structure of conventional DO is shown in Figure 3 [31]. In Figure 3(a), signals C(s), U(s), D_{ex}(s), N(s), and Y(s) are the command of the controller, the manipulated variable, the external disturbance, the measurement noise, and the controlled variable, respectively. Q(s) is the low-pass filter. Signals $\hat{D}(s)$ and $\hat{D}_f(s)$ are the estimates of the lumped disturbances before and after filtering. P(s) is the real object to be controlled, while G(s) is the model of the object. *e transfer function has the following form:

$$Y(s) = G_c(s)C(s) + G_d(s)D_{\text{ex}}(s) + G_n(s)N(s), \quad (6)$$

where

$$G_c(s) = \frac{P(s)G(s)}{G(s) + (P(s) - G(s))Q(s)},$$

$$G_d(s) = \frac{P(s)G(s)(1 - Q(s))}{G(s) + (P(s) - G(s))Q(s)},$$

$$G_n(s) = \frac{P(s)Q(s)}{G(s) + (P(s) - G(s))Q(s)}. \quad (7)$$

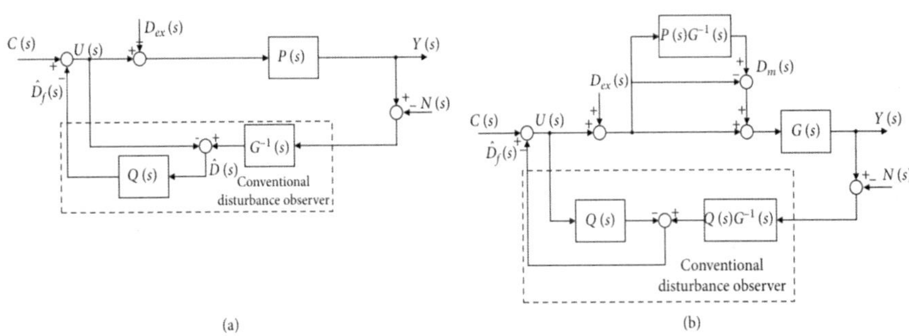

Figure 3. Block diagrams of conventional DO.

The design of DO is chiefly dependent on low-pass filter Q(s). Furthermore, as the anti-disturbance performance of the system is determined by time constants of Q(s), it is necessary to choose suitable time constants.

Case 1. If Q(s) ⟶ 1, then

$$G_c(s) \longrightarrow G(s),$$
$$G_d(s) \longrightarrow 0,$$
$$G_n(s) \longrightarrow 1. \quad (8)$$

Under this circumstance, the dynamic properties of the system resemble those of G(s).

Case 2. If Q(s) ⟶ 0, it can be deduced that

$$G_c(s) \longrightarrow P(s),$$
$$G_d(s) \longrightarrow P(s),$$
$$G_n(s) \longrightarrow 0. \quad (9)$$

Under the circumstance, the system is approximately open-loop. As a result, measurement noise is removed.

In low-frequency domain, Q(s) is approximately equal to 1, ensuring the estimated value of the lumped disturbance approach to the real value, which means the feed-forward compensation component can attenuate the influences of disturbances. Furthermore, in high-frequency domain, Q(s) approaches 0. It guarantees that the system is approximately open-loop and the high-frequency measurement noise can be removed by filtration.

In most cases, $G^{-1}(s)$ is physically unrealizable. In consideration of the internal disturbance $D_m(s)$ resulting from model mismatches, an equivalent conventional DO is illustrated in Figure 3(b). In order to ensure the realizability, $Q(s)G^{-1}(s)$ must be proper. *e relative degree of Q(s) ought to be no less than that of G(s).

Define D(s) as the lumped disturbances. It is obvious that D(s) include external disturbance $D_{ex}(s)$ and internal disturbance $D_m(s)$ resulting from model mismatches:

$$D(s) = D_{ex}(s) + D_m(s), \quad (10)$$

where $D_m(s) = [P(s)G^{-1}(s) - 1](U(s) + D_{ex}(s))$.

DO is able to deal with both types of disturbances without distinguishing. Therefore, the internal disturbance $D_m(s)$ aroused by model mismatching can be compensated.

In the process of grinding, as the inverse function of time delay part is physically unrealizable, conventional DO tends to be unavailable. Here modified disturbance observer [32] considering the time delay parts is used to deal with the problems mentioned above.

Take the two-input-two-output model for example:

$$\begin{bmatrix} Y_1(s) \\ Y_2(s) \end{bmatrix} = \begin{bmatrix} P_{11}(s) & P_{12}(s) \\ P_{21}(s) & P_{22}(s) \end{bmatrix} \begin{bmatrix} U_1(s) \\ U_2(s) \end{bmatrix} + \begin{bmatrix} D_{ex1}(s) \\ D_{ex2}(s) \end{bmatrix}, \quad (11)$$

where Y_i (s) (i = 1, 2) are the controlled variables, U_i (s) (i = 1,2) are the manipulated variables, P_{ij}(s)(i = 1,2;j = 1,2) are the models of process channels with P_{ij}(s) = $p_{ij}(s)e^{-\tau_{ij}s}\ (i=1,2;j=1,2)$, $p_{ij}(s)\ (i=1,2;j=1,2)$ are the minimum-phase parts of P_{ij}(s) and $e^{-\tau_{ij}s}\ (i=1,2;j=1,2)$ are the time delay parts of P_{ij}(s), D_{ex}(s) are the external disturbances with D_{exi}(s) = H_i(s)D_{ex}(s)(i = 1,2), D_{exi}(s) (i = 1,2) are the effects of D_{ex}(s) on Y_i(s), D_{exi}(s) = H_i(s)D_{ex}(s) (i = 1,2), and H_i(s)(i = 1,2) are the disturbance models.

The structure of modified DO for $u_1 - y_1$ loop is shown in Figure 4. The structure of $u_2 - y_2$ loop is similar to that of $u_1 - y_1$ loop. Obviously, the model P_{11}(s) is divided into two parts—a minimum-phase part p_{11} and a time delay part $e^{-\tau_{11}s}$:

$$P_{11}(s) = p_{11}(s)e^{-\tau_{11}s}. \quad (12)$$

The nominal model G_{11}(s) is

$$G_{11}(s) = g_{11}(s)e^{-\tau_{11n}s}. \quad (13)$$

It can be seen from Figure 4(a) that $e^{-\tau_{11}s}$ is inserted into the channel of U_1(s). Moreover, g_{11}(s) is inserted into the channel of Y_1(s) to substitute G(s).

In Figure 4(b), we take model mismatches into consideration. Obviously, D_1(s) consists of the external disturbance D_{ex1}(s) and the internal disturbance containing D_{m1}(s) resulting from model mismatching and D_{c1}(s) resulting from coupling, i.e.,

$$D_1(s) = D_{ex1}(s) + D_{m1}(s) + D_{c1}(s), \quad (14)$$

where $D_{m1}(s) = [p_{11}(s)e^{-\tau_{11}s} - g_{11}(s)e^{-\tau_{11n}s}]U_1(s)$, $D_{c1}(s) = p_{12}(s)e^{-\tau_{12}s}U_2(s)$.

From Figure 4(b), one obtains

$$Y_1(s) = g_{11}(s)e^{-\tau_{11n}s}U_1(s) + D_{m1}(s) + D_{ex1}(s) + D_{c1}(s). \quad (15)$$

Substituting (14) into (15) yields

$$Y_1(s) = g_{11}(s)e^{-\tau_{11n}s}U_1(s) + D_1(s). \tag{16}$$

If there is no sensor noise, i.e., $N_1(s) = 0$, it can be inferred from Figure 4(b) that

$$\hat{D}_{f1}(s) = Q(s)g_{11}^{-1}(s)Y_1(s) - Q(s)e^{-\tau_{11n}s}U_1(s). \tag{17}$$

Substituting (16) into (17) yields

$$\hat{D}_{f1}(s) = Q(s)g_{11}^{-1}(s)D_1(s). \tag{18}$$

Define $\tilde{D}(s)$ as the error between the real value and the estimated value of the lumped disturbance, i.e.,

$$\tilde{D}_1(s) = D_1(s) - g_{11}(s)e^{-\tau_{11n}s}\hat{D}_{f1}(s). \tag{19}$$

Substituting (18) into (19), one obtains

$$\tilde{D}_1(s) = [1 - Q(s)e^{-\tau_{11n}s}]D_1(s). \tag{20}$$

According to the final-value theorem, one has

$$\begin{aligned}\tilde{d}_1(\infty) &= \lim_{t\to\infty}\tilde{d}_1(t) = \lim_{s\to 0} s\tilde{D}_1(s)\\ &= \lim_{s\to 0}[1 - Q(s)e^{-\tau_{11n}s}]\lim_{s\to 0} sD_1(s)\\ &= \lim_{s\to 0}[1 - Q(s)e^{-\tau_{11n}s}]\lim_{t\to\infty} d_1(t)\\ &= \lim_{s\to 0}[1 - Q(s)e^{-\tau_{11n}s}]d_1(\infty).\end{aligned} \tag{21}$$

Obviously, if the steady-state gain of $Q(s)$ is 1, one obtains from (21) that

$$\tilde{d}_1(\infty) = 0. \tag{22}$$

Here $Q(s)$ is a first-order low-pass filter with the steady-state gain of 1, which has the following form:

$$Q(s) = \frac{1}{\lambda s + 1}. \tag{23}$$

Multiple DOs should be designed for the multi-input-multi-output system. The time constants of the filters in DOs must be debugged to compensate for different types of disturbances [33].

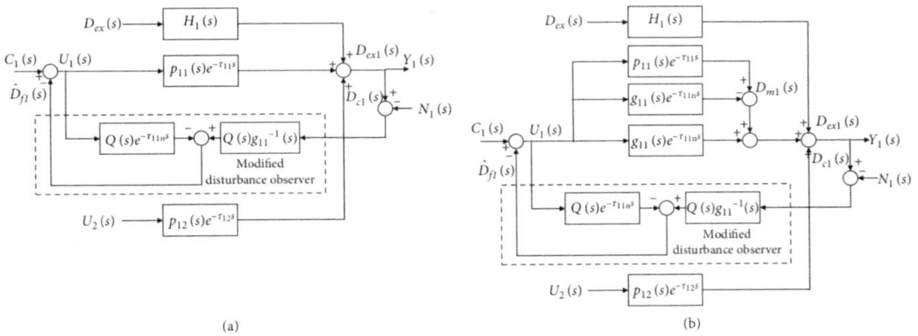

Figure 4. Block diagrams of modified DO.

SIMULATIONS

In this section, the three-input-three-output model is firstly introduced. Then, the schematic diagram of MPC-DO is illustrated. After that, simulation of tracking and anti-interference is given to verify the effectiveness of the proposed controller.

Model Description

As mentioned above, a two-input-two-output model [11] is generally applied to describe the grinding process of the ball mill. The manipulated variables of the second-order system are fresh ore feed rate u_1 (t/h) and dilution water feed rate u_2 (m^3/h), while the controlled variables are product particle size y_1 (%-200 mesh) and circulating load y_2 (t/h), respectively. However, this type of model cannot describe the system precisely as it does not take the factor of mill feed water rate into consideration. According to the analysis, the mill feed water rate ought to be another manipulated factor because it affects next series of operations. Moreover, on the basis of process description, mill solid concentration must be controlled within a reasonable range to avoid increasing the workload of subsequent operations. Therefore, it is another significant controlled variable.

In this paper, a three-input-three-output system is proposed to describe the grinding process of the ball mill. The manipulated variables are fresh ore feed rate u_1 (t/h), dilution water feed rate u_2 (m^3/h), and mill feed water feed rate u_3 (% solids), while the controlled variables are product particle size y_1 (%-200 mesh), circulating load y_2 (t/h), and mill solid concentration y_3 (% solids), respectively. The system is illustrated in Figure 5. Considering the standards of production, the scope of y_1 is 65%–75%, the scope of y_2 is 120 *t/h*–180 *t/h*, and the scope of y_3 is 40% solids~60% solids.

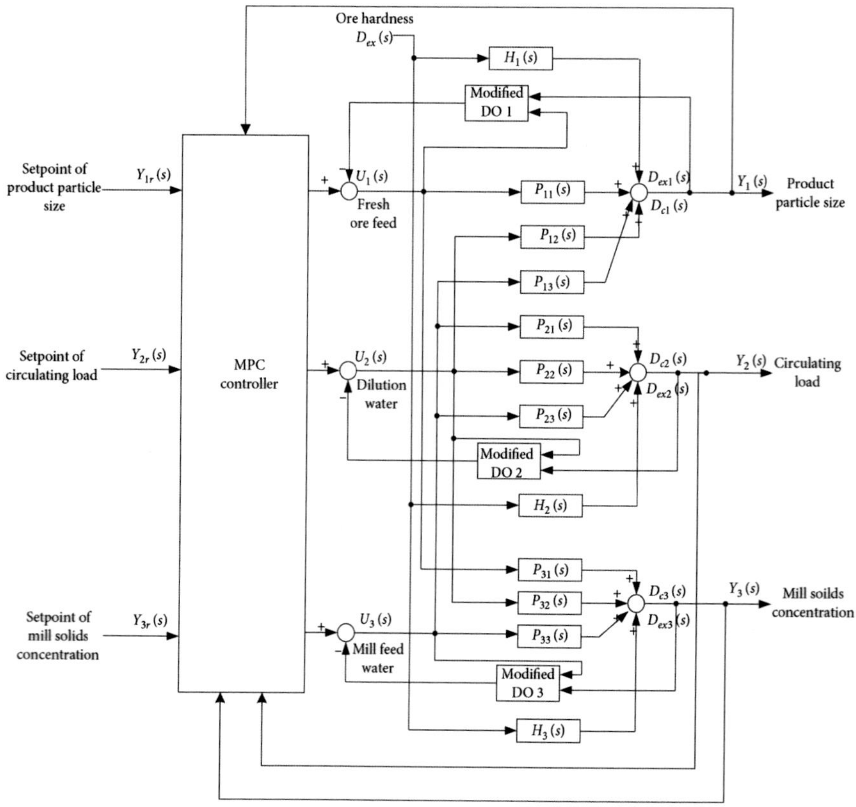

Figure 5. Schematic diagram of MPC-DO applied in the grinding process of ball mill.

Assume that the real grinding process of ball mill holds the following form:

$$\begin{bmatrix} Y_1(s) \\ Y_2(s) \\ Y_3(s) \end{bmatrix} = \begin{bmatrix} G_{11}(s) & G_{12}(s) & G_{13}(s) \\ G_{21}(s) & G_{22}(s) & G_{23}(s) \\ G_{31}(s) & G_{32}(s) & G_{33}(s) \end{bmatrix} \begin{bmatrix} U_1(s) \\ U_2(s) \\ U_3(s) \end{bmatrix}, \tag{24}$$

where

$$G_{11}(s) = g_{11}(s)e^{-\tau_{11n}s} = \frac{-2.75e^{-38s}}{89s+1},$$

$$G_{12}(s) = g_{12}(s)e^{-\tau_{12n}s} = \frac{1.85e^{-56s}\left(1-0.99e^{-260s}\right)}{(138s+1)(205s+1)},$$

$$G_{13}(s) = g_{13}(s)e^{-\tau_{13n}s} = \frac{8.2e^{-8s}\left(1-0.97e^{-220s}\right)}{(19s+1)(46s+1)},$$

$$G_{21}(s) = g_{21}(s) = \frac{1.54}{127s+1},$$

$$G_{22}(s) = g_{22}(s) = \frac{-3.23}{125s+1},$$

$$G_{23}(s) = g_{23}(s)e^{-\tau_{23n}s} = \frac{1.54e^{-84s}}{(83s+1)(187s+1)},$$

$$G_{31}(s) = g_{31}(s)e^{-\tau_{31n}s} = \frac{1.08e^{-48s}}{127s+1},$$

$$G_{32}(s) = g_{32}(s)e^{-\tau_{32n}s} = \frac{0.34e^{-94s}}{(127s+1)(264s+1)},$$

$$G_{33}(s) = g_{33}(s)e^{-\tau_{33n}s} = \frac{3.46e^{-7s}}{174s+1}. \quad (25)$$

It must be noted that the time constants are expressed in seconds. The establishment of the model is based on process testing. Specifically, a step disturbance is applied to one of the manipulated variables near the equilibrium point in turn to observe the dynamic changes of the three controlled variables. By observing the dynamic changes, the change curves can be obtained, and the transfer function models can be obtained by curve fitting. The transfer function consists of a minimum-phase part and a time delay part. The model holds the following form:

$$\begin{bmatrix} Y_1(s) \\ Y_2(s) \\ Y_3(s) \end{bmatrix} = \begin{bmatrix} P_{11}(s) & P_{12}(s) & P_{13}(s) \\ P_{21}(s) & P_{22}(s) & P_{23}(s) \\ P_{31}(s) & P_{32}(s) & P_{33}(s) \end{bmatrix} \begin{bmatrix} U_1(s) \\ U_2(s) \\ U_3(s) \end{bmatrix} + \begin{bmatrix} D_{ex1}(s) \\ D_{ex2}(s) \\ D_{ex3}(s) \end{bmatrix} + \begin{bmatrix} D_{m1}(s) \\ D_{m2}(s) \\ D_{m3}(s) \end{bmatrix} + \begin{bmatrix} D_{c1}(s) \\ D_{c2}(s) \\ D_{c3}(s) \end{bmatrix}, \quad (26)$$

where

$$P_{11}(s) = p_{11}(s)e^{-\tau_{11}s} = \frac{-2.85e^{-45s}}{95s+1},$$

$$P_{12}(s) = p_{12}(s)e^{-\tau_{12}s} = \frac{1.87e^{-60s}\left(1-1.09e^{-260s}\right)}{(140s+1)(210s+1)},$$

$$P_{13}(s) = p_{13}(s)e^{-\tau_{13}s} = \frac{8.32e^{-10s}\left(1-1.07e^{-220s}\right)}{(21s+1)(47s+1)},$$

$$P_{21}(s) = p_{21}(s) = \frac{1.57}{130s+1},$$

$$
\begin{aligned}
P_{22}(s) &= p_{22}(s) = \frac{-3.46}{130s+1}, \\
P_{23}(s) &= p_{23}(s)e^{-\tau_{23}s} = \frac{1.55e^{-90s}}{(85s+1)(190s+1)}, \\
P_{31}(s) &= p_{31}(s)e^{-\tau_{31}s} = \frac{1.14e^{-50s}}{140s+1}, \\
P_{32}(s) &= p_{32}(s)e^{-\tau_{32}s} = \frac{0.38e^{-100s}}{(130s+1)(270s+1)}, \\
P_{33}(s) &= p_{33}(s)e^{-\tau_{33}s} = \frac{3.59e^{-10s}}{180s+1}, \\
D_{ex1}(s) &= H_1(s)D_{ex}(s) = \frac{-6.4e^{-0.52s}}{4.8s+1}D_{ex}(s), \\
D_{ex2}(s) &= H_2(s)D_{ex}(s) = \frac{-8.6e^{-0.58s}}{5.2s+1}D_{ex}(s), \\
D_{ex3}(s) &= H_3(s)D_{ex}(s) = \frac{7.5e^{-0.46s}}{5.3s+1}D_{ex}(s), \\
D_{m1}(s) &= [p_{11}(s)e^{-\tau_{11}s} - g_{11}(s)e^{-\tau_{11n}s}]U_1(s), \\
D_{m2}(s) &= [p_{22}(s)e^{-\tau_{22}s} - g_{22}(s)e^{-\tau_{22n}s}]U_2(s), \\
D_{m3}(s) &= [p_{33}(s)e^{-\tau_{33}s} - g_{33}(s)e^{-\tau_{33n}s}]U_3(s), \\
D_{c1}(s) &= P_{12}(s)U_2(s) + P_{13}(s)U_3(s), \\
D_{c2}(s) &= P_{22}(s)U_2(s) + P_{23}(s)U_3(s), \\
D_{c3}(s) &= P_{32}(s)U_2(s) + P_{33}(s)U_3(s).
\end{aligned} \tag{27}
$$

Simulation of Tracking and Anti-Interference

The parameters set in the simulation are listed in Table 1.

Table 1. The parameters of the simulation.

Setpoint of u_1	Setpoint of u_2	Setpoint of u_3	Sampling time (s)
70%-200 mesh	150 t/h	50% solids	0.1
λ_1 of DO_1	λ_2 of DO_2	λ_3 of DO_3	—
1	0.8	0.7	—

Firstly, response curves of controlled variables and manipulated variables for tracking setpoint are listed as follows. The setpoints of u_1, u_2, and u_3 are selected as 70%-200 mesh, 150 *t/h*, and 50% solids, respectively. The whole period is selected as 500 seconds.

Figures 6(a)–6(c) illustrate the controlled variables during the process of ball mill grinding circuits. Furthermore, Figures 6(d)–6(f) show the variations of the manipulated variables.

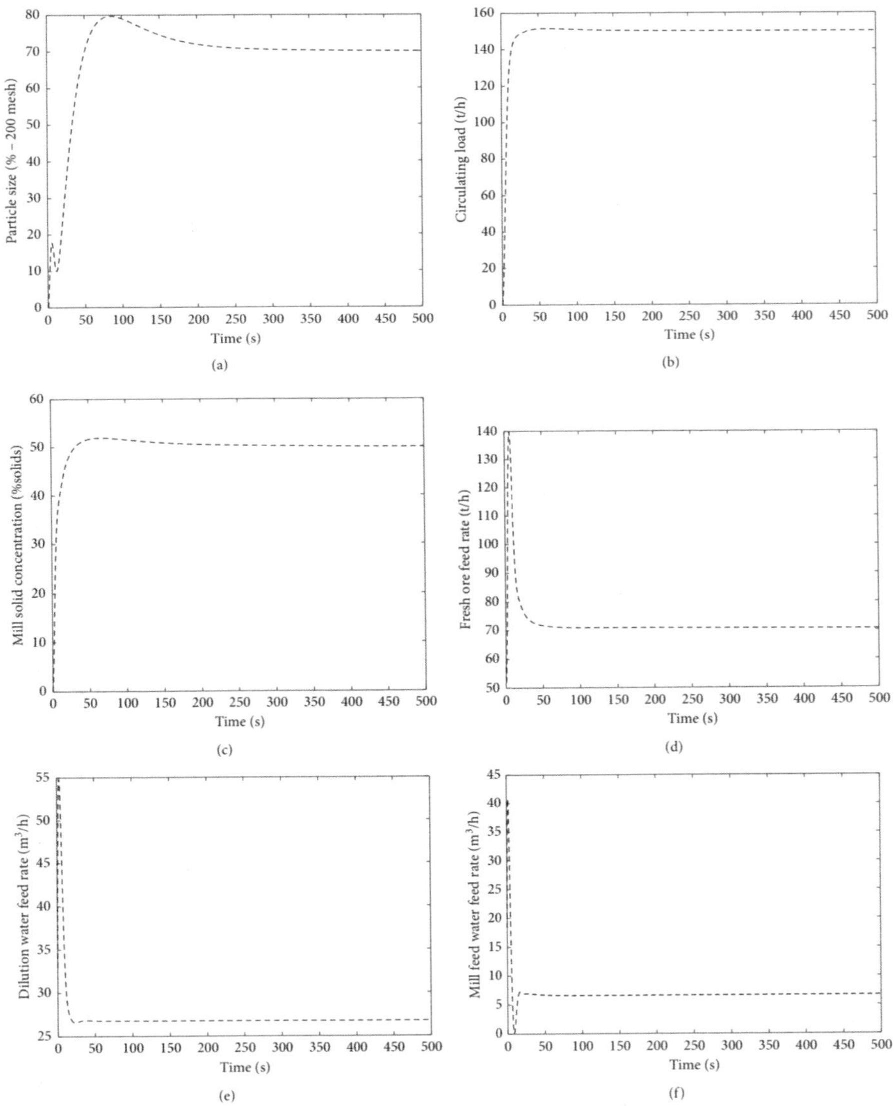

Figure 6. Response curves of controlled variables and manipulated variables for tracking setpoint: (a) product particle size; (b) circulating load; (c) mill solid concentration; (d) fresh ore feed rate; (e) dilution water feed rate; (f) mill feed water feed rate.

After that, a step signal is imposed as a strong external disturbance to verify the performance of anti-interference under the compound control strategy. The results are shown in Figure 7.

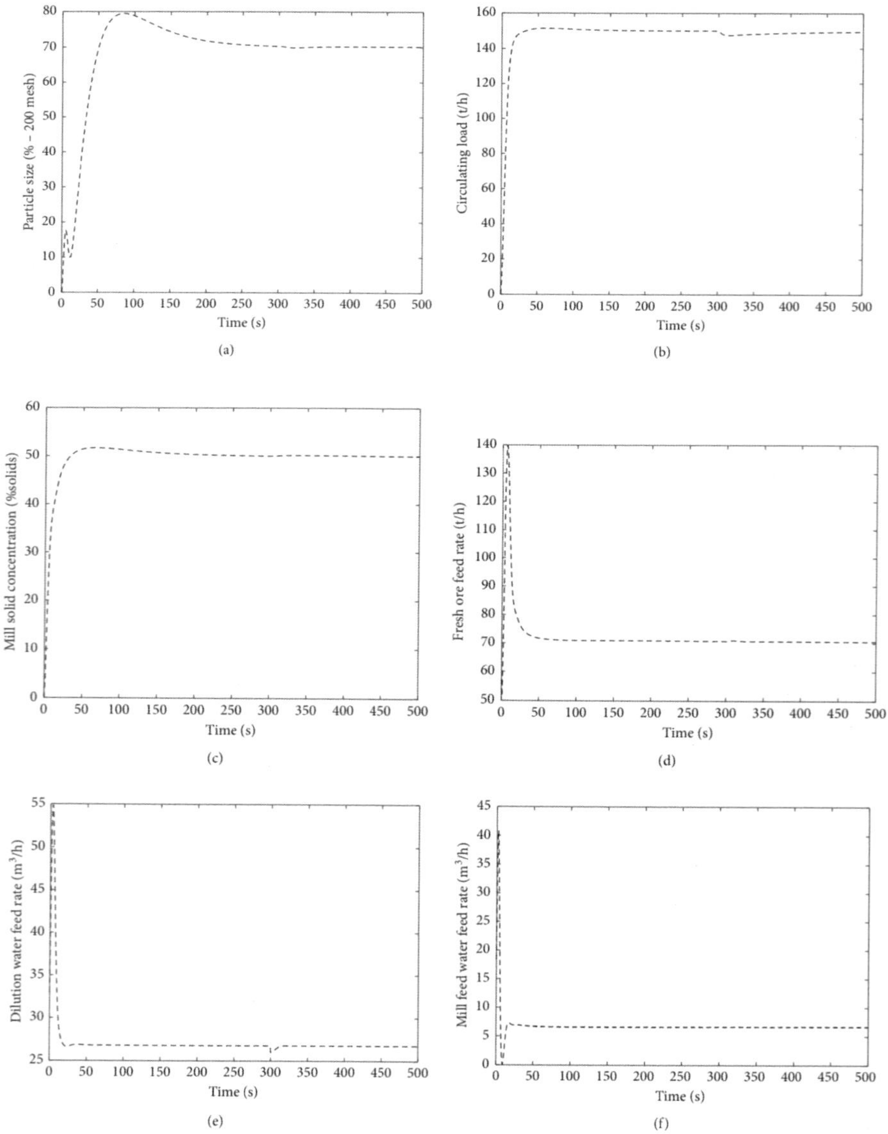

Figure 7. Response curves of controlled variables and manipulated variables in the presence of strong external disturbance: (a) product particle size; (b) circulating load; (c) mill solid concentration; (d) fresh ore feed rate; (e) dilution water feed rate; (f) mill feed water feed rate.

It can be seen from simulation that the system stabilizes after 25 s under the influence of manipulated variables. The composite control strategy based on MPC-DO shows a perfect anti-disturbance performance, achieving the goal of controller design.

To make a comparison between the proposed controller and MPC controller, the MPC controller is applied to the three-input-three-output system. The tracking performance is shown in Figure 8.

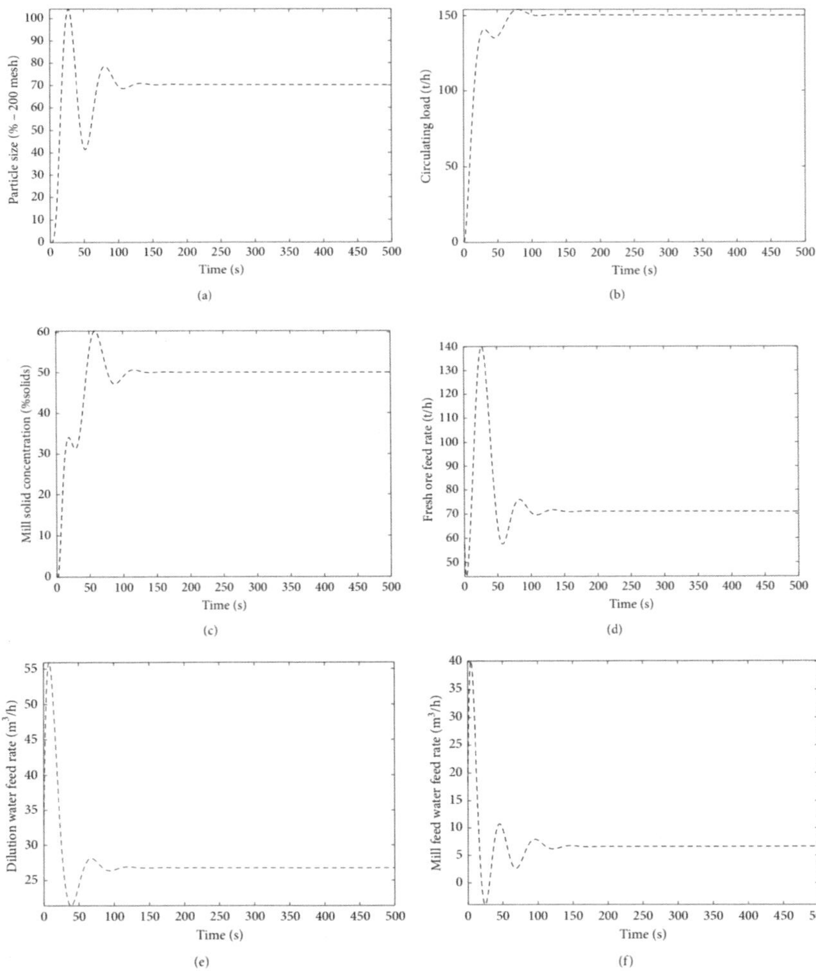

Figure 8. Response curves of controlled variables and manipulated variables for tracking setpoint: (a) product particle size; (b) circulating load; (c) mill solid concentration; (d) fresh ore feed rate; (e) dilution water feed rate; (f) mill feed water feed rate.

The performance indexes, including overshoot, peak time, and steady-state time, are employed here to estimate the control performance of setpoint tracking. Comparisons about the performance of MPC-DO controller and MPC controller are listed in Tables 2–4.

Table 2. The performance indexes of y_1.

Controller	Overshoot (%)	Peak time (s)	Steady-state time (s)
MPC-DO	13.52	84	335
MPC	45	104	320

Table 3. The performance indexes of y_2.

Controller	Overshoot (%)	Peak time (s)	Steady-state time (s)
MPC-DO	1	54	208
MPC	2.67	79	262

Table 4. The performance indexes of y_3.

Controller	Overshoot (%)	Peak time (s)	Steady-state time (s)
MPC-DO	3.4	51.7	298.5
MPC	20.66	58	228

From the performance index analysis above, it can be concluded that the MPC controller has the disadvantage of high overshoot when strong disturbances exist, especially product particle size y_1, which reaches 45%. It is likely to bring negative impact to the production process. With the assistance of MPC-DO controller, the problem of excessive overshoot is solved. It indicates that the proposed MPC-DO controller has excellent performance.

CONCLUSION

In GCP, complicated disturbances have negative effects on the ball mill grinding system. A three-input-three-output model is introduced in the paper to deal with the problem that traditional two-input-two-output model cannot describe the characteristics of the system precisely. Then, a compound control strategy including a feedback regulation component based on MPC and a feed-forward compensation component based on DO is proposed in order to improve the performance of anti-disturbance. Except for external disturbances, internal disturbances resulting from model mismatches are viewed as a part of the lumped disturbance. Simulation results demonstrate that the composite control strategy based on MPC-DO has achieved remarkable performance when applied to a three-input-three-output model of grinding process. In the follow-up research, advanced control methods ought to be studied further to meet the requirements of industrial production. Furthermore, the influence of other factors, such as the level of pump sump, the running speed of pump, and so on, ought to be studied to extend the model of the system to higher dimensions.

ACKNOWLEDGMENTS

This study was funded by the Open Foundation of State Key Laboratory of Process Automation in Mining & Metallurgy (BGRIMN-KZSKL-2020-11).

REFERENCES

1. L. Lv, Z. Deng, T. Liu, Z. Li, and W. Liu, "Intelligent technology in grinding process driven by data: a review," *Journal of Manufacturing Processes*, vol. 58, pp. 1039–1051, 2020.
2. X. Wang, Y. Wang, C. Yang, D. Xu, and W. Gui, "Hybrid modeling of an industrial grinding-classification process," *Powder Technology*, vol. 279, pp. 75–85, 2015.
3. X.-S. Chen, Q. Li, and S.-M. Fei, "Supervisory expert control for ball mill grinding circuits," *Expert Systems with Applications*, vol. 34, no. 3, pp. 1877–1885, 2008.
4. V. Chipakwe, P. Semsari, T. Karlkvist, J. Rosenkranz, and S. C. Chelgani, "A critical review on the mechanisms of chemical additives used in grinding and their effects on the downstream processes," *Journal of Materials Research and Technology*, vol. 9, no. 4, pp. 8148–8162, 2020.
5. J. Yu, Y. Qin, P. Gao, Y. Han, and Y. Li, "An innovative approach for determining the grinding media system of ball mill based on grinding kinetics and linear superposition principle," *Powder Technology*, vol. 378, pp. 172–181, 2021.
6. A. Pomerleau, D. Hodouin, A. Desbiens, and É. Gagnon, "A survey of grinding circuit control methods: from decentralized PID controllers to multivariable predictive controllers," *Powder Technology*, vol. 108, no. 2-3, pp. 103–115, 2000.
7. A. J. Niemi, L. Tian, and R. Ylinen, "Model predictive control for grinding systems," *Control Engineering Practice*, vol. 5, no. 2, pp. 271–278, 1997.
8. M. Ramasamy, S. S. Narayanan, and C. D. P. Rao, "Control of ball mill grinding circuit using model predictive control scheme," *Journal of Process Control*, vol. 15, no. 3, pp. 273–283, 2005.
9. X.-S. Chen, J.-Y. Zhai, S.-H. Li, and Q. Li, "Application of model predictive control in ball mill grinding circuit," *Minerals Engineering*, vol. 20, no. 11, pp. 1099–1108, 2007.
10. X.-S. Chen, Q. Li, and S.-M. Fei, "Constrained model predictive control in ball mill grinding process," *Powder Technology*, vol. 186, no. 1, pp. 31–39, 2008.
11. J. Yang, S. Li, X. Chen, and Q. Li, "Disturbance rejection of ball mill grinding circuits using DOB and MPC," *Powder Technology*, vol. 198,

no. 2, pp. 219–228, 2010.

12. A. Cortinovis, H. J. Ferreau, D. Lewandowski, and M. Mercangöz, "Experimental evaluation of MPC-based anti-surge and process control for electric driven centrifugal gas compressors," *Journal of Process Control*, vol. 34, pp. 13–25, 2015.
13. K. Najim, D. Hodouin, and A. Desbiens, "Adaptive control: state of the art and an application to a grinding process," *Powder Technology*, vol. 82, no. 1, pp. 59–68, 1995.
14. L. P. Xin, B. Yu, and L. Zhao, "Adaptive fuzzy backstepping control for a two continuous stirred tank reactors process based on dynamic surface control approach," *Applied Mathematics and Computation*, vol. 377, pp. 125–138, 2020.
15. J. J. Govindhasamy, S. F. McLoone, G. W. Irwin, J. J. French, and R. P. Doyle, "Neural modelling, control and optimisation of an industrial grinding process," *Control Engineering Practice*, vol. 13, no. 10, pp. 1243–1258, 2005.
16. O. Galán, G. W. Barton, and J. A. Romagnoli, "Robust control of a SAG mill," *Powder Technology*, vol. 124, no. 3, pp. 264–271, 2002.
17. É. M. Arruda, A. P. de Paiva, L. C. Brandão, and J. R. Ferreira, "Robust optimisation of surface roughness of AISI H13 hardened steel in the finishing milling using ball nose end mills," *Precision Engineering*, vol. 60, pp. 194–214, 2019.
18. H. Fujii, A. Iwanade, S. Kawada, and H. Kitaguchi, "The reduction of optimal heat treatment temperature and critical current density enhancement of ex situ processed MgB2 tapes using ball milled filling powder," *Cryogenics*, vol. 89, pp. 76–84, 2018.
19. X.-S. Chen, S.-H. Li, J.-Y. Zhai, and Q. Li, "Expert system based adaptive dynamic matrix control for ball mill grinding circuit," *Expert Systems with Applications*, vol. 36, no. 1, pp. 716–723, 2009.
20. J. Richalet, A. Rault, J. L. Testud, and J. Papon, "Model predictive heuristic control," *Automatica*, vol. 14, no. 5, pp. 413–428, 1978.
21. R. Rouhani and R. K. Mehra, "Model algorithmic control (MAC); basic theoretical properties," *Automatica*, vol. 18, no. 4, pp. 401–414, 1982.
22. X. S. Chen, J. Yang, and S. H. Li, "Disturbance observer based multivariable control of ball mill grinding circuits," *Journal of Process Control*, vol. 19, pp. 1205–1213, 2019.

23. K. Ohishi, M. Nakao, K. Ohnishi, and K. Miyachi, "Microprocessor-controlled DC motor for load-insensitive position servo system," *IEEE Transactions on Industrial Electronics*, vol. 34, no. 1, pp. 44–49, 1987.
24. S. Katsura, Y. Matsumoto, and K. Ohnishi, "Modeling of force sensing and validation of disturbance observer for force control," *IEEE Transactions on Industrial Electronics*, vol. 54, no. 1, pp. 530–538, 2007.
25. S. Ueki, T. Mouri, and H. Kawasaki, "Nonlinear disturbance observer for object grasping/manipulation by multi-fingered robot hand," *IFAC-PapersOnLine*, vol. 50, no. 1, pp. 12704–12709, 2017.
26. J. Ishikawa and M. Tomizuka, "A novel add-on compensator for cancellation of pivot nonlinearities in hard disk drives," *IEEE Transactions on Magnetics*, vol. 34, no. 4, pp. 1895–1897, 1998.
27. Y. Luo, T. Zhang, B. Lee, C. Kang, and Y. Chen, "Disturbance observer design with Bode's ideal cut-off filter in hard-disc-drive servo system," *Mechatronics*, vol. 23, no. 7, pp. 856–862, 2013.
28. H. Pan, W. Sun, H. Gao, and X. Jing, "Disturbance observer-based adaptive tracking control with actuator saturation and its application," *IEEE Transactions on Automation Science and Engineering*, vol. 13, no. 2, pp. 1–8, 2016.
29. D. Hodouin, S.-L. Jämsä-Jounela, M. T. Carvalho, and L. Bergh, "State of the art and challenges in mineral processing control," *Control Engineering Practice*, vol. 9, no. 9, pp. 995–1005, 2001.
30. D. Li, H. Cao, X. Zhang, X. Chen, and R. Yan, "Model predictive control based active chatter control in milling process," *Mechanical Systems and Signal Processing*, vol. 128, pp. 266–281, 2019.
31. W. R. Wang, S. X. Pan, and F. Wang, "A comparison study of two disturbance rejection control strategies for hydraulic position servo systems," in *Procceedings of the 30th Annual Conference of the IEEE Industrial Electronics Society, 2004 (IECON 2004)*, pp. 1993–1996, Busan, Republic of Korea, November 2004.
32. H. D. Zhu, G. H. Zhang, and H. H. Shao, "Control of the process with inverse response and dead-time based on disturbance observer," in *Procceedings of the 2005 American Control Conference*, pp. 4826–4831, Portland, OR, USA, June 2005.
33. T. Umeno and Y. Hori, "Robust speed control of DC servomotors using modern two degrees-of-freedom controller design," *IEEE Transactions on Industrial Electronics*, vol. 38, no. 5, pp. 363–368, 1991.

SECTION 2
PROCESS CONTROL AND AUTOMATION

CHAPTER 6

A Control Method for SMA Robotic Actuators

Tian Lyu

School of Mechatronics Engineering, Beijing Institute of Technology, Beijing, China.

ABSTRACT

Soft robots show remarkable benefits over conventional rigid robots due to their high energy density and other factors. We propose a circular soft robot integrating the control system, actuating it with several shape memory alloy (SMA) actuators. Our research methodology involved connecting the DC voltage supply and L298N module to provide uninterrupted power to the actuator and thence control the actuator via Arduino Uno MCU and TOF camera. We designed the controller to simultaneously complete the positioning

Citation: Lyu, T. (2022), "A Control Method for SMA Robotic Actuators". Journal of Computer and Communications, 10, 103-112. doi: 10.4236/jcc.2022.105007.

and manipulation tasks. A novel method utilizing visual servo and closed-loop control algorithm was proposed and integrated into the controller. This method involves the implementation of multi-gait locomotion using SMA actuators. Additionally, the development of closed-loop dynamic controllers for a continuous soft robot is also evaluated. The proposed control model is designed and simulated on the MATLAB tool. To verify the efficiency of the proposed forward-feedback controller, simulations and experiments were conducted in the current study. A new control method using PID control based on the Kalman filtering algorithm and visual servo for the SMA actuator designed in this research is introduced. We conclude that applying spike excitation voltage would benefit the actuating performance. Overall, the experimental results demonstrated a promising future for the purposed control method.

Keywords: Soft Robotics, Shape Memory Alloys, Visual Servo, PID Control

INTRODUCTION

Besides rigid robotics, soft robotics has shown greater efficiency in adapting to complex environments [1] [2] [3]. As soft robotics could be regarded as a complex rigid robot with infinite joints [4], it revealed a lack of appropriate control strategies, thus creating obstacles to the precise control of soft robotics [5]. Shape Memory Alloys (SMAs), as one of the novel materials, though found suitable for actuators of soft robotics, have a limitation of being short of control strategies [1] [2] [3] [6].

Fei *et al.* proposed a closed-loop control method embedding four curvature sensors as the feedback signals [1] [3]. However, their strategy involved a meticulous forging of the steel shell and affixing of the sensors, which increases the complexity of the system and scale. Cheng *et al.* developed an SMA-based soft robot, with a motion processing unit (MPU) for sending back the angular and velocity data [2]. Nevertheless, control methods based on integrated sensor feedback are feasible to operate larger-scale robots but not the centimetre-scale robot in this study. Consequently, a suitable control strategy for centimeter-scale soft robots is required. More measures can be found in controlling soft robots. Fan *et al.* applied Visual Servo to a type of soft robot and concluded this method's active regulation as being more efficient [7]. Instead of attaching peripheral sensors to the small-scale soft robots, the method of using a camera set to capture the position changes of the coiled robots should be considered. Nuchkrua T *et al.* proposed a contouring control of the robot, focusing on the error of the robot's end-factor [8]. This could be achieved by detecting the end terminals of the actuators via computer vision. Hence, it was implemented in the current research.

Therefore, a control system with a visual servo as the input signal and an integrated Proportional Integral Derivative (PID) and Kalman filter as the control algorithm is proposed.

DESIGN AND SIMULATION OF THE CONTROL SYSTEM

This control system aims to achieve the maximum position that an actuator can reach. Since the soft robot designed in this research gains mobility by extending its actuators to acquire friction force and reaction force, the position or deformation control of the actuators enables accurate control of this centimeter-scale robot.

To better simulate or construct a control system, a sufficient amount of data was obtained by recording the performance of a single actuator under different parameters. It showed the relationship among the diameter of the SMA coil wire, excitation voltage to generate the deformation of the SMAs, pulse width for excitation, pulse width for restoration, length of SMA coils, etc.

An optimized selection of the parameters is chosen. By altering its excitation pulse width duration, the maximum position of the actuator could be controlled. Afterward, a scattered map in relation to maximum deformation and excitation time is drawn using the data obtained. The Fourier fitting method is applied to obtain the input-output curve of the system [9].

Hence, the function of the curve can be determined using the following Equation (1):

$$O(t) = 37.31 - 8.766\cos(0.0427t) - 1.84\sin(0.0427t) \tag{1}$$

where, $O(t)$ denotes the deformation function of the actuator and signifies the duration of the excitation pulse. The graph shows a partial nonlinearity since its nonlinearity happens only in extreme conditions when excitation time is either too short or long. So, in a general control, we could still apply traditional control methods to this slightly nonlinear system.

Using Equation (1), we can obtain the transfer function in the frequency domain of the control system as:

$$TF(s) = \frac{133}{s^2 + 25s} \tag{2}$$

PID is commonly used in various linear control situations due to the deviation of the initial state and the uncertainty of the model, appropriately induced noise is conducive to a faster approximation to the real state. During the control process, noises or bias can be induced by several factors like manual measurement error, camera measurement error, etc. So, a PID control based on the Kalman filtering algorithm is suitable for the filtration of the controlled noise and the measuring noise [10].

Figure 1 shows the block diagram of the PID control based on the Kalman filtering algorithm control system. By cascading the two major control algorithms, we can obtain the following Equations (3-6):

From PID control [11], we can obtain:

$$u = k_p\left[e(t) + k_i\int_0^t e(t)\,\mathrm{d}t + k_d\frac{\mathrm{d}e(t)}{\mathrm{d}t}\right] \tag{3}$$

where, u is the output of the PID block.

From the Kalman filter, we have:

$$x_{k+1} = Ax_k + Bu_k \tag{4}$$

$$P_{k+1} = AP_kA^{\mathrm{T}} + Q \tag{5}$$

$$K_k = P_kH^{\mathrm{T}}\left(HP_kH^{\mathrm{T}} + R\right)^{-1} \tag{6}$$

$$x_k = x_k + K_k\left(z_k - Hx_k\right) \tag{7}$$

$$P_k = \left(I - K_kH\right)P_k \tag{8}$$

where, R and Q indicate the control error and measured error, respectively, in the diagram. Further, P_k denotes the anticipated actual position at present.

For y_d in diagram 2, we can predefine the demanded gait for the actuator. y in the diagram shows the data received from the camera and y_e is the anticipated actual gait same as the output value of Equation (8).

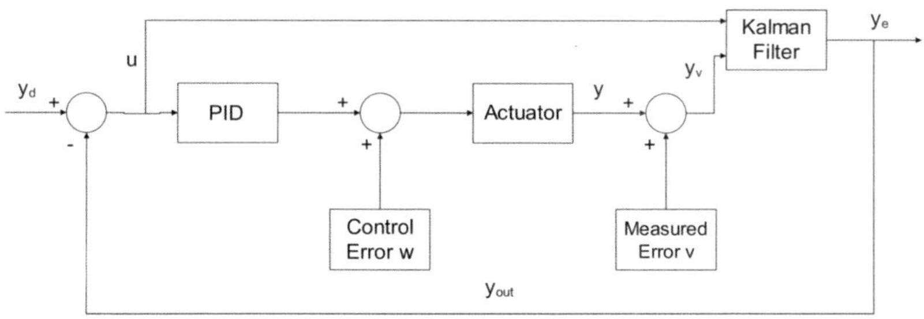

Figure 1. PID control based on the Kalman filtering algorithm block.

Accordingly, we can simulate the control process of the system via MATLAB software. The demanded gait was set to a constant 20 mm, and random noise was set to generate within a range of 1.

R, Q, k_p, k_I, k_D rely more on practical usage conditions, so in the simulation, they are set to be the value that does not affect the evaluation of this method and is calibrated in the actual experiment.

Table 1 shows the inputs variables and their values. As listed in the table, where *a* and *b* stands for the coefficients in the transfer function. y_d is the demanded gait length, which remains constant of 200 during the simulation. *M* is the number of steps of the simulation, which is 3000 steps in this case. *R*, *Q* are variables used in the Kalman filtering process.

In Figure 2, the simulation result is plotted, where the red curve indicates the ideal output gait after PID, the black line shows the signal under noise conditions, and the blue line demonstrates the demanded gait. The plot shows that the method is deemed plausible due to a gradual constraining of the simulated output signal to the demanded signal.

Till now, the design and simulation process has been briefly described. The following section focuses on experiment setups and approaches.

EXPERIMENT

To modulate the pulse width of the excitation pulse, Microcontroller Units (MCUs) are adequate for this task. Yet, the principle of SMAs to deform is contingent on the electric power it receives as springs are electrified and heat is generated [1] [6]. The MCUs could not provide such large electric power to generate heat on the SMA wires so a DC voltage supply and an L298N module are needed.

The MCU used in this experiment is an Arduino Uno due to its simplicity. The DC voltage supply and L298N module are connected, as shown in Figure 3. Thus, sufficient and constant power can be directly supplied to the actuator.

Table 1. Inputs of the MATLAB simulation.

Item	a	b	M	y_d	Q	R
Value	25	133	3000	200	1	1

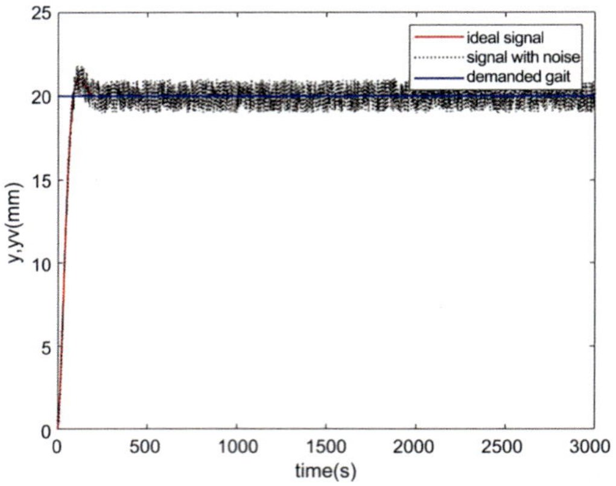

Figure 2. Result of the simulation.

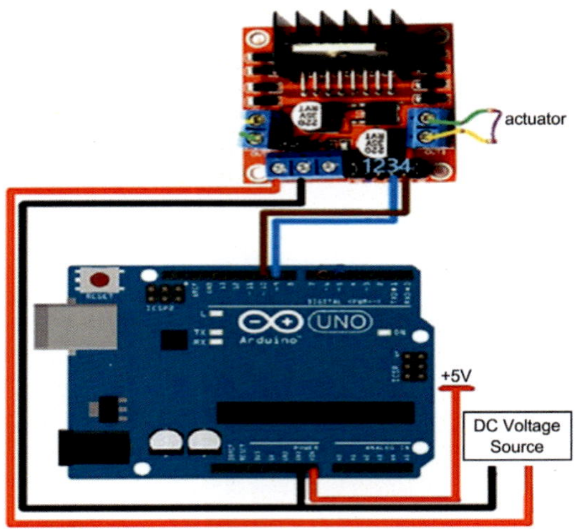

Figure 3. Hardware connection.

Figure 4 demonstrates a functional flow chart of MCU. The left part regarding receiving visual servo data from the computer is completed with a function block that is similar to a Universal Asynchronous Receiver-Transmitter (UART) interrupt. The UART block examines each package it receives from the computer and forsakes the invalid formatted packages. Maximum data are sorted and stored as a static variable in an appropriate package. This UART process is solitary from the main loop, thereby maintaining an optimal sampling rate despite the influence of the time delay function in the main loop. The maximum value is reset after the execution of the restoration process.

The visual servo in this experiment is aimed at detecting the distance between two marked points on the SMA wire and feedback on the real-time value.

A Time-of-Flight (TOF) camera (Axon Technology) is chosen. The M5 type camera has a detection range of 30 - 50 cm with 1 mm precision.

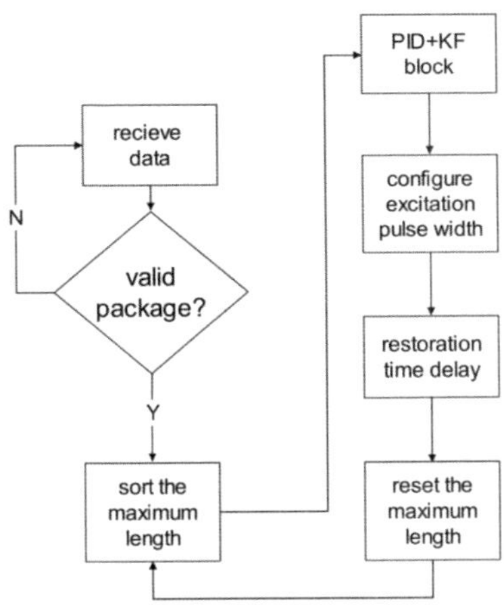

Figure 4. Functional flow chart of MCU.

TOF camera has some advantages over other depth cameras such as binocular cameras and RGB depth cameras; it maintains unaffected or slightly affected under various lighting conditions. Yet, a TOF camera can achieve only a 30 fps frame rate, leading to loss of data when tracking high-speed motion. Nevertheless, this type of depth camera still fits this experiment because the restoring procedure of the SMA wires is relatively slow though excitation occurs in milliseconds. The samples just after the excitation of SMAs can be regarded as the maximum deformation.

Two orange play dough points are affixed to each end of the actuator. The mark points ought to be reduced to their minimum size to optimize the detection performance. To obtain the distance between the points of interest, two main procedures are included, namely clarifying the image to detect the points and calculating the world coordinates of the points.

In the first part of the procedure, each frame of the video's color stream is taken and converted to Hue Saturation Value (HSV) stream. The threshold value of the mark points is selected in the HSV frame and thence to crop the area of interest placing two mark points at the center. The cropping process is relatively inevitable since environmental color may contaminate the mark points' location approximation. Image dilation followed by contour retrieval is executed. Then, the geometric centers of the contours are determined. Then, the coordinates of the centers could be conducted; the z coordinates of the centers can be obtained by mapping the depth stream and color stream, whereas the x and y positions could be converted from the image pixel coordinates after camera calibration (Qing) [12]. Still, only relative values are produced after these steps, so a precise gauge to measure the actual interval between the mark points is needed, in order to conduct the coefficient for each axis from image world coordinates to actual length separately. A general control system schematic diagram is shown in Figure 5.The coordinates' data of the SMA actuator are sampled, calculated, and converted to the approximate length on a laptop simultaneously. Meanwhile, the measured data from the visual servo along with demanded gait data, which can be pre-loaded on the MCU, are assigned to PID control based on the Kalman filtering algorithm. Eventually, a control signal is produced by MCU and the SMA robot is actuated.

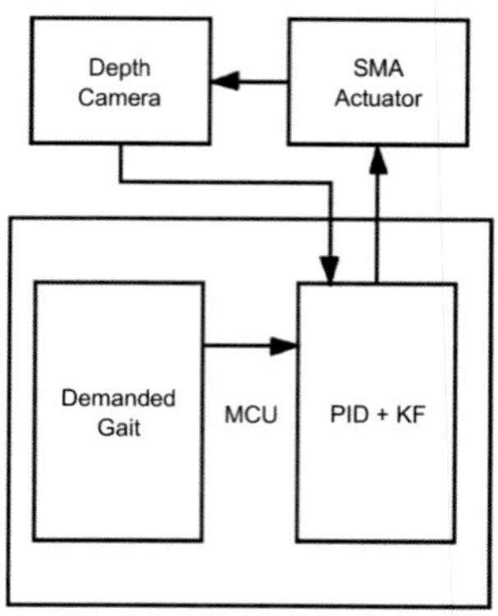

Figure 5. General control system scheme diagram.

Figure 6 demonstrates a general view of the test setups. The SMA actuator at one end is fixed on the white background with the other end free. Both terminals of the actuator are coated with white covers, and this is to eliminate the visual recognition error induced by the reflection of soldering material at the terminals of the SMA. Then, the SMA actuator is wired to the L298N module and the MCU.

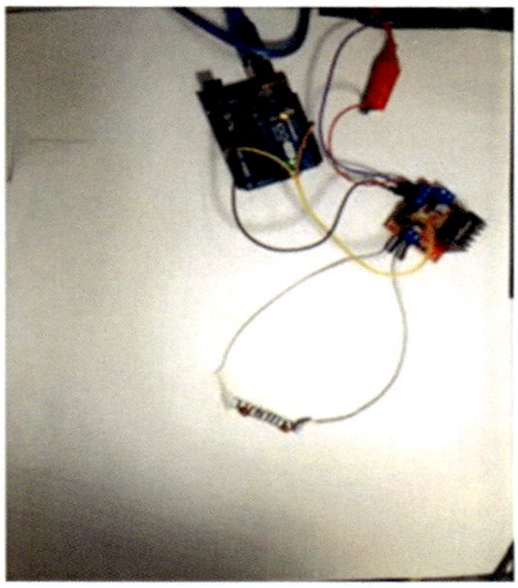

Figure 6. An overall view of an SMA actuator and its control system.

Actuating and controlling the SMA actuator in two conditions are tested; relatively low excitation voltage with more lasting pulse duration along with relatively high voltage with spike pulse excitation groups are recorded. By defining spike and gradual excitation, the excitation voltage lasting < 800 ms would be considered as a spike pulse, while 800 ms < voltage lasting < 4 s is considered as gradual excitation. Due to protecting the SMA wire from burning out, a upper limit of 4s excitation duration is selected. R, Q, k_p, k_I, k_D Parameters of PID control based on the Kalman filtering algorithm are carefully calibrated before data recording.

Lower excitation voltage requires a longer time for SMA to accumulate a sufficient amount of heat to deform, resulting in gradual distortion. This process is slow and conspicuous that the sampling rate of the TOF camera could fit the deformation procedure. To make the conjecture cogent, each test group uses different SMA actuators but is made to the approximately similar original length of 44 mm. Two groups of tests records are scattered in Figure 7(a). Both the test results show that the final maximum length has fallen into the margin of demand. The margin of demand is coded into MCU, defining 0.5 mm as its value, since the precision of the camera is 1 mm. On test 1, a gradual convergence can be seen according to the curve while on test 2, a maximum length regression of meeting the upper boundary of the range can be observed.

After several repetitions, our research proposes that applying spike excitation voltage would benefit the actuating performance. Such that a test sample of SMA actuator with an original length of 42 mm and demanded gait of 45 mm is plotted in Figure 7(b). The intercepted final convergence shows a satisfying result. Even though rapid deformation leads to the loss of sampling points on the excitation edge of the curve, the maximum length of the actuator still converges to set the target value.

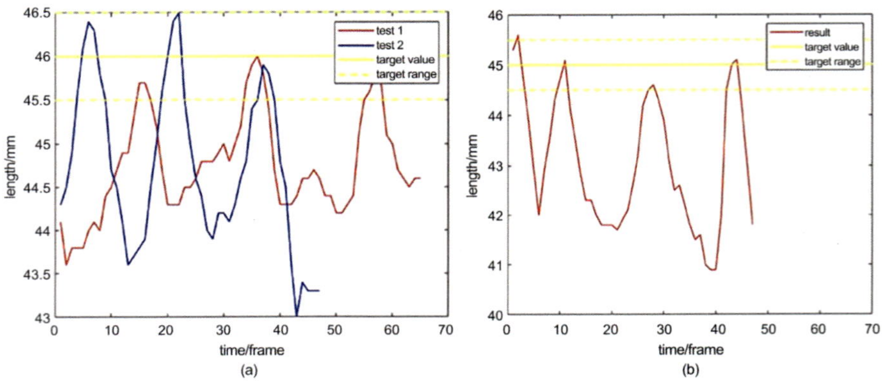

Figure 7. (a) Test data of gradual excitation, (Only final convergence interval intercepted), original length average: 44 mm targeted length: 46 0.5 mm; (b) Test data of spike excitation, (Only final convergence interval intercepted), original length average: 42 mm targeted length: 45 0.5 mm.

CONCLUSIONS

In this study, a control method using PID control based on the Kalman filtering algorithm and visual servo for the SMA actuator designed was introduced. Simulations with MATLAB and experiments are conducted, promising a prospecting future for the integral SMA soft robot and its control system.

Visual recognition stability and measurement precision are still technical barriers in this research. In future studies, we would embed a Reinforce Learning control of gait to this robotics system.

REFERENCES

1. Fei, Y. and Xu, H. (2017) Modeling and Motion Control of a Soft Robot. IEEE Industrial Electronics Society—Transactions on Industrial Electronics, 64, 1737-1742. https://doi.org/10.1109/TIE.2016.2572670
2. Cheng, C., Cheng, J. and Huang, W. (2019) Design and Development of a Novel SMA Actuated Multi-DOF Soft Robot. IEEE Access, 7, 75073-75080. https://doi.org/10.1109/ACCESS.2019.2920632
3. Li, J., Wang, J. and Fei, Y. (2019) Nonlinear Modeling on a SMA Actuated Circular Soft Robot with Closed-Loop Control System. Nonlinear Dynamics, 96, 2627-2635. https://doi.org/10.1007/s11071-019-04949-z
4. Wang, W.B., Xi, F.F., Tian, Y., Zhao, Y. and Li, Y. (2020) Modeling and Analysis of a Planar Soft Panel Continuum Mechanism. Journal of Mechanisms and Robotics, 12, Article ID: 044503. https://doi.org/10.1115/1.4046029
5. Della, S.C., Katzschmann, R.K., Bicchi, A. and Rus, D. (2020) Model-Based Dynamic Feedback Control of a Planar Soft Robot: Trajectory Tracking and Interaction with the Environment. International Journal of Robotics Research, 39, 490-513. https://doi.org/10.1177/0278364919897292
6. Mao, S., Dong, E., Jin, H., Xu, M., Zhang, S., Yang, J. and Low, K.H. (2014) Gait Study and Pattern Generation of a Starfish-Like Soft Robot with Flexible Rays Actuated by SMAs. Journal of Bionic Engineering, 11, 400-411. https://doi.org/10.1016/S1672-6529(14)60053-6
7. Xu, F., Wang, H., Liu, Z., Chen, W., and Wang, Y. (2021) Visual Servoing Pushing Control of the Soft Robot with Active Pushing Force Regulation. Soft Robotics. https://doi.org/10.1089/soro.2020.0178
8. Nuchkrua, T. and Chen, S. (2016) Precision Contouring Control of Five Degree of Freedom Robot Manipulators with Uncertainty. International Journal of Advanced Robotic Systems, 2017, 1-16. https://doi.org/10.1177/1729881416682703
9. Ge, L., Chen J. and Li, R. (2017) Feedforward Control Based on Fourier Series Trajectory Fitting Method for Industrial Robot. 2017 29th Chinese Control and Decision Conference (CCDC), Chongqing, 28-30 May 2017, 4890-4894. https://doi.org/10.1109/CCDC.2017.7979361

10. Hargrave, P.J. (1989) A Tutorial Introduction to Kalman Filtering. IEE Colloquium on Kalman Filters: Introduction, Applications and Future Developments, London, 21-21 February 1989, 1/1-1/6.
11. IEE Colloquium on Getting the Best Out of PID in Machine Control (Ref. No.1996/287). IEE Colloquium on Getting the Best Out of PID in Machine Control, London, 24-24 October 1996.
12. Qing (2019) Camera Calibration: Analysis of the Conversion Process from World Coordinate System to Image Pixel Coordinate System. https://blog.csdn.net/qq_27546529/article/details/89842373

CHAPTER 7

Wireless Communication for Robotic Process Automation Using Machine Learning Technique

C. Murugamani[1], Santosh Kumar Sahoo[2], Pravin R. Kshirsagar[3], Boppuru Rudra Prathap[4], Saiful Islam[5], Quadri Noorulhasan Naveed[6], Mohammad Rashid Hussain[6], Bui Thanh Hung[7], and Dawit Mamiru Teressa[8]

[1]HOD-IT, Bhoj Reddy Engineering College for Women, Hyderabad, India

[2]Department of Electronics & Instrumentation Engineering, CVR College of Engineering, Hyderabad, India

[3]Department of Artificial Intelligence, G.H. Raisoni College of Engineering, Nagpur, India

[4]Department of Computer Science and Engineering, CHRIST University, Bangalore, India

Citation: C. Murugamani, S.K. Sahoo, P. R. Kshirsagar, B.R. Prathap, S. Islam, Q.N. Naveed, M.R. Hussain, B.T. Hung, D.M. Teressa, "Wireless Communication for Robotic Process Automation Using Machine Learning Technique", Wireless Communications and Mobile Computing, vol. 2022, Article ID 4723138, 12 pages, 2022. https://doi.org/10.1155/2022/4723138.

[5]Civil Engineering Department, College of Engineering, King Khalid University, Abha 61411, Asir, Saudi Arabia

[6]College of Computer Science, King Khalid University, Abha 61413, Saudi Arabia

[7]Data Analytics & Artificial Intelligence Laboratory-DAAI Lab, Director of Artificial Intelligence & Information System Programme, Institute of Engineering-Technology, Thu Dau Mot University, 06 Tran Van On Street, Phu Hoa District, Thu Dau Mot City, Binh Duong Province, Vietnam

[8]Department of Chemical Engineering, College of Biological and Chemical Engineering, Addis Ababa Science and Technology University, Addis Ababa, Ethiopia

ABSTRACT

Machine intelligence is what has been generated by programming computers with certain aspects of human intellect, like training, solving problems, and priority setting. A machine can solve a number of complicated issues using these capabilities. In major industries, such as customer support and manufacturing, machine intelligence is now being employed. The growth and quick development of digital technology and artificial intelligence (AI) technologies are becoming more and more difficult. At now, sophisticated manufacturing, the world of invention, and broad acceptance are undergoing a fast transition. Robotics is much more vital as it may now be related to the human brain by the connection between machine and brain, as AI develops. The world's economy faces substantial difficulties by increasing productivity in the manufacturing industry. This study examines the present progress of robotic communication styles of artificial intelligence (AI). In many specific applications, communication between members of a robotic group or even people becomes vital. The paper solves the problem of implementation of an independent industry mobile robot in all fields in the major business, live interactive, planning, mobile robot technologies, and intending. In order to identify the best solution to this issue, a mixed integer robotic model has been developed.

INTRODUCTION

Artificial intelligence (AI) is the technology which allows robots and computers to learn, evaluate, and utilize their own reasoning. With the increasing complexity of technology, the need for artificial intelligence is expanding due to its capacity in a short time to handle difficult issues, with limited human resources and experience [1]. Artificial intelligence can be used to improve a robot's visual acuity and the accuracy of its image recognition. These are important for assembly, as robot welding or cutting can adapt to the smallest micro level tolerances. These really are artificial intelligence- (AI-) controlled robots. The majority of robots are not programmed to be smart. Prior lately, most industrial robots can only be configured to do a continuous series of movements that did not require AI technology, as we previously stated. AI is equipping technological skills and is able to enhance know-how in addition to learning and implement new techniques and techniques. The field of picture identification with machine intelligence is a major advancement, coupled with advancements in large datasets and GPU, which obviously have contributed to the growth of artificial information. An entity and its surroundings are the artificial intelligence (AI) system. The area is identified by an entity including such person or robotic using detectors and actuators [2]. It employs a technique called the searching and template comparison, which requires the machine to investigate the basis of the match obtained and to solve those problems if specified requirements are satisfied.

Robotics is a subject of integration between disciplines, which covers many fields, from mechatronic to management theories and software engineering, with new expansions to materials science, biotechnology, or the scientific method [3]. The interaction between AI and robotics is quite rich. It deals with problems like the following:

i. Intentional activity, organization, actions, observation, and rationalization of objectives
ii. Open settings to see, analyze, and comprehend
iii. Engage with clients as well as other robots
iv. Model of special needs by the methods mentioned

Robotics has always been a fruitful framework for AI technologies, which is often mentioned in its literary works, especially in the above-mentioned areas. AI has been abundant in innovative initiatives in the early days promoting a robotic platform that assumes research programme. The two areas were, nevertheless, evolved in different ways during the next years; robots largely grew beyond AI facilities. Ideally, there is a resurgence of cooperation between both the two areas [4]. This rebirth is mainly due to mature technologies in robotics and artificial intelligence, the emergence of low-cost robot systems with higher protection and actuation abilities, a lot of popular times, and a deeper grasp of the scientific problems of intelligent machines that we want to help with [4].

Manufacturers may envisage in combination with modern aid mechanization and mobile robots the best of the two the necessity for developing manufacturing system. Mobile robots can easily be operated for specific activities such as transport and machinery, feed supplies, preassembly, or quality costs at various manufacturing plants with arms to handle and battery to carry out.

These activities are sufficiently complicated that mobile robots can assume accountability [5]. Moreover, the usage of mobile robots means that the power usage or detailed design is decreased compared with conventional industrial ultimately deciding robots. These advantages make it possible to deploy robotic system in convertible production processes [6].

In this study, a specific issue is examined for a small digital robot, which executes numerous feeder operations autonomously by collection and reloading of container and feeders in the dispensers required. However, these feeding operations must be designed correctly in order to accurately employ mobile robots. Consequently, it is necessary to equip mobile robots for sequential feeding procedures to function well and fulfill various technological requirements [7]. Differing result techniques and algorithms have been used to identify the optimal schedule.

Many generally rigid robots are termed programme concept and function on certain routes and perform a limited sequence of actions numerous times. An unrestricted mobile robot can be planned in a manufacturing cell to perform several feeding activities involved in the collection, transport, and distribution of materials to the suppliers which is still unplanned [8].

REVIEW OF LITERATURE

In Zacharia et al.'s study [9], the robot outfitting technology was initialized to evaluate manufacturing facility effectiveness. Order collection is the activity to gather from storing area pieces required for the manufacturing of finished goods. Kitting is termed by producing a box or a component kit ready to buy. Adjustable mobile robots will substantially improve the efficiency of industry. In automation and artificial intelligence, the many elements and system features to be managed are diverse, assuming certain retaining restrictions. In this study, we are examining a robot suiting technique which works with a robot arm on a subway system and operates to choose pieces via a short aisle. The aim is to assess the effectiveness of robot kits in durations by modeling the basic kits carried out by the robot.

In Boudella et al.'s study [10], the CAS was developed to evaluate the conventional robotic arm. The CAS was implemented. Analysis took into consideration that the amount of production outputs efficiency and unit. In the case of partnership methods, estimating these performance results which in classic automatic production structures are easier becomes much more complex. The distribution of human and robot jobs and how they interact/cooperate during their assembling impact the CAS performance. Many system features and a mathematical formalism to assess the real efficiency of the commercial CAS installation have been created to take these aspects into consideration.

Kshirsagar et al. [6] created a number of computer simulations, along with an inhibitory in one layer, expectancies for the first layer, Rbf, Ffnn, Pnnn, Grnn, and so on. This human mind was recognized to be suited for a restricted number of modularly performed information and structures. But the number of data analysis and the duration of implementation have been lowered by new technologies [5]. Many respondents adjusted traditional machine learning to gather both a high number of data and a smaller duration. Neural networks could explain or create complicated and unreliable data outputs.

In Fernandez-Macias and Hurley's study [11], the most recent developments in robots, artificial intelligence, have to do with the ongoing discourse on mechanization, since the replacement of manual effort by mechanical inputs is becoming ever more common in some production processes and in the service industry. These technologies have made it possible for machines/robots to do actually observe to be too complex to automate activities. In addition, artificial intelligence still participates in tasks traditionally meant to be performed by human perception and expertise.

ARTIFICIAL INTELLIGENCE AND ROBOTICS

A system of development is essentially a randomized one. An operative utilizes his hands, wrists, senses, and mind to conduct activities such as gripping and keeping of work pieces of varied forms and sizes. Variable mechanization is used in large and micro manufacturing process. In additive manufacturing, robotics is crucial support. For accuracy, a robot must communicate in a similar way to the surroundings around it. So, if it is to emulate human talents, a robotic ought to be smart [12].

A smart robot has overcome both arms and end actuators and has computer-based sensing and adaptable controlling devices. Reactive control is needed to rectify faults in the work piece's and end-effectors' location and orientation.

The smart robots should identify the occurrences of causality [13]. This must thus identify and reduce the consequences of the defects. A machine carries out the mental processes like the brain activity. The body systems that could be accomplished with the fundamental principles and assumptions of software engineering are detecting and affecting. The mind and the body part must be synchronized in order to execute a job. There must thus be an intelligent robot with artificial intelligence that differentiates the robotic from some other engines [14].

Viewing, listening, touching, and measurement are included. The detectors collect the data and output it. You have little reasoning to do so. Functions can be performed.

Frauds can be done with the body, arm, hand, fingers, digits, legs, wheels, and diverse communication ways. In order to gain environmental information, the elements to understand, generate, and interpret are used [15]. These elements really detect, identify, and combine things and may guide changes in climate.

Data interpretation is a tool for environmental awareness [12]. Nevertheless, it is important to understand products in the correct environment. Creating a functionality is a tool for environmental impact. Unanticipated, inadequate, and unknown reasons are addressed, and knowledge is likely conflicting to act or adapt to the situation [16].

Robotic Communication Using AI Approaches

Due to its extraordinary capacity to cope with data analytics, intricacy, great precision, and high throughput, artificial intelligence is being used to improve technological development and operations. The familiar tools of IA, as illustrated in Figure 1, are artificial neural network, fuzzy logic, interfering network, genetic algorithm, pattern recognition, clustering, machine learning, profound teaching of particular swarm, etc. AI has worked in several fields including architecture, research, health, computers, economics, and finance. It was also utilized to increase the intelligibility of machinery [17].

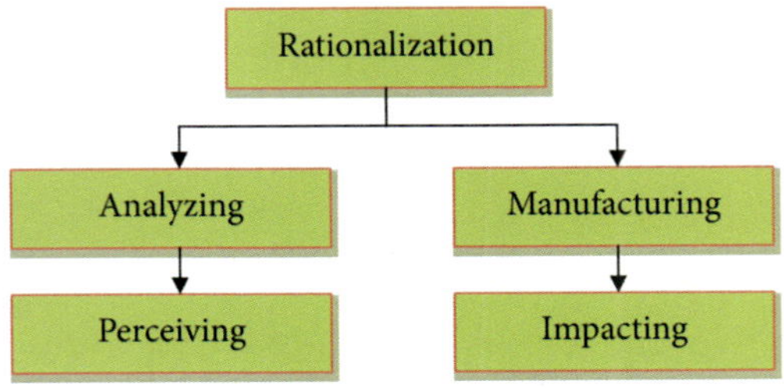

Figure 1. Standardized design of AI/robotics.

Intelligent machines are capable of understanding doing activities in a changing environment, like or near a human being. Thus, AI is a part of a computer science and is probably also a component of robot knowledge and choice [16]. To employ methods to IA and the robotics industry has become a powerful technology with a variety of automated capabilities in a variety of applications, including home care and space travel, medical treatments and intelligence actions, wind speed, weather, temperature, and wind. So, at not only work but also in homes and industries, we discover robotic capabilities that replace many activities that are harmful [18].

Classification of Robots

Two primary modes, service robots and camp robots, as indicated in Figure 2, may be categorized. In numerous social and economic elements of our civilization, robots have brought major changes.

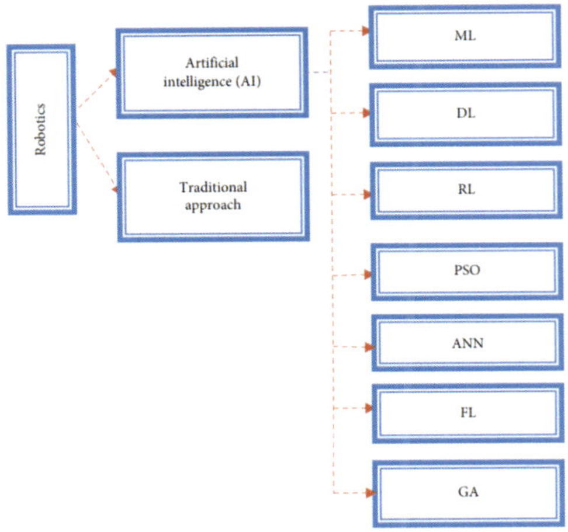

Figure 2. AI methods for robotic communication.

Intelligent robots are designed to build machines that can observe, act and behave as some human functions. In this light, smart robots have the capacity to personality, organize themselves and reproduce themselves. Robotics now becomes smart machines that use their AI skills and intelligence to do jobs swiftly and intelligently [16].

Robotics and AI can work together for a mechanical system. Robots will be omnipresent in the coming years and aid people omnipresent. Effectual interaction between robotics and people must be a key skill for future robotics in order to attain this aim. The basic values of many communications technologies including wireless technology have been addressed for robot networking [18].

For both study and practice, robotic technologies are commonly integrated with technology of the communication system. The integration of the two technologies will enable robotics to move independently and increase robot capabilities in order to execute a certain task effectively. In the exchange and transmission of data among robotics across the huge range, the wireless technology also communicates with others with people.

A robot network is a group of robot communications that are collaboratively conducted activities for a common purpose using mobile and wireless technologies. Teleoperated or independent communications can be possible. Figure 3 illustrates that robotics managed by telecommunications are robots that are managed entirely by human intervention. Autonomous robots, but at the other side, construction related or who have high degrees of maturity. Swarm robots are a huge decentralized set of robotic arms that carry out a shared job by detecting and sharing information through the internet or other methods independently, without even a person or centralized system control interference [19].

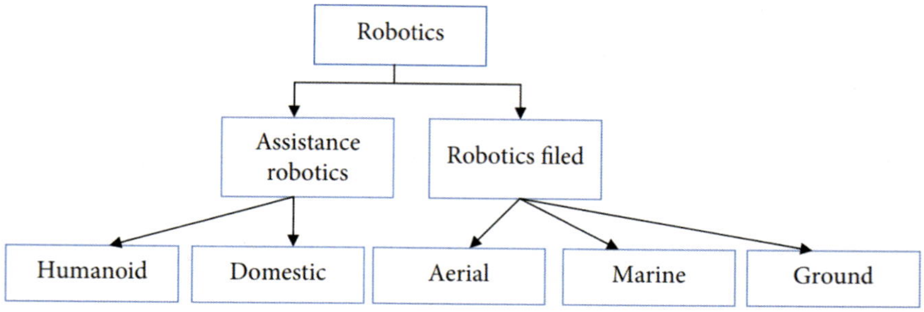

Figure 3. Classification of robotics.

In collaborative robotic group activities, good communication is handled. It is necessary to communicate among robotics to provide a barrier and prolong the life of interaction among them. The most widely utilized measures are categorized and provided as power and energy measures, service quality measurements, dependability, movement, catastrophe prevention, connection, and the cooperation among robotics [18]. There was much study in the field of technology among the robot team. Intelligent robotic communication is essential if it is in either sky, on the land, in moisture, or in another setting, as Figure 4 illustrates.

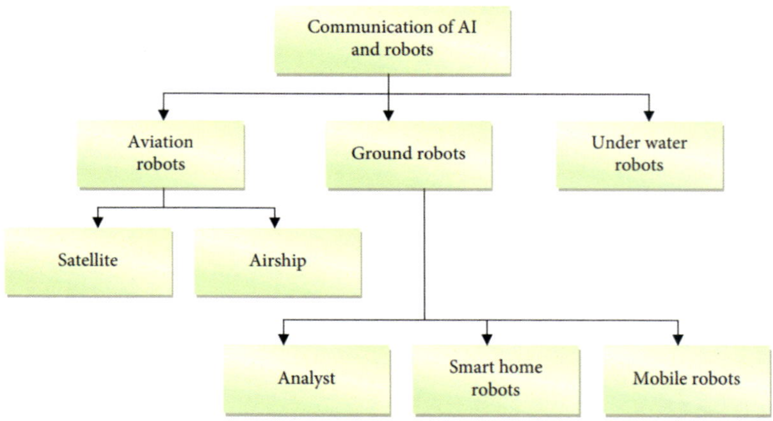

Figure 4. AI and robot communication in several fields.

Artificial Intelligence and Robotic Interface Assessment

Nanoantenna innovation provides a new modern communication model called nanowireless communications. For interaction between tiny entities, nanowireless transmission is necessary. Nanocommunication is one use of micro robot to communicate [20]. Robots usually employ the commercial networking devices due to independent nodes. Static and dynamic aggregate approaches are explored in the intelligent robotic system for a permanent communication platform among robotics for a wireless data transmission [21]. The static method allows the data transfer rate to be increased in an unsteady robotic system.

The flexible technology, though, enables the weight to be distributed among canals. For reliable channels of communication among robots, the advantage of these two approaches increased. AI is hard to accurately characterize. Nevertheless, the science and engineering design machinery described AI as "John McCarthy, regarded as AI's father." Consequently, AI refers to robot systems' capacity to process knowledge and to generate results similar to people's abilities in thinking, decision-making, and issue resolving. In addition, AI is always evolving to make presses more intelligent and intelligent [22]. Machine intelligence involves the capability of a technology to execute every mental job in any setting a human is going to do, as depicted in Figure 5. AI is being utilized to make machinery smarter. Machines that serve people or work with humans are supposed to know basic needs with little human dialogue. For this purpose, AI methods have been employed only by monitoring human bodily gestures to describe and comprehend human intentions [19].

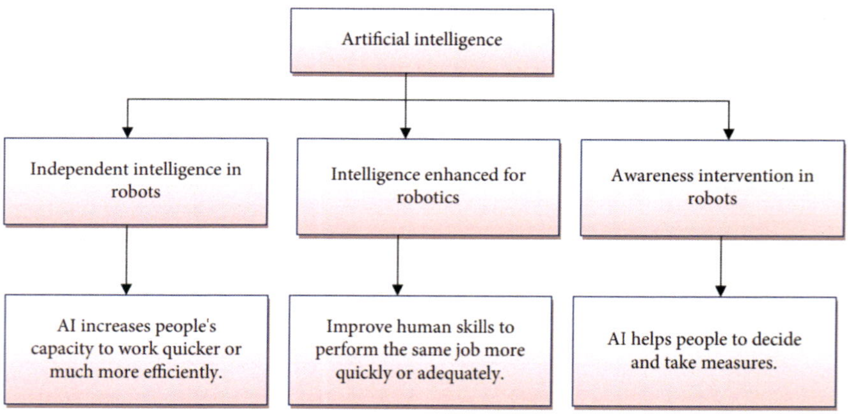

Figure 5. AI for robotic intelligence.

ADVANCED MANUFACTURING INDUSTRY

Manufactures fight to fulfill ever market change expectations. This requires production processes which are adaptable, smart, and adaptable enough to suit current requirements. Managers of companies and manufacturers finalized the integration of financial and political processes. Such convergence requires a major step in industry techniques and activities [23]. Furthermore, it will be able to integrate a range of elements—which include providers, manufacturing processes, and customers. AI systems were employed to assess sustainability throughout the previous decade. Self-employed people are attentive to the humanistic intention of informing them. People can work with robotics, not just without fear and satisfaction, and realize they have been appreciated and worked properly with the robot colleagues [21].

This will provide exceptionally high production standards, great trust, and cheaper prices. Robots are a programmable engine and a wonderful human worker who does repeated jobs under certain circumstances. A future modernization could begin a new era of robot manufacturing with collaborative robotics cognizant of the existing human condition and thus responsible for health risks [17]. A task is started by the employee, and a robot is using a camera. The robot is connected to a processing computer that takes the picture, does image transformation, and learns connections by a computer. It watches people, monitors the surroundings, and informs the controller what to do next using a profound evolution of human purpose. Figure 6 illustrates the fundamental premise of the sophisticated manufacturing sector.

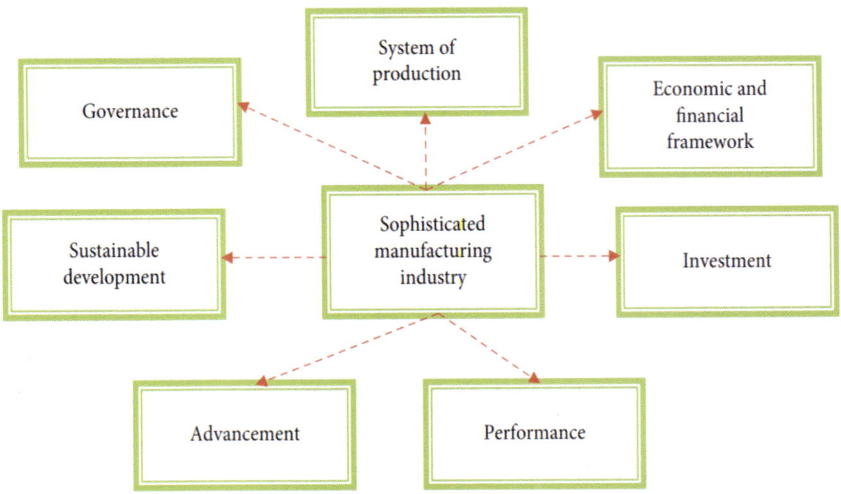

Figure 6. Sophisticated manufacture industry.

Intelligent Manufacturing Technology System

The system smart production process comprises largely of technologies in terms, fundamental information, smart foundation technological, abundant data networks, and smart manufacturing processes in the life cycle and technological support illustrated in Figure 7 [22].

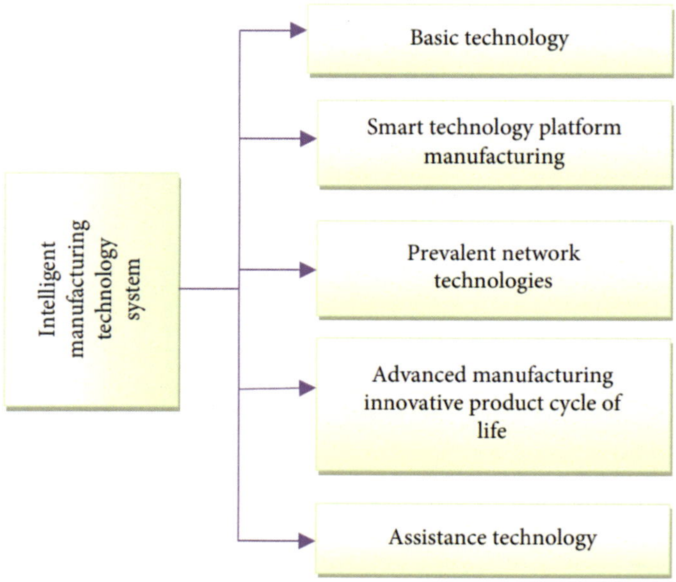

Figure 7. The Five Intelligent Manufacturing Systems Technologies.

Basic Technology

The innovation mainly includes smart architectural manufacturing technology, SDN, air-space systems innovations, business strategy innovations, marketing analysis and visualization innovations for industrial automation, systems engineering and management's decision.

Smart Technology Platform Manufacturing

In Figure 7, intelligent interface system involves, in particular, intelligent data analysis connectivity modern technologies for manufacturing-driven big data networks; intelligent commodity/capability and object information technology web; smart investment/data analysis customer service; smarter expertise/prototype; technical expertise for interactive human-computer/swarm-based technological advances; smart, research-driven creating new products; smart, human-computer industrial automation advanced technologies; a mixture of smart experimentation new tech; smart influence on the future; and smart internet digital remote access advanced technologies [19].

Prevalent Network Technologies

Prevalent communication network largely offers detailed fusing and aerial data networks.

Advanced manufacturing innovative product cycle of Life

Cycle of Product Life Smart manufacturing techniques comprises largely of cloud-based evolutionary theory, productive customer design, smart cloud production equipment technologies, smart cloud-based operational and managerial technologies, smart cloud analysis and empirical new tech, and smart cloud services.

Assistance Technology

The innovation supported largely includes new technologies, AI 2.0, new production technology, and manufacture application specialist technological advancements [21].

ARTIFICIAL INTELLIGENCE ENABLED ROBOTICS FOR MACHINE INTELLIGENCE

Designers presume that smart manufacturing is a new production prototype and the technological tool whereby the overall structure and cycle of life of product innovation include new information and communications technologies, smart technology and science, major manufacturing innovations, product quality, and based product new tech.

The production life cycle therefore utilizes independent detecting, interaction, collaborative effort, training, assessment, cognitive function, choice, regulation, and implementation of living thing, computer, substance, and ecological data to empower the incorporation and optimization of different aspects, such as three components and five streams, for a manufacturing company or corporation [24].

This simplifies manufacturing and provides users with an efficient, elevated, expense, and environmentally responsible product and so increases the manufacturing company or collective's competitive advantage. In the smart industrial sector, AI technology promotes the emergence of new models, methods and formats, system design, and technological platforms.

New Smart Production Methods, Techniques, and Manifestations

Web, service-based, participatory, personalizable, adaptable, and socialized industrial automation system allows consumers and offers user accounts as well as sentient incorporated, digitalized, web, virtualized, service-based, and cooperative services for the personalization, versatility, and intelligent manufacturing process [24]. Finally, ecology of intelligent production as illustrated in Figure 8 will develop a deep combination of the use of those models, methods, and patterns.

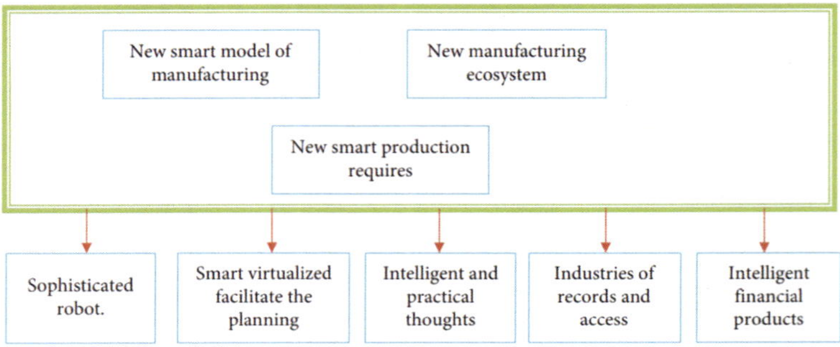

Figure 8. Artificial intelligence enabled robotics for machine intelligence.

METHODOLOGY

Robot-Assisted Mixed-Integer Programming Model (RA-MIPM)

Mathematical Model

A framework is designed to find the best sequencing for the robot manipulator that hasmouths for various feeding jobs. All feed duties are known to be supplied in preparation by tiny carriers. The smooth window paradigm, which enables upper boundary breakings [20], is considered for this model. The robot focuses on a main storage unit, limiting the energy requirements of the robot manipulator. Integer quantities make it very hard to resolve problems with nonconvex minimization. There is an only numeric value inside a set of selected parameters in the MIP framework. When the issue size is increased by adding extra rotation matrix, storage and execution times increase substantially. The MIP technique may thus be utilized generally only for modest problems with a limited percentage of feeds, i.e., with only a limited planned jobs, in the short design period of manufacturing lines. In order to evaluate the advantage range of typology heuristic analyses [25], the MIP system gives optimum information for different circumstances.

Pre Assumptions

i. All work is periodical, independent, and assigned to the very same robot

ii. A robotic arm can support one or maybe more SLC(s) at a single

iii. The minimum age as well as the period of a mobile robot and also the feeding component rate are specified, and the feeding mechanism is suitable

iv. The leading growth explanation is presented for all equipment feeds

v. In an area without disruption, an autonomous robotic IoT device is seen

Figure 9 shows the component functionalities of the robot manipulator platform. The design and independent diversity of the movable robot enable it to work well in cloud-based production contexts where a distinguishing aspect of cloud-based production is its dominant consideration flexibility [26]. Furthermore, the ability of a mobile robot to quickly reconfigure various jobs allows reduced downtime, high performance, and immediate reaction requirement in the cloud fabricating scenario.

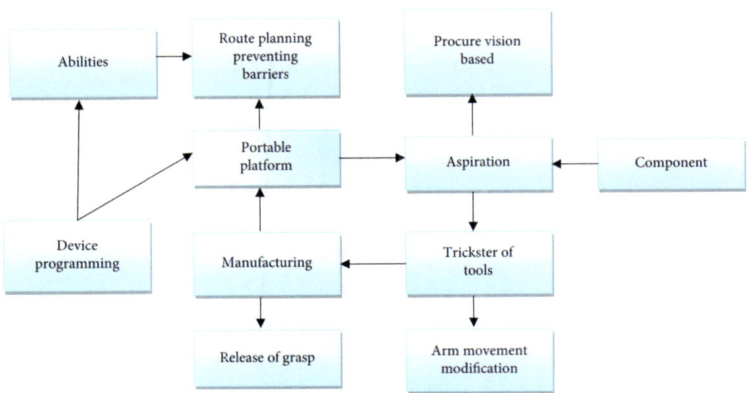

Figure 9. Mobile robot modular system.

The much more suitable process for the robot manipulator was previously investigated for the specially formatted feeding process, and that is to put several parts and materials in a feeder at a moment. The aim of this feeder is to provide machinery on each or more manufacturing lines with components mechanically in a normal factory. Periodically, several part feeds are made, manufacture jobs are introduced to the unvalued, and sometimes, the manufacturers are interrupted. Unless employees replenish feeders, manufacturing processes might be interrupted [27]. This also is a significant commercial promise for the management of multiplex feed. Using the robot arm decreases dependency more on multifaceted, human contact for feeder jobs and paves the path for developing and implementing cloud-based production processes.

Time Window

Time intervals enabling multipart robotic feed might be discovered in equations (1), (2), and (3)).

$$p_i = (b_i - \partial_i) f_i, \forall_i \in N \tag{1}$$

The job of feeding I is to accomplish the has said specified in Equation (1) with a particular amount of times and performances.

$$g_{it+1} = g_{it} + p_i, \forall_i \in N, t = 1/n_i \tag{2}$$

The releasing period for job j is set to some extent a i Equations (2), if the amount of components within the feeders I decreases.

$$d_{it} = g_{it} + (a_i - 0) f_i \tag{3}$$

When a job I is specified when the feeding I Formula does not provide a piece.

When necessary, I must execute a work demand l inside the timeframe period of that demand using autonomous vehicles. That indicates that after the maximum bound of the time span, the robotic arm enters Feed. The temporal limits, but at the other extreme, is called soft restrictions. The aim of the path length of jobs could therefore be modeled [4].

The late start of something like the robotic j feed leads the following job request to change key factors i. If the robot approaches before the lowest feed period, it is waiting for service to commence. Whenever a particular level a i falls into the feeder, the requested releasing period l of job i is established. If another feed I has no components, it specifies the reflect all available information l of job i.

$$d_{it} = g_{it},+(a_i - 0)f_i, t = (1, 2, 3 \cdots , n_i) \tag{4}$$

$$x_{it} \geq d_{it}, t = (1, 2, 3 \cdots , n_i) \tag{5}$$

Equation (5) ensures that an action proposal begins before it is published.

$$\sum_{o \in N\{0\}} \sum_{r=1}^{n} z_{01}^{\text{or}1} = 1 \tag{6}$$

The robot cannot retry the job at Equation (6).

$$\sum_{o \in N\{0\}} \sum_{r=1}^{n_0} \sum_{r=1}^{n} z_{01}^{\text{or}1} = 1 \tag{7}$$

Equation (7) eliminates the subtour among work performance:

$$\sum_{(i,t),(o,r) \in x} z_{it}^{ory} \leq |x| - 1\forall_s \in s,\forall x \subseteq xy \tag{8}$$

Equation (8) compels activities to be executed on one path perfectly.

$$\sum_{o \in N\{0\}} \sum_{r=1}^{n_0} \sum_{y \in s} z_{01}^{ory} \leq 1 \tag{9}$$

Equation (9) restricts robots to feed smaller load bearers than SLCs that are allowed to feed P, as seen in Figure 10.

A predictive genetic algorithm and computational model are suggested. In addition, in better operating situations, the robotic arm may interact and share data with various industrial smart phones and smart robotics. The technique offered provides insight into how robot technology moves towards the cloud manufacturing processes.

A hard part of the propeller manufacturing line at a plant and the computational tests are conducted to illustrate how efficient the proposed approach is. The GA systematic system may be translated to practically efficient outcomes for the recursive feeding issue of the robot manipulator.

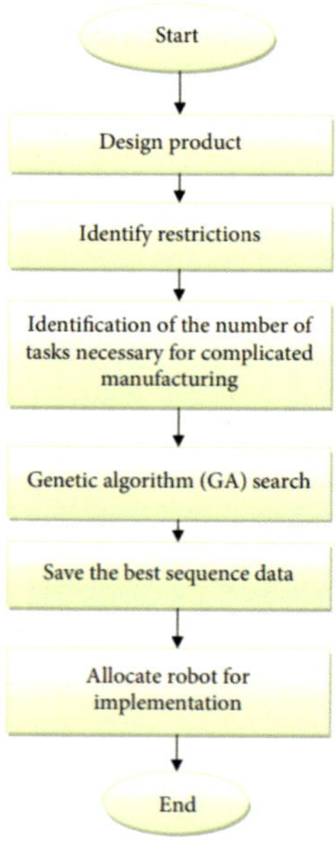

Figure 10. Displaying the whole description of the process.

Figure 10 shows the proposed technique as a flow chart. The information for the production is designed, and a connection pattern is constructed based on the linkage and order between pieces, i.e., physical restrictions. The relationships of importance are based on restrictions placed on an element during the maintenance work. When the component is dismantled in at least one assembly direction, a component may be either fully restricted or partly restricted. There are priority connections for an element that is totally restricted such that it needs to be placed before the first number of other elements. Next, provide the potential number of manufacturing sets that indicate predictive maintenance. In this way, create an initial population grid for the genetic algorithm. To alter the chromosome, a well-established crossing and recombination are employed as the approach.

RESULT AND DISCUSSION

Percentage of Productivity

The total productivity and earnings of robotic systems demonstrate an improvement in worker productivity. In addition, although automated machines have a substantial impact on total business time, there is proof that low-quality workers and in less deeper proportions are decreased to medium-sized workers. The suggested approach would improve operating efficiency and productivity, decrease manufacturing processes, and limit injuries to labor. The robot and man can generally operate with major components simultaneously. In this perfect situation, the production can achieve the greatest level. Small pieces can, nevertheless, prevent the robot arm for working and conversely through blockage. This results in optimum times and production decreases. The suggested feeding system enhancement fully understands in the production system.

Robotic Safety Analysis Effectiveness

The increased prevalence was conducted to evaluate the safety management of the business pre and post computerized technology was implemented. Overall, the automation concept has changed the development environment substantially and so this study suggests that the influence of automating security on the stages of the design is complicated and goes beyond just improving or decreasing the security. The test result is to minimize the entire journey duration of the robotic to a change mechanism range. Figure 11 illustrates the robot service's service quality.

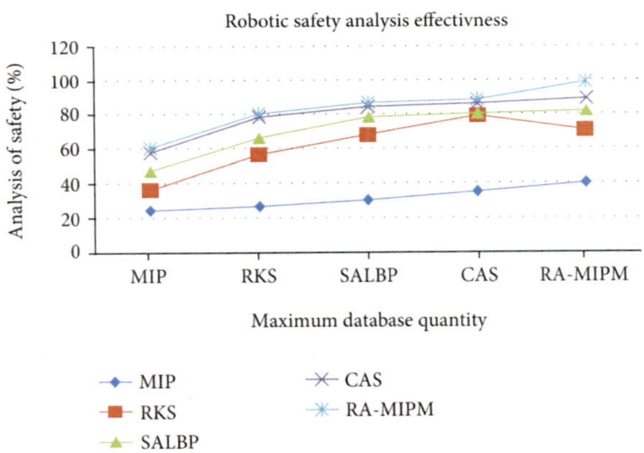

Figure 11. Robotic safety analysis effectiveness.

Table 1 shows genuine proof, and the analysis of the approach is done by water resource. The effectiveness made by mixing instruction set was particularly developed to measure two separate instances in the actual world. In conclusion, the effectiveness of the techniques suggested is demonstrated in several issue scenarios arbitrarily created and evaluated.

Table 1. Robotic safety analysis effectiveness.

Maximum database quantity	MIP	RKS	SALBP	CAS	RA-MIPM
15	25.3	27.8	31.5	36.3	41.4
25	36.9	57.5	69.3	80.4	72.2
35	47.0	66.4	78.2	80.3	81.8
45	58.0	79.0	84.4	85.8	88.7
55	60.8	80.2	87.0	88.2	98.8

Efficiency Estimation of the Ratio

Robots are important instruments for improving productivity. Utilities with effectiveness and accuracy than people are now produced in most of robotic adoptions. The confluence of electronic research and development has led to new platform characteristics: a highly voluminous, high-mix approach that generally allows manufacturers to more effectively fulfill farmers stores around the globe. In comparison to other current approaches, the presented scheme attains a high performance ratio. The operating lease of both the approaches presented is shown in Figure 12. Table 2 shows the RA-MIPM technique's operating efficiency. Robotics can increase and improve human capability and enable organizations to develop artificial intelligence (AI) better and more easily. It may cut expenses and improve efficiency, stability, effectiveness, and quality of work.

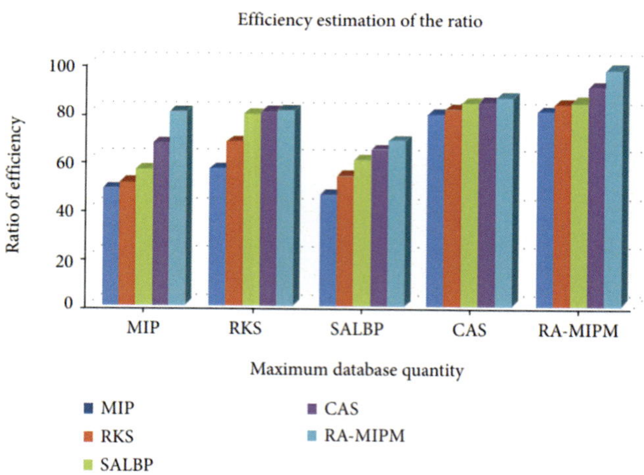

Figure 12. Efficiency estimation of the ratio.

Table 2. Efficiency estimation of the ratio.

Maximum database quantity	MIP	RKS	SALBP	CAS	RA-MIPM
15	49.2	57.3	46.6	80.2	81.2
25	51.4	68.3	54.5	81.9	84.4
35	56.7	79.8	60.8	84.7	85.6
45	67.6	80.9	65.2	85.5	91.8
55	79.9	80.4	68.4	86.3	98.0

Cost of Computation

The continual management of the industrial machinery for production plants is cost-effective and has a major impact on all operational activities linked to assets. The model allows for a significant decrease in costly unexpected breakdowns and extends to the remainder of the usable life of the manufacturing plant and equipment. Compared to current MIPs, RKS, SALBP-2, and CAS, the suggested RA-MIPM process offers minimal construction costs. The calculation cost of the suggested RA-MIPM technique can be seen in Figure 13 and Table 3.

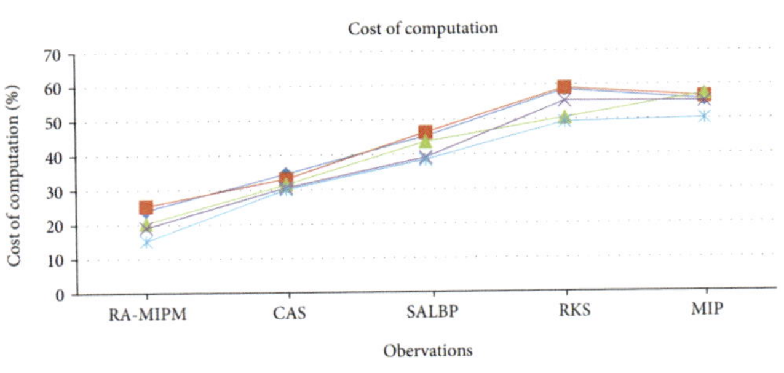

Figure 13. Cost of computation.

Table 3. Cost of computation.

Number of insinuations	RA-MIPM	CAS	SALBP	RKS	MIP
15	24.1	34.6	45.2	58.5	55.5
25	25.3	33.1	46.3	59.0	56.4
35	20.3	31.5	43.6	50.4	57.2
45	19.1	30.6	39.1	55.2	55.0
55	15.2	30.1	38.5	49.2	50.1

CONCLUSION

The essay explores the challenge of implementing an autonomously, movable automated machine in the real universe. In robotic arm scheduling and interactions, telecommunication, and autonomous vehicle technologies, the micro optimization issue is considered. The major news in this study is the simultaneous management of gentle working times and limited mobile robot capacity and the transfer from laboratory configuration to high-availability technologies. Moreover, robotic arm techniques, skills, and methods of someone using the robotic system constitute the backbone for developing cloud-based manufacturing techniques that have more energetic, adaptable configuration towards the existing resources and give fast instant access to capabilities, assets, and artificial intelligence. The integrity of all these responses may then be assessed as basic principles by applying MIP methods to extract value. The new type of robot for any kind of real-time application is tested using the latest and lightweight AI algorithm for cost reduction.

ACKNOWLEDGMENTS

The authors extend their appreciation to the Deanship of Scientific Research at King Khalid University for the funding and support for this work under research grant award number RGP. 1/184/42.

REFERENCES

1. V. Sharma and R. Kumar, "Cooperative frameworks and network models for flying ad hoc networks: a survey," *In Concurrencyand computation: practice and experience*, vol. 29, no. 4, 2017.
2. K. C. Morris, C. Schlenoff, and V. Srinivasan, "Guest editorial a remarkable resurgence of artificial intelligence and its impact on automation and autonomy," *IEEE Transactions on Automation Science and Engineering*, vol. 14, no. 2, pp. 407–409, 2017.
3. M. Wazid, A. K. Das, and J. H. Lee, "Authentication protocols for the internet of drones: taxonomy, analysis and future directions," *Journal of Ambient Intelligence and Humanized Computing*, 2018.
4. P. R. Kshirsagar, A. N. Rakhonde, and P. Chippalkatti, "MRI image based brain tumor detection using machine learning," *Test Engineering and Management*, vol. 81, pp. 3672–3680, 2020.
5. P. Kshirsagar, "Operational collection strategy for monitoring SMART waste management system using shortest PATH algorithm," *Journal of Environmental Protection and Ecology*, vol. 22, no. 2, pp. 566–577, 2021.
6. P. Kshirsagar, *Brain Tumor classification and Detection using Neural Network*, 2016.
7. V. Scilimati, A. Petitti, P. Boccadoro et al., "Industrial internet of things at work: a case study analysis in robotic-aided environmental monitoring," *In IET wireless sensor systems*, vol. 7, no. 5, pp. 155–162, 2017.
8. M. A. Messous, H. Sedjelmaci, and S. M. Senouci, "Implementing an emerging mobility model for a fleet of uavs based on a fuzzy logic inference system," *Pervasive and Mobile Computing*, vol. 42, pp. 393–410, 2017.
9. P. T. Zacharia, S. A. Tsirkas, G. Kabouridis, A. C. Yiannopoulos, and G. I. Giannopoulos, "Genetic-based optimization of the manufacturing process of a robotic arm under fuzziness," *Mathematical Problems in Engineering*, vol. 2018, 12 pages, 2018.
10. M. E. Boudella, E. Sahin, and Y. Dallery, "A mathematical model to assess the performance of a robotic kitting system in an assembly plant," in *In: 11th international conference on modeling, optimization and simulation-MOSIM'16*, 2016.

11. E. Fernandez Macias and J. Hurley, "Routine-biased technical change and job polarization in Europe," *Socio-Economic Review*, vol. 15, no. 3, pp. 563–585, 2016.
12. P. Roy and P. Dickinsion, *How Robotic Process Automation and Artificial Intelligence will Change Outsourcing*, Seminar on June 7, 2016.
13. P. R. Kshirsagar, P. P. Chippalkatti, and S. M. Karve, "Performance Optimization of Neural Network using GA Incorporated PSO," *Journal of Adv Research in Dynamical & Control Systems*, vol. 10, pp. 156–169, 2018.
14. A. Burgess, "Robotic Process Automation & Artificial Intelligence," *handbook posted on 7th june*, 2017.
15. B. Zhang, C. H. Liu, J. Tang, Z. Xu, J. Ma, and W. Wang, "Learning-based energy-efficient data collection by unmanned vehicles in smart cities," *IEEE Transactions on Industrial Informatics*, vol. 14, no. 4, pp. 1666–1676, 2018.
16. S. Akojwar and P. Kshirsagar, "A novel probabilistic-PSO based learning algorithm for optimization of neural networks for benchmark problems," *WSEAS Transactions on Electronics*, vol. 7, pp. 79–84, 2016.
17. L. Zhang, X. Diao, and O. Ma, "A preliminary study on a robot's prediction ofhuman intention," in *In: 2017 IEEE 7th annualinternational conference on CYBER technology in automation,control, and intelligent systems (CYBER)*, pp. 1446–1450, Honolulu, HI, USA, 2017.
18. P. Valente Klaine, M. A. Imran, O. Onireti, and R. D. Souza, "A survey of machine learning techniques applied to Self-Organizing cellular networks," *IEEE Communications Surveys andTutorials.*, vol. 19, no. 4, pp. 2392–2431, 2017.
19. O. Vermesan, A. Bröring, E. Tragos, M. Serrano, D. Bacciu, and S. Chessa, "Internet of robotic things: converging sensing/actuating, hypoconnectivity, artificial intelligence and IoT platforms," *Cognitivehyperconnected digital transformation: Internet of things intelligenceevolution*, River Publishers, pp. 1–35, 2017.
20. S. Oza, "IoT: The Future for Quality of Services," *Lecture notes in electrical engineering*, Springer, Singapore, vol. 570, 2020.

21. P. Kshirsgar, V. More, V. Hendre, P. Chippalkatti, and K. Paliwal, "IOT Based Baby Incubator for Clinic," *Lecture Notes in Electrical Engineering*, Springer, Singapore, vol. 570, 2019.
22. J. Lee, B. Bagheri, and H.-A. Kao, "A Cyber-Physical Systems architecture for Industry 4.0-based manufacturing systems," *Manufacturing Letters*, vol. 3, pp. 18–23, 2015.
23. S. Sundaramurthy, C. Saravanabhavan, and P. Kshirsagar, "Prediction and classification of rheumatoid arthritis using ensemble machine learning approaches," in *2020 International Conference on Decision Aid Sciences and Application (DASA)*, pp. 17–21, Sakheer, Bahrain, 2020.
24. X. W. Wang, *Industry 4.0: Road to the Future Industry of German Manufacturing Industry 2025 Diagram*, China Machine Press, Beijing, 2015.
25. M. R. Pedersen, L. Nalpantidis, R. S. Andersen et al., "Robot skills for manufacturing: from concept to industrial deployment," *Robotics and Computer-Integrated Manufacturing*, vol. 37, no. 37, pp. 282–291, 2016.
26. A. Keisner, J. Raffo, and S. Wunsch-Vincent, "Robotics: breakthrough technologies, innovation, intellectual property," *Foresight and STI Governance*, vol. 10, no. 20, pp. 7–27, 2016.
27. M. Faccio, M. Bottin, and G. Rosati, "Collaborative and traditional robotic assembly: a comparison model," *The International Journal of Advanced Manufacturing Technology*, vol. 102, no. 5–8, pp. 1355–1372, 2019.

CHAPTER 8

Voice Control for an Industrial Robot as a Combination of Various Robotic Assembly Process Models

Svitlana Maksymova[1], Rami Matarneh[2], Vyacheslav V. Lyashenko[3], and Nataliya V. Belova[3]

[1]Department of Computer-Integrated Technologies, Automation and Mechatronics, Kharkiv National University of Radio Electronics, Kharkiv, Ukraine.

[2]Department of Computer Science, Prince Sattam Bin Abdulaziz University, Al-Kharj, Saudi Arabia.

[3]Department of Informatics, Kharkiv National University of Radio Electronics, Kharkiv, Ukraine.

Citation: Maksymova, S. , Matarneh, R. , Lyashenko, V. and Belova, N. (2017), "Voice Control for an Industrial Robot as a Combination of Various Robotic Assembly Process Models". Journal of Computer and Communications, 5, 1-15. doi: 10.4236/jcc.2017.511001.

ABSTRACT

Models for the design of assembly processes are considered. Various models for the voice control of an industrial robot are considered: a logical model, semantic networks, a frame model and Petri nets. It is shown that this set of models allows describing the process of designing the technological process for an industrial robot. The logical model of the technological process allows you to define logical relationships. A model based on semantic networks describes the relationship between assembly units in a detail. This allows you to determine the order and method of registration, as well as the mutual orientation of assembly units in the product. The frame model provides the ability to streamline the execution of the build process. A model based on Petri nets allows one to describe the type and sequence of technological transitions. Based on the proposed models, a method of voice control for an industrial robot is developed. The basic principles of voice control for an industrial robot are considered.

Keywords: Robot, Voice Control, Frame Model, Petri Nets, Logic Model, Semantic Nets, Robotic Assembly Processes

INTRODUCTION

Assembly as technological process is a part of production process. It is associated with qualitative transformation (changing) of production object. We have to notice assembly realization task is multi-variant. It is explained by the fact that for the same operation and all the process functions realization systems with different structures and functional elements (modules) may be used. Such systems have different technical and economic parameters. For this task solving it is expediently to use simulation methods and graph theory and search and optimization mathematical methods main positions. As a result, we can design optimal robotic assembly technological process and identify errors in robot control program before its using in industrial conditions. Using modeling we can develop collision free paths.

In [1] Zixiang Li, Qiuhua Tang, Li Ping Zhang propose to minimize energy consumption and cycle time using restarted simulated annealing algorithm. Marco Faber, SinemKuz, Marcel Ph. Mayer, Christopher M. Schlick in [2] use simulations in order to evaluate parameters, new assembly or planning strategies or the assembly of new products. In [3] authors present Taguchi optimization method in conjunction with simulation modeling in a new application for dynamic scheduling problems in robotic flexible assembly cells, in order to minimize total tardiness and number of tardy jobs. George Michalos, Konstantinos Kaltsoukalas, Panagiotis Aivaliotis, PlatonSipsas, Andreas Sardelis, George Chryssolouris [4] use semantic model representing the robot/resource skills along with the operations that a resource has to execute. In [5] authors write simulation-based technologies constitute a focal point of digital manufacturing solutions, since they allow for the experimentation and validation of different product, process and manufacturing system configurations. In [6] SongliangNiea, Yuwen Lib, GuoShuaia, Song Taoa, Fengfeng Xi propose method for modeling and analyzing the fatigue life of robots with flexible joints. Hod Lipson [7] uses simulation for soft robots. In [8] authors use frame model to facilitate as-built Building Information Model generation. Arne Wahrburg, Stefan Zeiss, Björn Matthias, Jan Peters, Hao Ding [9] propose to use modeling in order to create primitives that encapsulates the capabilities to coordinate, control and supervise an elementary robot task. At last in [10] autorspropese even powerful general-purpose robot simulation framework called V-REP that renders simulations and simulation models more accessible to a general-public, by reducing the simulation model deployment complexity.

In general case robotic assembly process includes next operations:

- robot has orientation with required accuracy on the specified assembly details surfaces;
- mating detail clamping, its move in working space to assembly device or base detail;
- mating detail orientation with required accuracy on base detail or device surface;
- details conjugation that are mutually oriented with required accuracy;
- completed details fixation.

Currently it is expediently to use robot voice control as computer-aided design (CAD) interactive interface implementation mean. It allows representing specific technological task features, to take into account equipment and robot functioning workspace characteristics, to provide corrections into technological transitions set, robotic system state operational control.

In [11] audio information is used for robot motion control. Authors in [12] integrate the industrial robot ABB IRB140 with the system for automatic speech recognition that has a task to recognize human speech and use it as a command to the robot, so the robot can manipulate the objects. Panagiota

Tsarouchi, Sotiris Makris, George Chryssolouris [13] discuss process of learning by demonstration as well as the instructive systems is reviewed, focusing mainly on programming through visual guidance and imitation, voice commands and haptic interaction. In [14] authors propose approach that is distributed, fault tolerant and scalable, such that any recognition algorithm or language support can be added and used without any changes to the existing system. Alberto Poncela and Leticia Gallardo-Estrella [15] represent a human-robot interface to teleoperate a robotic platform via the user's voice. In [16] Somnath Kar, Ankit Jana, Debarshi Chatterjee, Dipayan Mitra use voice control technique to navigate the robot in various directions in the Cartesian plane. In [17] authors research industrial robot voice control in noisy environment. Carlos Mateo, Alberto Brunete, Ernesto Gambao, Miguel Hernando [18] use voice control to make easier to program tasks for industrial robots like polishing, milling or grinding. In [19] H. C. Fang, S. K. Ong, A. Y. C. Nee to use voice control for novel augmented reality based interface for interactive robot path and end-effector (EE) orientation planning. Adam Rogowski [20] proposes even Web-based remote robotized cells voice control. So we can say industrial robot voice control has plenty of advantages and may be used in different industrial fields.

MODELING AS A DECISION-MAKING TOOL

A model is any image, analog, description, scheme, drawing, map, etc. of any object, process or phenomenon used as its substitute or representative. Each model reflects certain selected characteristics and neglects non-essential for this model. Work not with the object itself, but with its model allows to quickly and without material costs investigate its properties and behavior in certain situations. There are four main presentation forms: analytical, tabular, graphical and graphical. The analytical form represents a model in the form of a formula, an analytical expression, a set of analytical expressions. It is applied under certain model or object behavior laws. The graphical form uses models or data set representation in the form of curves, graphs, diagrams. The tabular form gives model or its characteristics representation in the form of one or a set of interrelated tables. The graph form is based on the representation of the model in the form of a graphical scheme called a graph.

Robotic assembly production processes design is difficult to be mathematically modeled, because it is based on technological rules using and often depends on subjective factors. But scientists try use different model for their goals achievement [4] [8] [21] [22] [23] [24] [25] . In [21] modeling is used for making logic and for evaluation of the control logic performance and the results. K. Abd, K. Abhary, R. Marian [22] used simulation for results evaluation. Models are often used to represent in a formalized way the structure, the behavior, and the requirements of the assembly systems [23] . In [24] authors propose to use fuzzy logic based model in order to predict surface roughness. George Pintzos, Nikolaos Nikolakis, Kosmas Alexopoulos, George Chryssolouris [26] , Yanru Zhonga, Yuchu Qinb, Meifa Huangc, Wenlong Lub, Liang Chang [27] , Mehmet Dogar, Ross A. Knepper, Andrew Spielberg, Changhyun Choi, Henrik I. Christensen, Daniela Rus [28] , Q. Yanga, D. L. Wuc, H. M. Zhuc, J. S. Baoc, Z. H. Wei [29] and George Michalos, Konstantinos Kaltsoukalas, Panagiotis Aivaliotis, PlatonSipsas, Andreas Sardelis, George Chryssolouris [4] use semantic models for assembly simulation. Gong-dong Wang, Yao Yang, Wei Wang, L. V. Si-Chao [30] , Jiaxiang Luo, Jiyin Liu, Yueming Hu [31] and L. Westover, J. Olearczyk, U. Hermann, S. Adeeb, Y. Mohamed [32] use frame models for assembly simulation. Hesuan Hu, Meng Chu Zhou [33] , Nai-Chieh Wei, I-Ming Chao, Chin-Jung Liu and, Hong Long Chen [34] , Hesuan Hu, Mengchu Zhou, Zhiwu Li, Ying Tang [35] , Yan Yang, Hesuan Hu, Yang Liu [36] and Hao Yue, Hesuan Hu, Weimin Wu, Hongye Su, Jihui Zhang [37] use Petri Nets for assembly simulation.

Robotic Assembly Processes Logic Model

As a result of mathematical models analysis assembly technological process design logical model is proposed.

In developing of assembly technological process design logical model we have to take into account next constructions features: all parts are limited in displacements in all directions, each item of the product is oriented towards other parts of this product and all parts of the product are interconnected. Any product construction may be assembled. Let us consider hinged elements mounting on a board. Let the circuit board require the installation of chips that can vary from minimum to maximum values: capacitors in the range from 0 to i, 0 to j transistors, 0 to 1 resistors, 0 to k inductors and 0 to m diodes. For realization of technological processes implementations, it is suggested to formulate a set of basic technological rules, so we propose a form the main rules set, where each detail is characterized by certain parameters, which can be represented as follows:

parameters (name, square, dimension, mount_variant)

Mount variant takes into account which outputs contain a hinged element (planar or core). For radio electronic products assembly the feature of each element can be written as follows:

Feature (name, [control_points_list])

where [control_points_list]=[cp1,cp2], cp1, cp2—the point of orientation of the element on the board.

Then assembly unit may be represented in next way:

assembly_unit (number, board, [hinged_elements_list])

For radio electronic products assembly, let us formulate our proposed rules as follows:

Rule 1. Assembly unit may include different assembly units types (type1, type2, ···):

$$\exists x \exists y \left(\text{type}(x, \text{board}) \wedge \text{type}\left(y, [\text{hinged\_elements\_list}]\right)\right)$$
$$\rightarrow \exists u \left(\text{type}(u, \text{assembly\_unit}) \wedge \text{contains}\left(u, [x, y]\right)\right).$$

Rule 2. Base details have similar features (for example, holes), fastening elements-appropriate dimensions, that allow to connect base details (B) with other details (D):

$$\exists u \left(\text{type}(u, \text{assembly\_unit}) \wedge \text{feature}\left(u, \{B, D\}\right)\right)$$
$$\rightarrow \text{assembly\_unit\_found}(u, B, D)$$

Rule 3. Product is considered to be assembled when all its assembly rules are assembled. Assembly unit is considered to be assembled when all details included in it are founded.

From the model point of view we can consider assembly technological process. Assembly technological process start stage is base detail finding:

$$\exists B \exists D \exists u \left(\text{base\_found}(B) \wedge \text{compatible\_detail}(B, D)\right.$$
$$\left. \wedge \text{assembly\_unit\_found}(u, B, D) \wedge \text{assembled}(\text{и})\right) \rightarrow \text{process\_completed}$$

Then we find detail that will be connected with the base detail (compatible part) and assembly unit that includes base and compatible detail.

Base detail is the detail with the biggest dimensions.

$$\exists B \left(\text{base\_found}(B) \vee \text{мах\_dimension}(B)\right) \rightarrow \text{base\_found}(B)$$

Detail compatible with base must have compatible features:

$$\exists B \exists D \exists S(\text{detail}(B) \wedge \text{detail}(D) \wedge \text{feature}(B,S) \wedge \text{feature}(D,S))$$
$$\rightarrow \text{compatible\_detail}(B,D)$$

Then it is necessary to determine assembly process sequence and details which will connect base and compatible details.

Semantic Nets in Robotic Assembly Process Simulation

For connotations between assembly units in product analysis we propose to represent assembly technological process using semantic net. Semantic net and its elements describe relations between assembly units. It is a directed graph with marked verticles and arcs. Verticles are mapped to objects (in our case details) and arcs are mapped to semantic relations between them. By the definition scheme, or relations intensional Ri, is next pairs set:

$$\text{INT}(\text{Ri}) = \{\cdots,[\text{Aj}, \text{DOM}(\text{Aj})],\cdots\},$$

where Ri—relations name;

DOM(Aj) —domain Aj, i.e. attribute values set Aj of relation Ri

All domains union is an extensional of relation Ri, i.e.

$$\text{EXT}(\text{Ri}) = \{\text{F1},\cdots,\text{Fp}\}$$

where Fk(k = 1, p)—fact of relation Ri.

Fact is set by pairs "attribute-value" totality. They are called attributive pairs. In graphic form graph is a semantic net subgraph that has a star like structure. Subgraph root is a predicate type verticle. From subgraph verticle edges start marked as fact attributes names. They are directed to base set verticles that are these attributes values. A fragment of the semantic product description network is shown in Figure 1. A model of a semantic network can be used at the initial stage of the product analysis and will determine the order and method of alignment in addition to the mutual orientation of assembly units in the product.

Thus, the developed semantic model of the manufacturing process assures the description of the relations between assembly units in the assembled product. Although the semantic network is designed to display the existence of semantic links between assembly units, the full description of the product seems rather cumbersome, even for simple products. Therefore, it seems advisable to use the semantic network model only to display individual nodes, including standardized in composition and technological execution.

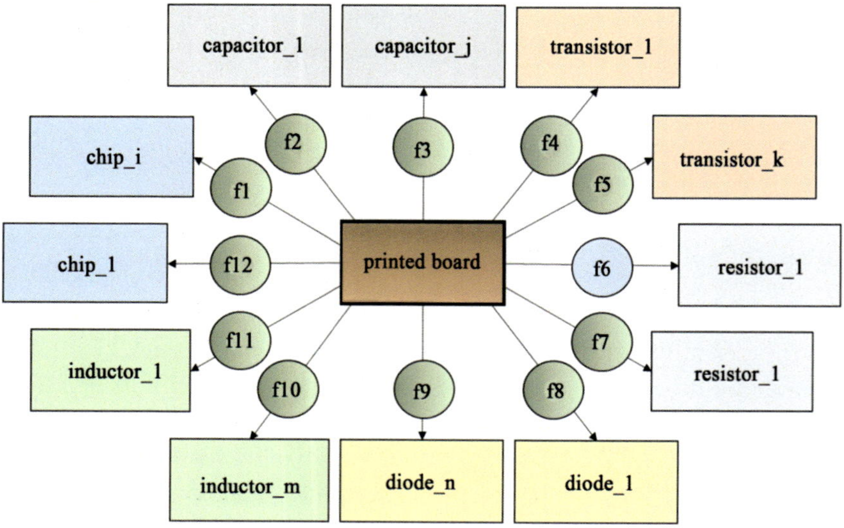

Figure 1. Semantic net fragment describing relation "Orientation".

Robotic Assembly Processes Frame Model

Since in assembly process there are often separate units assembly processes typical situations it is expediently to consider assembly technological process frame model.

For each assembly unit there are plenty of assembly schemes CD.

In particular, CD may be represented in next way:

$$CD = \langle C(D)1, C(D)2, \cdots, C(D)N \rangle$$

where CD—assembly process schemes plurality,

C(D)i—possible assembly process scheme,

N—possible assembly process scheme number.

In its turn C(D) may be written in next form:

$$C(D) = \langle da, db, \cdots, dm \rangle$$

where D—necessary assembly operations plurality;

di D—separate assembly operation;

m—necessary assembly operations number.

Each separate assembly operation is performed by specific instrument. It is also described by start and finish positions of assembled detail. It can be written in next way:

$$di = \langle I, ndet1, ndet2, kdet1, kdet2 \rangle$$

where I—instrument;

ndet1—detail1 start position;

ndet2—detail2 start position;

kdet1—detail1 finish position;

kdet2—detail2 finish position.

In scheme C(D)I may be typical assembly process sub-sequence , that will represent typical assembly technological operation. So, it is a technological operation frame.

Example of frame of typical technological operations accentuation in technological process structure is shown on Figure 2 (sequences (di, dj, dk) and (da, db)).

Let us notice for every elements type mounting on printed board it is necessary to implement actions totality-technological process sub-sequences. Frame "Diodes_mounting" contains operations that are necessary for diodes mounting: operation of taking the diode from the store, placing the diode on the necessary place on the printed circuit board, orienting it relative to the control points, and the installation itself. Similarly, frames are allocated for each type of mounted elements: chips, transistors, capacitors, resistors, inductors.

Frame	Diode_mounting	
instrument	Robot1, grippe1	
Detil1	Diode1	
DetilN	DiodeN	

Frame	resistors_mounting	
instrument	Robot1, grippe1	
Detil1	Resistor1	
DetilN	DiodeN	

Figure 2. Example of frame of typical technological operations accentuation in technological process structure.

When analyzing the technological processes of the assembly, it is necessary to find typical sub-sequences and streamline the assembly process according to the criteria of saving time, materials, equipment. To fulfill these criteria, it is advisable to select sub-sequences that can be executed in parallel, minimize the need for reorientation of the robot and parts, as well as tool changes during the assembly process.

Examination of the frame model allows us to make the following conclusion: when assembly technological process is designing, the frame model is suitable for describing typical sub-sequences of technological processes, individual technological operations and transitions. Even in a completely new product, it is possible to have typical assembly units for which there will be an appropriate frame in the structure of the process planning system that describes assembly process element—a kind of micro-know-how for an individual node, transition or operation.

Petri Nets in Robotic Assembly Processes Modeling

It is convenient to use Petri nets when robotic production technological processes are designing. The application of Petri nets can be viewed in two ways. On the one hand, with the help of the Petri net, it is possible to represent the detailed structure of technological transitions of the designed assembly technological process. It is assumed that the sequence of the assembly process has already been determined and can be developed either with the help of Petri nets or with the help of other models. On the other hand, the model of Petri nets can be the basis for technological process design with initially uncertain character of the transitions. In this case, all possible assembly process variants may be described by the sequence of nodes—assembly elements and subassemblies. The choice of a particular sequence can be determined by means of a logical model and evaluated using definiteness coefficients.

Let's consider the simulation of the assembly robotic complex work using a Petri net. Assembly robotic complex work begins with industrial robot turn on. Further, under the condition of the performed turning on, the manipulator is checked in the start position (at point 1'). At the next step, the manipulator is moved to the location of detail 2 (details store). After that, the detail is clamped with a grip, after which it is transferred to the installation point (destination 1). The next step is to install the detail in the required position and release the gripping device. The manipulator is moved to the destination 1. The number of installed details is incremented, all details are checked, and if not, the manipulator moves to the required point 2 (details store). The cycle is repeated until all details are mount, after which the manipulator arm is moved to the starting point 1’, and the robot turns off. Petri net can be used in determining the order of the assembly process in the robotic complex (Figure 3).

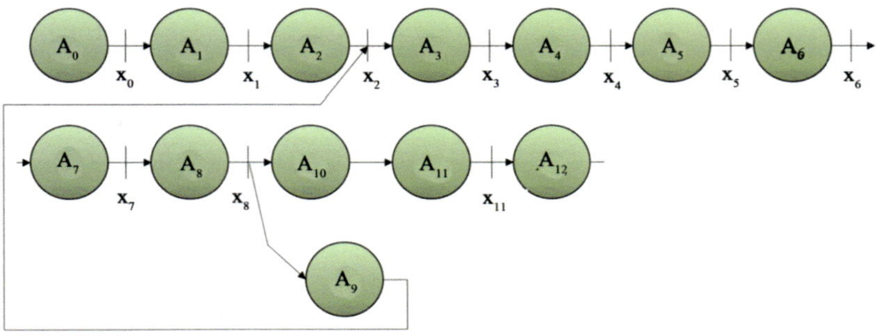

Figure 3. Petri net for assembly robotic complex functioning.

In Figure 3 variables take the following values:

x0—robotic complex on, x1—industrial robot manipulator at the starting point 1’.

x2—industrial robot manipulator at the required point 2.

x3—the part is clamped by industrial robot gripper device.

x4—industrial robot manipulator at the destination 1.

x5—detail.

x6—the detail is unclenched by industrial robot gripper device.

x7—industrial robot manipulator at the destination 1.

x8—i = i + 1.

x9—i<=””” style=”box-sizing: border-box; padding: 0px; margin: 0px;”>

x10—i = n + 1.

x11—industrial robot manipulator at the starting point1’.

A0—is robotic complex turn on.

A1—is the movement of industrial robot manipulator to the start point 1’.

A2—is the movement of industrial robot manipulator PR to the required point 2.

A3—is work piece capture by robot.

A4—moving the manipulator PR to the point 1.

A5—setting the part.

A6—expanding the workpiece by the robot.

A7—moving industrial robot manipulator to the point 1.

A8—i = i + 1.

A9—i<=”” style=”box-sizing: border-box; padding: 0px; margin: 0px;”>

A10—i = n + 1.

A11—move industrial robot manipulator to the starting point 1’.

A12—turn off the robot. It is proposed to use the Petri net to analyze the robot’s technological process to detect errors, the presence of looping and non-rational movements of the robot’s manipulator.

In the event that these situations are detected, it is necessary to correct the process of the robot directly to avoid errors in the robot application program development.

Thus, the model allows to describe detailed the nature of technological transitions in assembly robotic complex technology design. It is also reasonable to use the Petri net in determining the order of the assembly process in the robotic complex, eliminating possible errors, looping and inefficient movements of the robot manipulator.

Given the multivariance of the assembly technology design process and the need to evaluate each option, we should also talk about the applicability of the approach of fuzzy Petri nets.

VOICE CONTROL METHOD FOR AN INDUSTRIAL ROBOT

In the CAD (computer-aided design) development for various purposes, the necessary stage is technological processes models selection, which depends on the modeled object structure, its properties, and the application of the various data processing facilities. The nature of the models determines the essence of design methods.

Robotic production technological processes design is most often performed in off-line mode.

At the initial stages of the design, the sets of working points of robot trajectories are memorized. Next, each set of trajectories is associated with a set of movement commands. Between the move commands, the manipulation of working objects is usually placed. Models set presented above must be used as a sequence. At first the product is analyzed using rules and positions represented in logical model. Then we use semantic model in order to understand connotations between details. After analyzing connotations we have to check standard situations presence using frame. This stage may simplify developing technological process. And next we have to form robot's technological operations and transitions. For this it is expediently to use model based on Petri nets. So all these models are used in order to form sequence of industrial robot actions.

Let us consider voice control method used after product analysis based on models sequence presented above. It is a description of the set of techniques and operations used in robotic production assembly technological processes automated design.

In particular, it provides that the design of the assembly process using voice control for an industrial robot is based on the following principles.

1. Assembly units placement in the robot workspace determines the nature and sequence of technological and auxiliary transitions. In terms of predicate logic, this expression can be written in next way:

$$\forall x\left(\text{type}\left(x,\text{assembly\_unit}\right)\right)\exists y\left(\text{type}\left(y,\text{technological\_transition}\right)\right)$$
$$\exists r\left(\text{type}\left(r,\text{rule}\right)\right)\rightarrow\exists\vec{t}\left(\text{type}\left(\vec{t},\text{technological\_transitions\_sequence}\right)\right).$$

2. Coordinates of any point, any command for moving a manipulator is given by a voice in a limited natural language. This expression can be represented in the following form:

$$\forall p\left(\text{type}\left(p,\text{point}\right)\right)\forall c\left(\text{type}\left(c,\text{command}\right)\right)$$
$$\exists vp\left(\text{type}\left(vp,\text{voice\_presentation}\right)\right).$$

3. Commands, the sequence of which provides the fulfillment of the goals of technological or auxiliary transitions, are combined into meta-commands. This expression can be written in next way:

$$\exists c\left(\text{type}\left(c,\text{command}\right)\right)c\in y\left(\text{type}\left(y,\text{technological\_transition}\right)\right)$$
$$\wedge y\in T\left(\text{type}\left(T,\text{typical\_technological\_transition}\right)\right)$$
$$\rightarrow\exists mc\left(\text{type}\left(mc,\text{meta}-\text{command}\right)\right).$$

4. Sequences of control commands, given by voice, realize separate technological or auxiliary transitions, separate technological operations and form the technological process. This provision is represented by the following group of formal expressions:

$$\exists vp(type(vp, voice\_presentation)) \rightarrow \exists \vec{VP}\left(type\left(\vec{VP}, voice\_commands\_sequence\right)\right),$$

$$\exists \vec{VP} \rightarrow \exists \vec{t}\left(type\left(\vec{t}, technological\_transitions\_sequence\right)\right),$$

$$\exists \vec{t} \rightarrow \exists \vec{to}\left(type\left(\vec{to}, technological\_operations\_sequence\right)\right),$$

$$\exists \vec{to} \rightarrow \exists \vec{tp}\left(type\left(\vec{tp}, technological\_process\right)\right).$$

5. Commands voice input application order is established on the basis of the assembly drawing analysis and is determined by the above models of assembly products and technological processes representation.
6. Information voice input ensures the work for information input subsystem in the CAD of robotic production control programs.

CONCLUSIONS

In the article, robotic production assembly technological processes design models are considered. The formal representation of assembly processes is based on the general representation of the assembly detail as an ordered set of assembly units. The nature of ordering is determined by the design intent of the assembly detail, taking into account the interaction of assembly units, their dimensional relationships, formal and informal logical patterns of technological processes design.

Based on the properties of assembly technological processes design, technological process logical model was developed. Technological process model based on semantic networks is proposed, which describes the relationship between assembly units in a detail. To identify typical situations in assembly process sequence, technological process frame model was developed, which makes it possible to order the assembly process execution. Assembly technological process model based on Petri nets is proposed, which allows describing in detail the type and sequence of technological transitions.

The specified set of models will allow describing technological process design process from different points of view and taking into account voice control using, provide a holistic description and prepare the basis for the practical implementation of the voice control subsystem in the CAD robot technological process system. Based on the proposed models, a voice control method for an industrial robot is developed, the main principles of which are given in the paper.

It should also be noted that using the ideology of image processing or wavelet analysis methodology [38] - [45] together with voice control of the robot will create new applications for technological processes.

REFERENCES

1. Li, Z., Tang, Q. and Zhang, L. (2016) Minimizing Energy Consumption and Cycle Time in Two-Sided Robotic Assembly Line Systems Using Restarted Simulated Annealing Algorithm. Journal of Cleaner Production, 135, 508-522. https://doi.org/10.1016/j.jclepro.2016.06.131
2. Faber, M., Kuz, S., Mayer, M.P. and Schlick, C.M. (2013) Design and Implementation of a Cognitive Simulation Model for Robotic Assembly Cells. International Conference on Engineering Psychology and Cognitive Ergonomics, Springer, Berlin, Heidelberg, 205-214.
3. Abd, K., Abhary, K. and Marian, R. (2014) Simulation Modelling and Analysis of Scheduling in Robotic Flexible Assembly Cells Using Taguchi Method. International Journal of Production Research, 52, 2654-2666. https://doi.org/10.1080/00207543.2013.867082
4. Michalos, G., Kaltsoukalas, K., Aivaliotis, P., Sipsas, P., Sardelis, A. and Chryssolouris, G. (2014) Design and Simulation of Assembly Systems with Mobile Robots. CIRP Annals-Manufacturing Technology, 63, 181-184. https://doi.org/10.1016/j.cirp.2014.03.102
5. Mourtzis, D., Papakostas, N., Mavrikios, D., Makris, S. and Alexopoulos, K. (2015) The Role of Simulation in Digital Manufacturing: Applications and Outlook. International Journal of Computer Integrated Manufacturing, 28, 3-24. https://doi.org/10.1080/0951192X.2013.800234
6. Nie, S., Li, Y., Shuai, G., Tao, S. and Xi, F. (2016) Modeling and Simulation for Fatigue Life Analysis of Robots with Flexible Joints under Percussive Impact Forces. Robotics and Computer-Integrated Manufacturing, 37, 292-301. https://doi.org/10.1016/j.rcim.2015.04.001
7. Lipson, H. (2014) Challenges and Opportunities for Design, Simulation, and Fabrication of Soft Robots. Soft Robotics, 1, 21-27. https://doi.org/10.1089/soro.2013.0007
8. Feng, C., Xiao, Y., Willette, A., McGee, W. and Kamat, V.R. (2015) Vision Guided Autonomous Robotic Assembly and As-Built Scanning on Unstructured Construction Sites. Automation in Construction, 59, 128-138. https://doi.org/10.1016/j.autcon.2015.06.002
9. Wahrburg, A., Zeiss, S., Matthias, B., Peters, J. and Ding, H. (2015). Combined Pose-Wrench and State Machine Representation for

Modeling Robotic Assembly Skills. 2015 IEEE/RSJ International Conference on Intelligent Robots and Systems (IROS), 852-857. https://doi.org/10.1109/IROS.2015.7353471

10. Rohmer, E., Singh, S.P. and Freese, M. (2013) V-REP: A Versatile and Scalable Robot Simulation Framework. 2013 IEEE/RSJ International Conference on Intelligent Robots and Systems (IROS), 1321-1326. https://doi.org/10.1109/IROS.2013.6696520
11. Pleshkova, S., Bekiarski, A., Dehkharghani, S.S. and Peeva, K. (2015) Perception of Audio Visual Information for Mobile Robot Motion Control Systems. In: Computer Vision in Control Systems-2, Springer International Publishing, 135-167. https://doi.org/10.1007/978-3-319-11430-9_6
12. Tasevski, J., Nikolic, M. and Miskovic, D. (2013) Integration of an Industrial Robot with the Systems for Image and Voice Recognition. Serbian Journal of Electrical Engineering, 10, 219-230. https://doi.org/10.2298/SJEE1301219T
13. Tsarouchi, P., Makris, S. and Chryssolouris, G. (2016) Human-Robot Interaction Review and Challenges on Task Planning and Programming. International Journal of Computer Integrated Manufacturing, 29, 916-931. https://doi.org/10.1080/0951192X.2015.1130251
14. Gupta, M., Verma, P., Bhattacharya, T. and Das, P.K. (2015) A Mobile Agents Based Distributed Speech Recognition Engine for Controlling Multiple Robots. In: Proceedings of the 2015 Conference on Advances in Robotics, ACM, New York, 28. https://doi.org/10.1145/2783449.2783477
15. Poncela, A. and Gallardo-Estrella, L. (2015) Command-Based Voice Teleoperation of a Mobile Robot via a Human-Robot Interface. Robotica, 33, 1-18. https://doi.org/10.1017/S0263574714000010
16. Kar, S., Jana, A., Chatterjee, D., Mitra, D., Banerjee, S., Kundu, D. and Gupta, S.D. (2015) Image Processing Based Customized Image Editor and Gesture Controlled Embedded Robot Coupled with Voice Control Features. Image, 6, 91-96. https://doi.org/10.14569/IJACSA.2015.061113
17. Silaghi, H., Rohde, U., Spoiala, V., Silaghi, A., Gergely, E. and Nagy, Z. (2014) Voice Command of an Industrial Robot in a Noisy Environment. 2014 International Symposium on Fundamentals of Electrical Engineering, 1-5. https://doi.org/10.1109/ISFEE.2014.7050596

18. Mateo, C., Brunete, A., Gambao, E. and Hernando, M. (2014) Hammer: An Android Based Application for End-User Industrial Robot Programming. 2014 IEEE/ ASME 10th International Conference on Mechatronic and Embedded Systems and Applications, 1-6. https://doi.org/10.1109/MESA.2014.6935597
19. Fang, H.C., Ong, S.K. and Nee, A.Y.C. (2014) A Novel Augmented Reality-Based Interface for Robot Path Planning. International Journal on Interactive Design and Manufacturing, 8, 33-42. https://doi.org/10.1007/s12008-013-0191-2
20. Rogowski, A. (2013) Web-Based Remote Voice Control of Robotized Cells. Robotics and Computer-Integrated Manufacturing, 29, 77-89.
21. Michalos, G., Sipsas, P., Makris, S. and Chryssolouris, G. (2016) Decision Making Logic for Flexible Assembly Lines Reconfiguration. Robotics and Computer-Integrated Manufacturing, 37, 233-250.
22. Abd, K., Abhary, K. and Marian, R. (2013) A Methodology for Scheduling Robotic Flexible Assembly Cells Using Fuzzy Logic and Simulation. Proceedings of the World Congress on Engineering, Vol. 1, London, 3-5 July 2013.
23. Thramboulidis, K. (2016) An Open Distributed Architecture for Flexible Hybrid Assembly Systems: A Model-Driven Engineering Approach. The International Journal of Advanced Manufacturing Technology, 85, 1449-1460. https://doi.org/10.1007/s00170-015-8064-4
24. Barzani, M.M., Zalnezhad, E., Sarhan, A.A., Farahany, S. and Ramesh, S. (2015) Fuzzy Logic Based Model for Predicting Surface Roughness of Machined Al-Si-Cu-Fe Die Casting Alloy Using Different Additives-Turning. Measurement, 61, 150-161.
25. Fattahi, A., Elaoud, S., SadeqiAzer, E. and Turkay, M. (2014) A Novel Integer Programming Formulation with Logic Cuts for the U-Shaped Assembly Line Balancing Problem. International Journal of Production Research, 52, 1318-1333. https://doi.org/10.1080/00207543.2013.832489
26. Pintzos, G., Nikolakis, N., Alexopoulos, K. and Chryssolouris, G. (2016) Motion Parameters Identification for the Authoring of Manual Tasks in Digital Human Simulations: An Approach Using Semantic Modelling. Procedia CIRP, 41, 752-758.
27. Zhong, Y., Qin, Y., Huang, M., Lu, W. and Chang, L. (2014) Constructing a Meta-Model for Assembly Tolerance Types with a Description Logic

Based Approach. Computer-Aided Design, 48, 1-16.

28. Dogar, M., Knepper, R.A., Spielberg, A., Choi, C., Christensen, H.I. and Rus, D. (2016) Towards Coordinated Precision Assembly with Robot Teams. In: Experimental Robotics, Springer International Publishing, 655-669. https://doi.org/10.1007/978-3-319-23778-7_43
29. Yang, Q., Wu, D.L., Zhu, H.M., Bao, J.S. and Wei, Z.H. (2013) Assembly Operation Process Planning by Mapping a Virtual Assembly Simulation to Real Operation. Computers in Industry, 64, 869-879.
30. Wang, G.D., Yang, Y., Wang, W. and Si-Chao, L.V. (2016) Variable Coefficients Reciprocal Squared Model Based on Multi-Constraints of Aircraft Assembly Tolerance Allocation. The International Journal of Advanced Manufacturing Technology, 82, 227-234. https://doi.org/10.1007/s00170-015-7299-4
31. Luo, J., Liu, J. and Hu, Y. (2017) An MILP Model and a Hybrid Evolutionary Algorithm for Integrated Operation Optimisation of Multi-Head Surface Mounting Machines in PCB Assembly. International Journal of Production Research, 55, 145-160. https://doi.org/10.1080/00207543.2016.1200154
32. Westover, L., Olearczyk, J., Hermann, U., Adeeb, S. and Mohamed, Y. (2014) Analysis of Rigging Assembly for Lifting Heavy Industrial Modules. Canadian Journal of Civil Engineering, 41, 512-522. https://doi.org/10.1139/cjce-2013-0192
33. Hu, H. and Zhou, M. (2015) A Petri Net-Based Discrete-Event Control of Automated Manufacturing Systems with Assembly Operations. IEEE Transactions on Control Systems Technology, 23, 513-524. https://doi.org/10.1109/TCST.2014.2342664
34. Wei, N.C., Chao, I.M., Liuand, C.J. and Chen, H.L. (2014) Applying Branch-and-Bound and Petri Net Methods in Solving the Two-Sided Assembly Line Balancing Problem. World Academy of Science, Engineering and Technology, International Journal of Mechanical, Aerospace, Industrial, Mechatronic and Manufacturing Engineering, 8, 136-141.
35. Hu, H., Zhou, M., Li, Z. and Tang, Y. (2013) Deadlock-Free Control of Automated Manufacturing Systems with Flexible Routes and Assembly Operations Using Petri Nets. IEEE Transactions on Industrial Informatics, 9, 109-121. https://doi.org/10.1109/TII.2012.2198661

36. Yang, Y., Hu, H. and Liu, Y. (2015) A Petri Net-Based Distributed Control of Automated Manufacturing Systems with Assembly Operations. 2015 IEEE International Conference on Automation Science and Engineering, 1090-1097. https://doi.org/10.1109/CoASE.2015.7294244
37. Yue, H., Hu, H., Wu, W., Su, H. and Zhang, J. (2017) Polynomial-Complexity Supervisory Control for Flexible Assembly Systems Based on Petri Nets. International Journal of Computer Integrated Manufacturing, 1-16. https://doi.org/10.1080/0951192X.2017.1335436
38. Lyashenko, V., Matarneh, R., Kobylin, O. and Putyatin, Y. (2016) Contour Detection and Allocation for Cytological Images Using Wavelet Analysis Methodology. International Journal of Advance Research in Computer Science and Management Studies, 4, 85-94.
39. Lyashenko, V., Matarneh, R. and Kobylin, O. (2016) Contrast Modification as a Tool to Study the Structure of Blood Components. Journal of Environmental Science, Computer Science and Engineering & Technology, 5, 150-160.
40. Lyashenko, V.V., Matarneh, R. and Deineko, Z.V. (2016) Using the Properties of Wavelet Coefficients of Time Series for Image Analysis and Processing. Journal of Computer Sciences and Applications, 4, 27-34.
41. Ahmad, M.A., Lyashenko, V.V., Deineko, Z.V., Baker, J.H. and Ahmad, S. (2017) Study of Wavelet Methodology and Chaotic Behavior of Produced Particles in Different Phase Spaces of Relativistic Heavy Ion Collisions. Journal of Applied Mathematics and Physics, 5, 1130. https://doi.org/10.4236/jamp.2017.55100
42. Lyashenko, V., Babker, A. and Lyubchenko, V. (2017) Wavelet Analysis of Cytological Preparations Image in Different Color Systems. Open Access Library Journal, 4, 1-9.
43. Lyashenko, V.V., Ahmad, M.A., Deineko, Z.V. and Rasool, M.H. (2017) Methodology of Wavelets in Relativistic Heavy Ion Collisions in One Dimensional Phase Space. Journal of High Energy Physics, 3, 254-266.
44. Lyashenko, V.V., AMAaa, B. and Kobylin, O.A. (2016) Using the Methodology of Wavelet Analysis for Processing Images of Cytology Preparations. National Journal of Medical Research, 6, 98-102.

45. Lyashenko, V.V., Babker, A.M.A.A. and Kobylin, O.A. (2016) The Methodology of Wavelet Analysis as a Tool for Cytology Preparations Image Processing. Cukurova Medical Journal, 41, 453-463. https://doi.org/10.17826/cukmedj.237468

CHAPTER 9

Hamiltonian Servo: Control and Estimation of a Large Team of Autonomous Robotic Vehicles

Vladimir Ivancevic[1] and Peyam Pourbeik[2]

[1]Joint and Operations Analysis Division, Defence Science & Technology Group, Canberra, Australia.

[2]CEWD, Assured Communications, Survivable Networks, Defence Science & Technology Group, Adelaide, Australia.

ABSTRACT

This paper proposes a novel Hamiltonian servo system, a combined modeling framework for control and estimation of a large team/fleet of autonomous robotic vehicles. The Hamiltonian servo framework represents high-dimensional, nonlinear and non-Gaussian generalization of the classical

Citation: Ivancevic, V. and Pourbeik, P. (2017). "Hamiltonian Servo: Control and Estimation of a Large Team of Autonomous Robotic Vehicles". Intelligent Control and Automation, 8, 175-197. doi: 10.4236/ica.2017.84014.

Kalman servo system. After defining the Kalman servo as a motivation, we define the affine Hamiltonian neural network for adaptive nonlinear control of a team of UGVs in continuous time. We then define a high-dimensional Bayesian particle filter for estimation of a team of UGVs in discrete time. Finally, we formulate a hybrid Hamiltonian servo system by combining the continuous-time control and the discrete-time estimation into a coherent framework that works like a predictor-corrector system.

Keywords: Team of UGVs, Kalman Servo, Hamiltonian Control, Bayesian Estimation

INTRODUCTION

Today's military forces face an increasingly complex operating environment, whether dealing with humanitarian operations or in the theater of war. In many situations one has to deal with a wide variety of issues from logistics to lack of communications infrastructure [1] . In many situations one may be even denied satellite access and hence unable to utilize services such as GPS for Geo-location and the ability reconnect back to HQ [2] .

Given the increasing success of artificial intelligence (AI) techniques in recent times and the advancements made in unmanned vehicle (UV) design, autonomous systems in particular teams or swarms of autonomous vehicles have become a potential attractive solution to solving the complexity issues facing the military today. Teams or swarms of intelligent autonomous vehicles would also introduce a level of survivability to the systems which would increase its utility in such complex and hostile environments [3] .

A problem which arises in utilizing large teams or swarms of autonomous vehicles is one of control. Recently, the US Defence Advanced Research Projects Agency (DARPA) has sought to address some of the control issues associated with large swarm through its Offensive Swarm-Enabled Tactics (OFFSET) [4] . Therefore, there is a clear need for both local and global control of large teams or swarms of UV. This would allow for autonomous or semi-autonomous operation of large teams of UV which would be easily manageable.

This paper proposes a Hamiltonian servo system which is a combined modeling framework for the control and estimation of such autonomous teams of UV's. This paper will demonstrate the ability of this framework to handle the high-dimensional, nonlinear and non-Gaussian nature of the problem at hand. The authors will also show this to be a generalization of the classic Kalman Servo System. Finally results of a number of simulations will be presented to demonstrate the validity of the framework through the authors implementation using Mathematica & C++ code (which is provided in the Appendix).

Slightly more specifically, we recall here that traditional robotics from the 1970s-80s were mainly focused on building humanoid robots, manipulators and leg locomotion robots. Its modeling framework was linear dynamics and linear control, that is, linearized mechanics of multi-body systems (derived using Newton-Euler, Lagrangian, Gibs-Appel or kinetostatics equations of motion) and controlled by Kalman's linear quadratic controllers (for a comprehensive review, see [5] - [11]). The pinnacle of this approach to robotics in the last decade has been the famous Honda robot ASIMO (see [12]), with a related Hamiltonian biomechanical simulator [13] .

In contrast, contemporary robotics research has mainly been focused on field robotics, based on estimation rather than control, resulting in a variety of self-localization and mapping (SLAM) algorithms (see [14] [15] [16] [17] [18] and the OpenSLAM web-site [19]).

Instead of following one of these two sides of robotics, in the present paper we will try to combine them. From a robotics point-of-view, control and estimation are two complementary sides, or necessary components, of efficient manipulation of autonomous robotic vehicles. In this paper we attempt to develop a computational framework (designed in Wolfram Mathematica[Ò] and implemented in C++) that unifies both of these components. From a mathematical point-of-view, in this paper we attempt to provide nonlinear, non-Gaussian and high-dimensional generalization of the classical Kalman servo system, outlined in the next section.

MOTIVATION: KALMAN SERVO

In 1960s, Rudolf Kalman developed concept of a controller-estimator servo system (see Figure 1; for original Kalman's work see [20] [21] [22]), which consists of the following three main components, written in classical boldface matrix notation:

- **System state dynamics**: linear, low-dimensional, multi-input multi-output process cascade (as implemented in Matlab[Ò] Control and Signal Processing toolboxes):

$$\dot{\boldsymbol{x}} = \boldsymbol{A}\boldsymbol{x} + \boldsymbol{B}\boldsymbol{u}, \quad \boldsymbol{y} = \boldsymbol{C}\boldsymbol{x} + \boldsymbol{D}\boldsymbol{u}, \quad \left[\boldsymbol{x}(0) = \boldsymbol{x}_0\right], \tag{1}$$

where we have three vector spaces: state space $\mathbb{X} \subset \mathbb{R}^n$, input space $\mathbb{U} \subset \mathbb{R}^m$ and output space $\mathbb{Y} \subset \mathbb{R}^k$, such that $x(t) \in \mathbb{X}$ is an n-vector of state variables, $u(t) \in \mathbb{U}$ is an m-vector of inputs and $y(t) \in \mathbb{Y}$ is a k-vector of outputs; x_0 is the initial state vector. The matrix variables are the following maps: i) $n \times n$ dimensional state dynamics $A = A(t) : \mathbb{X} \to \mathbb{X}$; ii) $n \times m$ dimensional input map $B = B(t) : \mathbb{U} \to \mathbb{X}$; iii) $k \times n$ dimensional output map $C = C(t) : \mathbb{X} \to \mathbb{Y}$; and (iv) $k \times m$ dimensional input-output transform $D = D(t) : \mathbb{U} \to \mathbb{Y}$. The first equation in (1) is called the state or dynamics equation and the second equation in (1) is called the output or measurement equation.

- **Kalman feedback controller**: optimal LQR-controller:[1]

$$\boldsymbol{y} = \boldsymbol{K}\left[\boldsymbol{x}_{ref}(t) - \boldsymbol{x}(t)\right], \tag{2}$$

with the controller gain matrix: $\boldsymbol{K} = \boldsymbol{K}(t) : \mathbb{X} \to \mathbb{Y} = \boldsymbol{R}_c^{-1}\boldsymbol{B}^{\mathrm{T}}\boldsymbol{P}_c$ determined by the matrix P_c governed by the controller Riccati ordinary differential equation (ODE):

$$\dot{\boldsymbol{P}}_c = \boldsymbol{A}^{\mathrm{T}}\boldsymbol{P}_c + \boldsymbol{P}_c\boldsymbol{A} + \boldsymbol{Q}_c - \boldsymbol{P}_c\boldsymbol{B}\boldsymbol{R}_c^{-1}\boldsymbol{B}^{\mathrm{T}}\boldsymbol{P}_c, \quad \left[\boldsymbol{P}_c(0) = \boldsymbol{P}_0^c\right],$$

where $\boldsymbol{P}_0^C$ is the initial controller matrix. $Q_c : \mathbb{X} \to \mathbb{X}$ and $R_c : \mathbb{U} \to \mathbb{U}$ are matrices from the quadratic cost function:

$$J = \int_{t_0}^{t_1} \left(\boldsymbol{x}^{\mathrm{T}} Q \boldsymbol{x} + \boldsymbol{u}^{\mathrm{T}} R \boldsymbol{u}\right) \mathrm{d}t. \tag{3}$$

- **Kalman filter**: optimal LQG-estimator[2] with the (additive, zero-mean, delta- correlated) Gaussian white noise $\eta(t)$:

$$\dot{\boldsymbol{x}} = \boldsymbol{A}\boldsymbol{x} + \boldsymbol{B}\boldsymbol{u} + \boldsymbol{L}\left[\boldsymbol{y} - (\boldsymbol{C}\boldsymbol{x} + \boldsymbol{D}\boldsymbol{u})\right] + \boldsymbol{\eta}, \quad \left[\boldsymbol{x}(0) = \boldsymbol{x}_0\right], \tag{4}$$

with the filter gain matrix: $\boldsymbol{L} = \boldsymbol{L}(t) : \mathbb{X} \to \mathbb{Y} = \boldsymbol{R}_f^{-1}\boldsymbol{P}_f\boldsymbol{C}^{\mathrm{T}}$ determined by the matrix P_f governed by the filter Riccati ODE:

$$\dot{\boldsymbol{P}}_f = \boldsymbol{A}^{\mathrm{T}}\boldsymbol{P}_f + \boldsymbol{P}_f\boldsymbol{A} + \boldsymbol{Q}_f - \boldsymbol{P}_f\boldsymbol{C}^{\mathrm{T}}\boldsymbol{R}_f^{-1}\boldsymbol{C}\boldsymbol{P}_f, \left[\boldsymbol{P}_f(0) = \boldsymbol{P}_0^f\right],$$

where $\boldsymbol{P}_0^f$ is the initial filter matrix. $Q_f : \mathbb{X} \to \mathbb{X}$ and $R_f : \mathbb{U} \to \mathbb{U}$ are covariance matrices from the noisy η-generalization of the quadratic cost function (3).

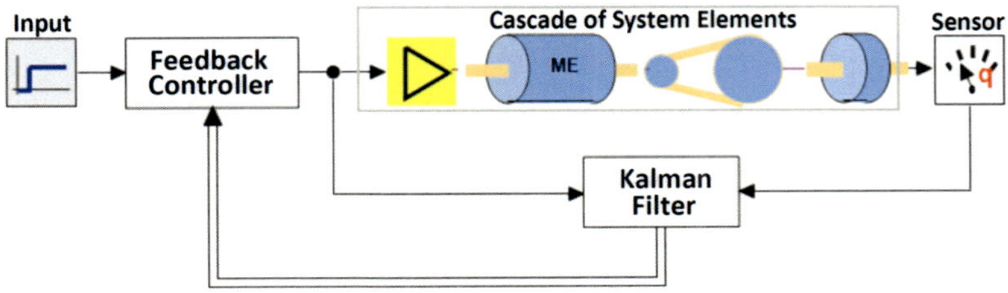

Figure 1. Kalman's fundamental controller-estimator servo system, including state-space dynamics, Kalman filter and feedback controller.

Defined in this way, Kalman's controller-estimator servo system has a two-cycle action, similar to the predictor-corrector algorithm: in the first time-step the servo controls the plant (e.g., a UGV), in the second step it estimates the plants (x, y) coordinates, in the next step it corrects the control, then estimates again, and so on.

In the remainder of this paper we will develop a nonlinear, non-Gaussian and high-dimensional generalization of the Kalman servo.

CONTROL OF AN AUTONOMOUS TEAM OF UGVS: AFFINE HAMILTONIAN NEURAL NETWORK

Now we switch from matrix to ($\alpha, \beta = 1, \cdots, n$)-index notation,[3] to label the position of $n \times n$ individual UGVs within the swarm's global plane coordinates, longitude and latitude, respectively. The first step in the nonlinear generalization of the Kalman servo is replacing the linear, low-dimensional, state dynamics (1) and control (2)-(3) with nonlinear, adaptive and high-dimensional dynamics and control system, as follows.

In the recent paper [23] , we have proposed a dynamics and control model for a joint autonomous land-air operation, consisting of a swarm of unmanned ground vehicles (UGVs) and swarm of unmanned aerial vehicles (UAVs). In the present paper we are focusing on a swarm of UGVs only, from both control and estimation perspectives. It can be modeled as an affine Hamiltonian control system (see [24]) with many degrees-of-freedom (DOFs), reshaped in the form of a (q, p)-pair of attractor matrix equations with nearest-neighbor couplings defined by the matrix of affine Hamiltonians:

$$H_{\alpha\beta} = \frac{1}{2}\left(q_{\alpha\beta}q_{\alpha-1\beta-1} + p_{\alpha\beta}p_{\alpha-1\beta-1}\right).$$

The following pair of attractor matrix equations defines the activation dynamics of a bidirectional recurrent neural network:

$$\dot{q}_{\alpha\beta} = \varphi\left(A^q - q_{\alpha\beta} - \omega^q_{\alpha\beta} q^2_{\alpha\beta} p_{\alpha\beta}\right) - \sum_{j,k} \partial_{p_{\alpha\beta}} H_{\alpha\beta} u^q_{\alpha\beta}, \tag{5}$$

$$\dot{p}_{\alpha\beta} = \varphi\left(A^p - p_{\alpha\beta} - \omega^p_{\alpha\beta} p^2_{\alpha\beta} q_{\alpha\beta}\right) + \sum_{j,k} \partial_{q_{\alpha\beta}} H_{\alpha\beta} u^p_{\alpha\beta}, \tag{6}$$

where superscripts (q, p) denote the corresponding Hamiltonian equations and the partial derivative is written as: $\partial_u H \equiv \partial H/\partial u$. In Equations ((5), (6)) the matrices $q_{\alpha\beta} = q_{\alpha\beta}(t)$ and $p_{\alpha\beta} = p_{\alpha\beta}(t)$ define time evolution of the swarm's coordinates and momenta, respectively, with initial conditions: $q_{\alpha\beta}(0)$ and $p_{\alpha\beta}(0)$. The field attractor lines A^q and A^p (with the field strength φ) are defined in the simulated urban environment using the A-star algorithm that finds the shortest Manhattan distance from point A to point B on the environmental digital map (i.e., street directory). The adaptive synaptic weights matrices: $\omega^q_{\alpha\beta} = \omega^q_{\alpha\beta}(t)$ and $\omega^p_{\alpha\beta} = \omega^p_{\alpha\beta}(t)$ are trained by Hebbian learning (11) defined below. Input matrices $u^q_{\alpha\beta} = u^q_{\alpha\beta}(t)$ and $u^p_{\alpha\beta} = u^p_{\alpha\beta}(t)$ are Lie-derivative based controllers (explained below).

Equations ((5), (6)) are briefly derived as follows (for technical details, see [23] and the references therein). Given the swarm configuration manifold M (as a product of Euclidean groups of motion for each robotic vehicle), the affine Hamiltonian function $H_a(q, p, u) : T_*M \to \mathbb{R}$ is defined on its cotangent bundle T_*M by (see [24]):

$$H_a(q,p,u) = H_0(q,p) - \sum_{j<n} H_j(q,p)u_j(t,q,p), \tag{7}$$

where $H_0(q, p) : T_*M \to \mathbb{R}$ is the physical Hamiltonian function (kinetic plus potential energy of the swarm) and control inputs $u_j(t, q, p)$ are defined using recurrent Lie derivatives:[4]

$$u_j = \frac{\dot{o}_R^{(r)j} - L_f^{(r)} H_j + \sum_{s=1}^{r} c_{s-1}\left(o_R^{(s-1)j} - L_f^{(s-1)} H_j\right)}{L_g L_f^{(r-1)} H_j}. \tag{8}$$

Using (7) and (8), the affine Hamiltonian control system can be formulated (for i = 1, ⋯ , n; j < i) as:

$$\dot{q}_i = V_i + \partial_{p_i} H_0 - \sum_{j<i} \partial_{p_i} H_j u_j + \partial_{p_i} D, \tag{9}$$

$$\dot{p}_i = F_i - \partial_{q_i} H_0 + \sum_{j<i} \partial_{q_i} H_j u_j + \partial_{q_i} D. \tag{10}$$

where D is Rayleigh's dissipative function and $V_i = V_i(t, q, p)$ and $F_i = F_i(t, q, p)$ are velocity and force controllers, respectively.

The affine Hamiltonian control system (9)-(10) can be reshaped into a matrix form suitable for a UGV-swarm or a planar formation of a UAV-swarm: $q_i \to q_{\alpha\beta}$, $p_i \to p_{\alpha\beta}$, $u_i \to u_{\alpha\beta}$, by evaluating dissipation terms into cubic terms in the brackets, and replacing velocity and force controllers with attractors A^q and A^p, respectively (with strength φ). If we add synaptic weights $\omega_{\alpha\beta}$ we come to our recurrent neural network Equations ((5), (6)).

Next, to make Equations ((5), (6)) adaptive, we use the abstract Hebbian learning scheme:

New Value = Old Value + Innovation

which in our settings formalizes as (see [25]):

$$\dot{\omega}^q_{\alpha\beta} = -\omega^q_{\alpha\beta} + \Phi^q_{\alpha\beta}\left(q_{\alpha\beta}, p_{\alpha\beta}\right),$$

$$\dot{\omega}^p_{\alpha\beta} = -\omega^p_{\alpha\beta} + \Phi^p_{\alpha\beta}\left(q_{\alpha\beta}, p_{\alpha\beta}\right), \tag{11}$$

where the innovations are given in the matrix signal form (generalized from [26]):

$$\Phi^q_{\alpha\beta} = S^q_{\alpha\beta}\left(q_{\alpha\beta}\right)S^q_{\alpha\beta}\left(p_{\alpha\beta}\right) + \dot{S}^q_{\alpha\beta}\left(q_{\alpha\beta}\right)\dot{S}^q_{\alpha\beta}\left(p_{\alpha\beta}\right),$$

$$\Phi^p_{\alpha\beta} = S^p_{\alpha\beta}\left(q_{\alpha\beta}\right)S^p_{\alpha\beta}\left(p_{\alpha\beta}\right) + \dot{S}^p_{\alpha\beta}\left(q_{\alpha\beta}\right)\dot{S}^p_{\alpha\beta}\left(p_{\alpha\beta}\right), \tag{12}$$

with sigmoid activation functions $S(\cdot) = tanh(\cdot)$ and signal velocities: $\dot{S}(\cdot) = 1 - tanh(\cdot)$.

In this way defined affine Hamiltonian control system (5)-(6), which is also a recurrent neural network with Hebbian leaning (11)-(12), represents our nonlinear, adaptive and high-dimensional generalization of Kalman's linear state dynamics (1) and control (2)-(3).

Using Wolfram Mathematica$^{\grave{O}}$ code given below, the Hamiltonian neural net was simulated for 1 sec, with the matrix dimensions n = 10 (i.e., with 100 neurons in both matrices q_{ij} and p_{ij}, which are longitudes and latitudes of a large fleet of 100 UGVs, see Figures 2-4). The Figures illustrate both stability and convergence of the recurrent Hamiltonian neural net.

C++ code for the the Hamiltonian neural net is given in Appendix 1.1.

Mathematica® code

Dimensions:

$n = 10; \text{Tfin} = 1; \varphi = 30;$

Coordinates and momenta:

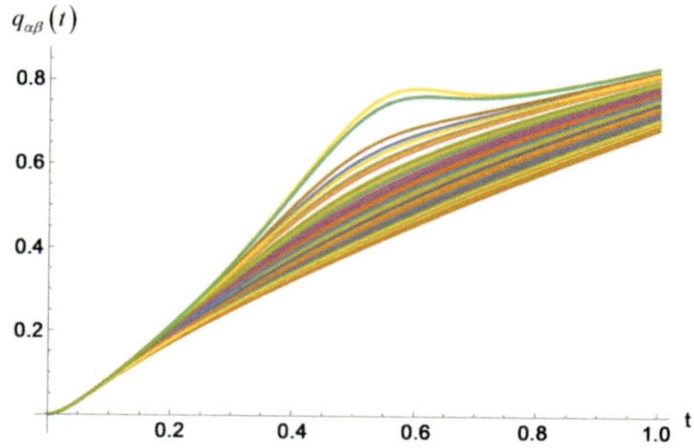

Figure 2. Adaptive control for a large fleet (a 10 × 10 matrix) of 100 UGVs: time evolution of coordinates $q_{\alpha\beta}(t)$, where $\alpha, \beta = 1, \cdots, 10$. As expected, we can see the exponential-type growth of coordinates $q_{\alpha\beta}(t)$ together with their spreading from zero initial conditions. This plot is the simulation output of the recurrent activation dynamics given by Equations ((5), (6)) with Hebbian learning dynamics given by Equations ((11), (12)), starting from zero initial coordinates and momenta and random (−1,1) initial weights.

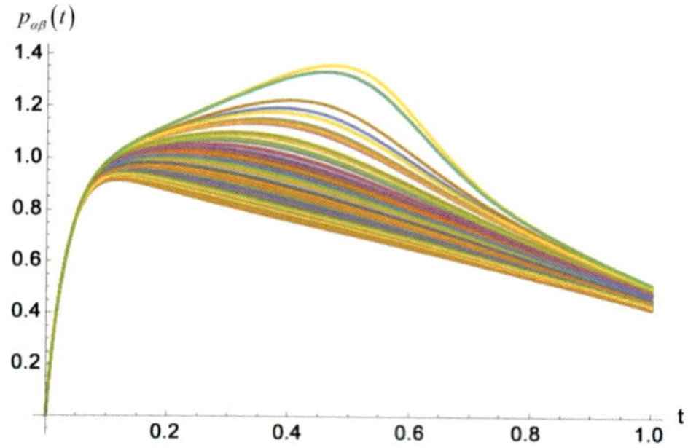

Figure 3. Adaptive control for a large fleet (a 10 × 10 matrix) of 100 UGVs: time evolution of momenta $p_{\alpha\beta}(t)$. As expected, we can see the exponential-type

decay of momenta $p_{\alpha\beta}(t)$ together with their spreading, after the initial convergent growth from zero initial conditions. This plot is the simulation output from the same recurrent neural network as in Figure 2.

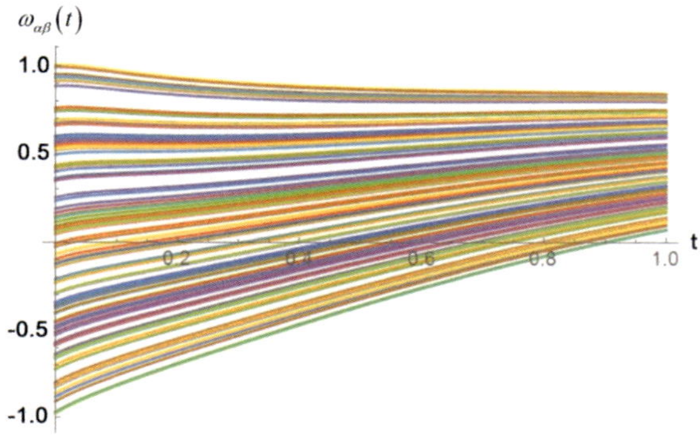

Figure 4. Adaptive control for a large fleet (a 10 × 10 matrix) of 100 UGVs: time evolution of synaptic weights $\omega_{\alpha\beta}(t)$. As expected, we can see the converging behavior of the weights $\omega_{\alpha\beta}(t)$, starting from their initial random distribution. Simulating the same recurrent neural network from Figure 2.

$$\text{Table}\left[q_{\alpha,\beta}[t],\{\alpha,n\},\{\beta,n\}\right];$$
$$\text{Table}\left[p_{\alpha,\beta}[t],\{\alpha,n\},\{\beta,n\}\right];$$

Hyperbolic tangent activation functions:

$$\text{Table}\left[\text{Sq}_{\alpha,\beta}=\text{Tanh}\left[q_{\alpha,\beta}[t]\right],\{\alpha,n\},\{\beta,n\}\right];$$
$$\text{Table}\left[\text{Sp}_{\alpha,\beta}=\text{Tanh}\left[p_{\alpha,\beta}[t]\right],\{\alpha,n\},\{\beta,n\}\right];$$

Derivatives of activation functions:

$$\text{Table}\left[\text{Sdq}_{\alpha,\beta}=1-\text{Sq}^2_{\alpha,\beta},\{\alpha,n\},\{\beta,n\}\right];$$
$$\text{Table}\left[\text{Sdp}_{\alpha,\beta}=1-\text{Sp}^2_{\alpha,\beta},\{\alpha,n\},\{\beta,n\}\right];$$

Control inputs:

$$\text{Table}\left[a_{\alpha,\beta}=\text{RandomReal}\left[\{-1,1\}\right],\{\alpha,n\},\{\beta,n\}\right];$$
$$\text{Table}\left[b_{\alpha,\beta}=\text{RandomReal}\left[\{-1,1\}\right],\{\alpha,n\},\{\beta,n\}\right];$$
$$\text{Table}\left[c_{\alpha,\beta}=\text{RandomReal}\left[\{-1,1\}\right],\{\alpha,n\},\{\beta,n\}\right];$$
$$\text{Table}\left[u_{\alpha,\beta}[t]=a_{\alpha,\beta}\text{Sin}\left[b_{\alpha,\beta}+c_{\alpha,\beta}t\right],\{\alpha,n\},\{\beta,n\}\right];$$

Affine Hamiltonians (nearest-neighbor coupling):

$$\text{Table}\left[H_{\alpha,\beta}=\frac{1}{2}\left(q_{\alpha,\beta}[t]q_{\alpha-1,\beta-1}[t]+p_{\alpha,\beta}[t]p_{\alpha-1,\beta-1}[t]\right),\{\alpha,2,n\},\{\beta,2,n\}\right];$$

Attractor lines (figure-8 attractor):

$$\text{Aq}=\text{Sin}[t];\text{Ap}=\text{Cos}[t];$$

Equations of motion:

$$\text{Eqns}=\text{Flatten}\Big[\text{Join}\Big[$$

$$\text{Table}\Big[\Big\{q'_{\alpha,\beta}[t]==\varphi\left(\text{Aq}-q_{\alpha,\beta}[t]-\omega_{\alpha,\beta}[t]q_{\alpha,\beta}[t]^{2}\,p_{\alpha,\beta}[t]\right)$$

$$-\sum_{\alpha=1}^{n}\sum_{\beta=1}^{n}\partial_{p_{\alpha,\beta}[t]}H_{\alpha,\beta}u_{\alpha,\beta}[t],q_{\alpha,\beta}[0]==0\Big\},\{\alpha,n\},\{\beta,n\}\Big],$$

$$\text{Table}\Big[\Big\{p'_{\alpha,\beta}[t]==\varphi\left(\text{Ap}-p_{\alpha,\beta}[t]-\omega_{\alpha,\beta}[t]p_{\alpha,\beta}[t]^{2}\,q_{\alpha,\beta}[t]\right)$$

$$+\sum_{\alpha=1}^{n}\sum_{\beta=1}^{n}\partial_{q_{\alpha,\beta}[t]}H_{\alpha,\beta}u_{\alpha,\beta}[t],p_{\alpha,\beta}[0]==0\Big\},\{\alpha,n\},\{\beta,n\}\Big],$$

$$\text{Table}\Big[\{\omega'_{\alpha,\beta}[t]==-\omega_{\alpha,\beta}[t]+\text{Sq}_{\alpha,\beta}\text{Sp}_{\alpha,\beta}+\text{Sdq}_{\alpha,\beta}\text{Sdp}_{\alpha,\beta},$$

$$\omega_{\alpha,\beta}[0]==\text{RandomReal}[\{-1,1\}]\},\{\alpha,n\},\{\beta,n\}\Big]\Big]\Big];$$

Numerical solution:

$$\text{sol}=\text{NDSolve}\Big[\text{Eqns},\text{Flatten}\Big[\text{Join}\Big[\text{Table}\left[q_{\alpha,\beta},\{\alpha,n\},\{\beta,n\}\right],$$

$$\text{Table}\left[p_{\alpha,\beta},\{\alpha,n\},\{\beta,n\}\right],\text{Table}\left[\omega_{\alpha,\beta},\{\alpha,n\},\{\beta,n\}\right]\Big]\Big],\{t,0,\text{Tfin}\}\Big];$$

Plots of coordinates:

$$\text{Plot}\Big[\text{Evaluate}\Big[\text{Table}\left[q_{\alpha,\beta}[t]/.\text{sol},\{\alpha,n\},\{\beta,n\}\right]\Big],\{t,0,\text{Tfin}\}\Big],$$

$$\text{PlotRange}\to\text{All},\text{PlotStyle}\to\text{AbsoluteThickness}[1.5],\text{Frame}\to\text{True}\Big]$$

Plots of momenta:

$$\text{Plot}\Big[\text{Evaluate}\Big[\text{Table}\left[p_{\alpha,\beta}[t]/.\text{sol},\{\alpha,n\},\{\beta,n\}\right]\Big],\{t,0,\text{Tfin}\}\Big],$$

$$\text{PlotRange}\to\text{All},\text{PlotStyle}\to\text{AbsoluteThickness}[1.5],\text{Frame}\to\text{True}\Big]$$

Plots of weights:

$$\text{Plot}\Big[\text{Evaluate}\Big[\text{Table}\left[\omega_{\alpha,\beta}[t]/.\text{sol},\{\alpha,n\},\{\beta,n\}\right]\Big],\{t,0,\text{Tfin}\}\Big],$$

$$\text{PlotRange}\to\text{All},\text{PlotStyle}\to\text{AbsoluteThickness}[1.5],\text{Frame}\to\text{True}\Big]$$

ESTIMATION OF AN AUTONOMOUS TEAM OF UGVS: HIGH-DIMENSIONAL BAYESIAN PARTICLE FILTER

The second step in the nonlinear generalization of the Kalman servo is to replace the linear/Gaussian Kalman filter (4) with a general nonlinear/non-Gaussian Monte-Carlo filter, as follows.

Recursive Bayesian Filter

Recall that the celebrated Bayes rule gives the relation between the three basic conditional probabilities: i) a Prior P(A) (i.e., an initial degree-of-belief in event A, that is, Initial Odds), ii) Likelihood P(B|A) (i.e., a degree-of-belief in event B, given A, that is, a New Evidence); and Posterior: P(A|B) (i.e., a degree-of-belief in A, given B, that is, New Odds). Provided P(B) ≠ 0, Bayes rule reads:

$$\overset{\text{Posterior}}{P(A\mid B)}=\frac{\overset{\text{Prior}}{P(A)}\times\overset{\text{Likelihood}}{P(B\mid A)}}{P(B)}.$$

In statistical settings, Bayes rule (13) can be rewritten as:

$$\overset{\text{Posterior}}{p(H\mid D)}=\frac{\overset{\text{Prior}}{p(H)}\times\overset{\text{Likelihood}}{p(D\mid H)}}{p(D)},$$

where p(H) is the prior probability density function (PDF) that the hypothesis H is true, p(D|H) is the likelhood PDF for the data D given a hypothesis H, p(H|D) is the posterior PDF that the hypothesis H is true given the data D, and the normalization constant: $p(D)=\sum_{H_i}p(D\mid H)p(H)$ is the PDF for the data D averaged over all possible hypotheses H_i.

When Bayes rule (13)-(14) is applied iteratively/recursively over signal distributions evolving in discrete time steps, in such a way that the Old Posterior becomes the New Prior and New Evidence is added, it becomes the recursive Bayesian filter, the formal origin of all Kalman and particle filters. The Bayesian filter can be applied to estimate the hidden state x_t of a nonlinear dynamical system evolving in discrete time steps (e.g., a numerical solution of a system of ODEs, or discrete-time sampling measurements) in a recursive manner by processing a sequence of observations $Y_T=\{y_t\}_{t=1}^{T}$ dependent on the state x_t within a dynamic noise (either Gaussian, or non-Gaussian) η_t measured as μ_t. The state dynamics and measurement, or the so-called state-space model (SSM) are usually given either by Kalman's state Equation (1), or by its nonlinear generalization. Instead of the Kalman filter Equation (4) combined with the cost function (3), the Bayesian filter includes (see, e.g. [27] [28]):

Time-update Equation

$$\underbrace{p(x_t \mid Y_{t-1})}_{\text{Predictor}} = \int_{\mathbb{R}^n} \underbrace{p(x_t \mid x_{t-1})}_{\text{Prior}} \, \underbrace{p(x_{t-1} \mid Y_{t-1})}_{\text{Old Posterior}} \mathrm{d}x_{t-1}, \tag{15}$$

and

Measurement-Update Equation

$$\underbrace{p(x_t \mid Y_t)}_{\text{New Posterior}} = Z_t^{(-1)} \underbrace{p(x_t \mid Y_{t-1})}_{\text{Predictor}} \, \underbrace{l(y_t \mid x_t)}_{\text{Likelihood}}, \tag{16}$$

where Z_t is the normalizing constant, or partition function, defined by:

$$Z_t = \int_{\mathbb{R}^n} p(x_t \mid Y_{t-1}) l(y_t \mid x_t) \mathrm{d}x_t.$$

In the Bayesian filter (15)-(16), given the sequence of observations Y_t, the posterior fully defines the available information about the state of the system and the noisy environment—the filter propagates the posterior (embodying time and measurement updates for each iteration) across the nonlinear state-space model; therefore, it is the maximum a posteriori estimator of the system state. In a special case of a linear system and Gaussian environment, the filter (15)-(16) has optimal closed-form solution, which is the Kalman filter (4). However, in the general case of a nonlinear dynamics and/or the non-Gaussian environment, it is no longer feasible to search for closed-form solutions for the integrals in (15)-(16), so we are left with suboptimal, approximate, Monte Carlo solutions, called particle filters.

Sequential Monte Carlo Particle Filter

Now we consider a general, discrete-time, nonlinear, probabilistic SSM, defined by a Markov process $\{x_t\}_{t\geq 1} \subseteq \mathbb{R}^{n_x}$, which is observed indirectly via a measurement process $\{y_t\}_{t\geq 1} \subseteq \mathbb{R}^{n_y}$. This SSM consists of two probability density functions: dynamics f$(\cdot)$ and measurement $h(\cdot)$. Formally, we have a system of nonlinear difference equations, given both in Bayesian probabilistic formulation (left-hand side) and nonlinear state-space formulation (right-hand side):

$$\text{Dynamics}: x_{t+1} \mid x_t \sim f(x_{t+1} \mid x_t) \Leftrightarrow x_{t+1} = f(x_t) + \epsilon_t, \tag{17}$$

$$\text{Measurement}: y_t \mid x_t \sim h(y_t \mid x_t) \Leftrightarrow y_t = h(x_t) + \nu_t, \tag{18}$$

$$\text{with initial state}: x_1 \sim \mu(x_1),$$

where f and h are the state and output vector-functions, ϵ_t and ν_t are non-Gaussian process and measurement noises and μ is a given non-Gaussian distribution. The state filtering problem means to recover information about the current state x_t in (17) based on the available measurements $y_{1:t}$ in (18) (see, e.g. [29] ; for a recent review, see [30] and the references therein).

The principle solution to the nonlinear/non-Gaussian state filtering problem is provided by the following recursive Bayesian measurement and time update equations:

$$\overbrace{p(x_t \mid y_{1:t})}^{\text{New Filter}} = \frac{\overbrace{h(y_t \mid x_t)}^{\text{Measure}} \overbrace{p(x_t \mid y_{1:t-1})}^{\text{Predictor}}}{p(y_t \mid y_{1:t-1})}, \tag{19}$$

where

$$\overbrace{p(x_t \mid y_{1:t-1})}^{\text{Predictor}} = \int \overbrace{f(x_t \mid x_{t-1})}^{\text{Dynamics}} \overbrace{p(x_{t-1} \mid y_{1:t-1})}^{\text{Old Filter}} \mathrm{d}x_{t-1}. \tag{20}$$

In a particular case of linear/Gaussian SSMs (17)-(18), the state filtering problem (19)-(20) has an optimal closed-form solution for the filter PDF $p(x_t \mid y_{1:t})$, given by the Kalman filter (4). However, in the general case of nonlinear/non-Gaussian SSMs—such an optimal closed-form solution does not exist.

The general state filtering problem (19)-(20) associated with any nonlinear/ non-Gaussian SSMs (17)-(18) can be only numerically approximated using the sequential Monte Carlo (SMC) methods. The SMC-approximation to the filter PDF, denoted by $\hat{p}(x_t \mid y_{1:t})$, is an empirical distribution represented as a weighted sum of Dirac-delta functions:

$$\hat{p}(x_t \mid y_{1:t}) = \sum_{i=1}^{N} w_t^i \delta_{x_t^i}(x_t), \tag{21}$$

where the samples $\{x_t^i\}_{i=1}^N$ are called particles (point-masses 'spread out' in the state space); each particle x_t^i represents one possible system state and the corresponding weight w_t^i assigns its probability. In this way defined particle filter (PF) plays the role of the Kalman filter for nonlinear/non-Gaussian SSMs. PF approximates the filter PDF $p(x_t \mid y_{1:t})$ using the SMC delta-approximation (21).

Briefly, a PF can be interpreted as a sequential application of the SMC technique called importance sampling (IS, see e.g. [31]). Starting from the initial approximation: $\hat{p}(x_1 \mid y_1) \propto h(y_1 \mid x_1)\mu(x_1)$, at each time step the IS is used to approximate the filter PDF $p(x_t \mid y_{1:t})$ by using the recursive Bayesian Equations ((19), (20)) together with the already generated IS approximation of $p(x_{t-1} \mid y_{1:t-1})$. The particles $\{x_t^i\}_{i=1}^N$ are sampled independently from some proposal distribution $r(x_1)$. To account for the discrepancy between the proposal distribution and the target distribution, the particles are assigned importance weights, given by the ratio between the target and the proposal: $w_1^i \propto h(y_1 \mid x_1^i)\mu(x_1^i)/r(x_1^i)$, where the weights are normalized to sum to one (for a recent technical review, see e.g., [30]).

The IS proceeds in an inductive fashion:

$$\hat{p}(x_{t-1} \mid y_{1:t-1}) = \sum_{i=1}^{N} w_{t-1}^i \delta_{x_{t-1}^i}(x_{t-1}),$$

which, inserted as an old/previous filter into the time-update Equation (20), gives a mixture distribution, approximating $p(x_t \mid y_{1:t-1})$ as:

$$\hat{p}(x_t \mid y_{1:t-1}) = \int \overbrace{f(x_t \mid x_{t-1})}^{\text{Dynamics}} \sum_{i=1}^{N} w_{t-1}^i \delta_{x_{t-1}^i}(x_{t-1})\mathrm{d}x_{t-1} = \sum_{i=1}^{N} w_{t-1}^i f(x_t \mid x_{t-1}^i). \tag{22}$$

Subsequent insertion of (22) as a predictor into the measurement-update Equation (19), gives the following approximation of the filter PDF:

$$\overbrace{p(x_t \mid y_{1:t})}^{\text{Filter}} \approx \frac{\overbrace{h(y_t \mid x_t)}^{\text{Measure}}}{p(y_t \mid y_{1:t-1})} \overbrace{\sum_{i=1}^{N} w_{t-1}^i f(x_t \mid x_{t-1}^i)}^{\text{Weighted Dynamics}},$$

which needs to be further approximated using the IS. The proposal density is pragmatically chosen as: $r(x_t \mid y_{1:t}) = \sum_{j=1}^{N} w_{t-1}^j f(x_t \mid x_{t-1}^j)$. The N ancestor indices $\{a_t^i\}_{i=1}^N$ are resampled into a new set of particles $\{a_{t-1}^i\}_{i=1}^N$

which is subsequently used to propagate the particles to time t.

The final step is to assign the importance weights to the new particles as:

$$\overline{w}_t^i = \frac{h\left(y_t \mid x_t^i\right)\sum_{j=1}^{N} w_{t-1}^j f\left(x_t^i \mid x_{t-1}^j\right)}{\sum_{j=1}^{N} w_{t-1}^j f\left(x_t \mid x_{t-1}^j\right)}.$$

By evaluating $\overline{w}_t^i$ for $i = 1,\cdots,N$ and normalizing the weights, we obtain a new set of weighted particles $\left\{x_t^i, w_t^i\right\}_{i=1}^N$, constituting an empirical approximation of the filter PDF $p(x_t \mid y_{1:t})$. Since these weighted particles $\left\{x_t^i, w_t^i\right\}_{i=1}^N$ can be used to iteratively approximate the filter PDF at all future times, this completes the PF-algorithm with an overall computational complexity of O(N) (as pioneered by [31] ; for more technical details, see e.g., [30] and the references therein).

We remark that the pinnacle of the PF-theory is the so-called Rao-Blackwellized (or, marginalized) PF, which is a special kind of factored PF, where the state-space dynamics (17) is split into a linear/Gaussian part and a nonlinear/non-Gaussian part, so that each particle has the optimal linear-Gaussian Kalman filter associated to it (see [32] - [40]). However, in our case of highly nonlinear and adaptive dynamics, the use of Rao-Blackwellization does not give any advantage, while its mathematics is significantly heavier than in an ordinary PF.

HAMILTONIAN SERVO FRAMEWORK FOR GENERAL CONTROL AND ESTIMATION

The Hamiltonian servo system is our hybrid[5] global control-estimation framework for a large team/fleet of UGVs (e.g., $10 \times 10 = 100$), consisting of the following four components:

Adaptive Control, which is nonlinear and high-dimensional, defined by:

$$\dot{q}_{\alpha\beta} = \varphi\left(A^q - q_{\alpha\beta} - \omega_{\alpha\beta}^q q_{\alpha\beta}^2 p_{\alpha\beta}\right) - \sum_{j,k} \partial_{p_{\alpha\beta}} H_{\alpha\beta} u_{\alpha\beta}^q,$$

$$\dot{p}_{\alpha\beta} = \varphi\left(A^p - p_{\alpha\beta} - \omega_{\alpha\beta}^p p_{\alpha\beta}^2 q_{\alpha\beta}\right) + \sum_{j,k} \partial_{q_{\alpha\beta}} H_{\alpha\beta} u_{\alpha\beta}^p, \qquad (23)$$

$$\dot{\omega}^q_{\alpha\beta} = -\omega^q_{\alpha\beta} + \Phi^q_{\alpha\beta}\left(q_{\alpha\beta}, p_{\alpha\beta}\right), \quad \dot{\omega}^p_{\alpha\beta} = -\omega^p_{\alpha\beta} + \Phi^p_{\alpha\beta}\left(q_{\alpha\beta}, p_{\alpha\beta}\right),$$
$$\Phi^q_{\alpha\beta} = S^q_{\alpha\beta}\left(q_{\alpha\beta}\right)S^q_{\alpha\beta}\left(p_{\alpha\beta}\right) + \dot{S}^q_{\alpha\beta}\left(q_{\alpha\beta}\right)\dot{S}^q_{\alpha\beta}\left(p_{\alpha\beta}\right), \ (\alpha,\beta = 1,\cdots,n)$$
$$\Phi^p_{\alpha\beta} = S^p_{\alpha\beta}\left(q_{\alpha\beta}\right)S^p_{\alpha\beta}\left(p_{\alpha\beta}\right) + \dot{S}^p_{\alpha\beta}\left(q_{\alpha\beta}\right)\dot{S}^p_{\alpha\beta}\left(p_{\alpha\beta}\right). \tag{24}$$

Measurement process, given in discrete-time, probabilistic and state-space Hamiltonian (q, p)-form:

$$p_t^{\alpha\beta} \mid q_t^{\alpha\beta} \sim h\left(p_t^{\alpha\beta} \mid q_t^{\alpha\beta}\right) \Leftrightarrow p_t^{\alpha\beta} = h\left(q_t^{\alpha\beta}\right) + v_t, \ \text{with: } q_1^{\alpha\beta} \sim \mu\left(q_1^{\alpha\beta}\right), \tag{25}$$

where v_t is a non-Gaussian measurement noise. Note that here we use the inverse dynamics (given coordinates, calculate velocities, accelerations and forces)—as a bridge between robot's self-localization and control: measuring coordinates $q_t^{\alpha\beta}$ at discrete time steps t and from them calculating momenta $p_t^{\alpha\beta}$ and control forces $\dot{p}_t^{\alpha\beta}$;

Bayesian recursions, given in discrete-time Hamiltonian (q, p)-form:

$$\overbrace{p\left(q_t^{\alpha\beta} \mid p_{1:t}^{\alpha\beta}\right)}^{\text{New Filter}} = \frac{\overbrace{h\left(p_t^{\alpha\beta} \mid q_t^{\alpha\beta}\right)}^{\text{Measure}} \overbrace{p\left(q_t^{\alpha\beta} \mid p_{1:t-1}^{\alpha\beta}\right)}^{\text{Predictor}}}{p\left(p_t^{\alpha\beta} \mid p_{1:t-1}^{\alpha\beta}\right)},$$

where

$$\overbrace{p\left(q_t^{\alpha\beta} \mid p_{1:t-1}^{\alpha\beta}\right)}^{\text{Predictor}} = \int \overbrace{f\left(q_t^{\alpha\beta} \mid q_{t-1}^{\alpha\beta}\right)}^{\text{Dynamics}} \overbrace{p\left(q_{t-1}^{\alpha\beta} \mid p_{1:t-1}^{\alpha\beta}\right)}^{\text{Old Filter}} \mathrm{d}x_{t-1}; \tag{26}$$

and

$$\overbrace{p\left(q_t^{\alpha\beta} \mid p_{1:t}^{\alpha\beta}\right)}^{\text{Filter PDF}} \approx \frac{\overbrace{h\left(p_t^{\alpha\beta} \mid q_t^{\alpha\beta}\right)}^{\text{Measure}}}{p\left(p_t^{\alpha\beta} \mid p_{1:t-1}^{\alpha\beta}\right)} \overbrace{\sum_{(i)=1}^{N} w_{t-1}^{(i)} f\left(q_t^{\alpha\beta} \mid q_{t-1}^{\alpha\beta(i)}\right)}^{\text{Weighted Dynamics}}, \tag{27}$$

with IS weights:

$$w_t^{(i)} = \frac{h\left(p_t^{\alpha\beta} \mid q_t^{\alpha\beta(i)}\right)\sum_{n=1}^{N} w_{t-1}^{(n)} f\left(q_t^{\alpha\beta(i)} \mid q_{t-1}^{\alpha\beta(n)}\right)}{\sum_{n=1}^{N} w_{t-1}^{(n)} f\left(q_t^{\alpha\beta} \mid q_{t-1}^{\alpha\beta(n)}\right)},$$

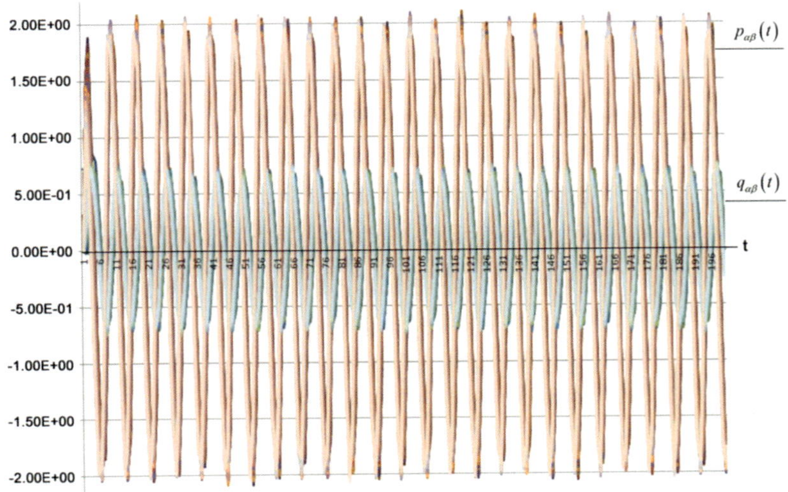

Figure 5. Sample output of the Hamiltonian servo framework applied to a large fleet (a 9 × 9 matrix) of 81 UGVs. As expected, the plot shows almost periodic time evolution of coordinates $q_{\alpha\beta}(t)$ and momenta $p_{\alpha\beta}(t)$. Slight variation of both coordinates $q_{\alpha\beta}(t)$ and momenta $p_{\alpha\beta}(t)$ from the intended regular harmonic motion is due to adaptive control with initial random distribution of synaptic weights $\omega_{\alpha\beta}(t)$. This plot is the simulation output of the Hamiltonian neural network (23)-(24) and recurrent estimation (larger amplitudes, governed by the Bayesian particle filter (25)-(27)).

where bracketed superscripts (i, n) label IS weights.

This Hamiltonian servo framework has been implemented in C++ language (using Visual Studio 2015; see Appendix 1.2). A sample simulation output of the servo system is given in Figure 5. The Figure demonstrates the basic stability and convergence of the servo system.

CONCLUSION

In this paper we have outlined a novel control and estimation framework called the Hamiltonian Servo system. The framework was shown to be a generalization of the Kalman Servo System. Using an example of a large team/fleet of 81 (9 × 9) unmanned ground vehicles (UGVs), it was shown that the framework described provided the control and estimation as required. This approach encompassed a three tired system of adaptive control, measurement process and Bayesian recursive filtering which was demonstrated to enable control and estimation of multidimensional, non-linear and non-Gaussian systems.

ACKNOWLEDGEMENTS

The authors are grateful to Dr. Martin Oxenham, JOAD, DST Group, Australia for his constructive comments which have improved the quality of this paper.

APPENDIX: C++ CODE

Affine Hamiltonian Neural Network

```
// Affine Hamiltonian neural network
// Using Runge-Kutta-4 fixed ODE integrator
// Simulating UGV-swarm

#include <stdio.h>
#include <stdlib.h>
#include <math.h>
#include <time.h>

// attractor: (Aq(t),Ap(t))
#define Aq(t) ( sin(t) )
#define Ap(t) ( 3.*cos(t) )
#define M 2
#define N 10

double k1[M], f1[M], k2[M], f2[M], k3[M], f3[M], k4[M];


double h = 0.001;          // Time step
double t = 0.;             // Initial time
double Tfin = 30.;         // Final time
double fld = 50.;
double eta = 0.5;
double q[N][N], p[N][N], w[N][N], qc[2], wt[3];
double ranf();             // calling random generator

// Affine Hamiltonian equations
void Eqs(double wt[], double t, double y[], double dy[]) {
dy[0] = fld*(Aq(t) - y[0] - wt[0] * y[0] * y[1] * y[0] - y[0] * wt[1]);
dy[1] = fld*(Ap(t) - y[1] - wt[0] * y[1] * y[0] * y[1] - y[1] * wt[2]);
}

// RK4 integrator
void RK4(double *wt, double t, double *x) {
double h2;  int i;
```

```
Eqs(wt, t, x, k1);  h2 = h / 2.;
for (i = 0; i < M; i++)  f1[i] = x[i] + h2*k1[i];
Eqs(wt, t + h2, f1, k2);
for (i = 0; i < M; i++)  f2[i] = x[i] + h2*k2[i];
Eqs(wt, t + h2, f2, k3);
for (i = 0; i < M; i++)  f3[i] = x[i] + h2*k3[i];
Eqs(wt, t + h, f3, k4);
for (i = 0; i < M; i++)
x[i] += h*(k1[i] + 2.0*k2[i] + 2.0*k3[i] + k4[i]) / 6.;
}

//initializing q[i][j] and p[i][j]
void Initialise() {

 for (int i = 0; i < N; i++)
 for (int j = 0; j < N; j++) {
 q[i][j] = 0.3*(ranf() - 0.5);
 p[i][j] = 0.3*(ranf() - 0.5);
 w[i][j] = 0.3*(ranf() - 0.5);
 }
  q[-1][-1] = 0.3*(ranf() - 0.5);   // bounds for q[i][j] and p[i][j]
 p[N][N] = 0.3*(ranf() - 0.5);
 }

 // better quality random generator
 double ranf() {
 const int ia = 16807, ic = 2147483647, iq = 127773, ir = 2836;
 int il, ih, it;  double rc;  int iseed = 0;
 ih = iseed / iq;  il = iseed%iq;  it = ia*il - ir*ih;
 if (it > 0) { iseed = it; }
 else { iseed = ic + it; }
 rc = ic;  return iseed / rc;
 }

 // main code
 int main() {
 int ticks;  double ms;
 clock_t start, stop;
  FILE* pOut;                          // file pointer to output file
 FILE* wOut;
  fopen_s(&pOut, "swarmRK4.csv", "w");  // opening files
  start = clock() * CLK_TCK;            // clock setup
 Initialise();
  while (t <= Tfin)                     // main time loop
 for (int i = 0; i < N; i++)
 for (int j = 0; j < N; j++) {
 // representing q[i][j] and p[i][j] using array qc[]
 qc[0] = q[i][j];  qc[1] = p[i][j];


 // representing other constant terms using array wt[]
 wt[0] = w[i][j]; wt[1] = p[i + 1][j + 1]; wt[2] = q[i - 1][j - 1];
 RK4(wt, t, qc);
                  t += h;        // updating time

 //Hebb's rule: covariance hypothesis
     w[i][j] =-w[i][j]+ eta*(q[i][j] -
         (q[i][j] + p[i][j]) / M)*(qc[0] - (qc[0] + qc[1]) / M);

 // updating q[i][j] and p[i][j]:
     q[i][j] = qc[0];  p[i][j] = qc[1];

 // printouts to files
    fprintf(pOut, ",%.5f,%.5f", qc[0], qc[1]);
    fprintf(wOut, ",%.5f", w[i][j]);
    } fprintf(pOut, "\n"); fprintf(wOut, "\n");
 }
```

```
    stop = clock() * CLK_TCK;             // timing
    ticks = stop - start;
    ms = (float)ticks / 1000.0f;
    printf("finished... (ms = %f)\n", ms);
    getch();  return 0;
}
```

High-Dimensional Bayesian Particle Filter

The following C++ code uses the Armadillo matrix library [41].

```
// Integration of the swarm control with the particle filter

#include <iostream>
#include <armadillo>        // using Armadillo (64-bit)

using namespace std;
using namespace arma;

// Define a float matrix type.
typedef Mat<float> matf;
float dt = 0.01, t = 0.;
int N = 9;           //Number of vehicles
int b = 2*N*N;       //Number of equations

struct m_struct
{
 int nx;            // State dimension
 int ny;            // Measurement dimension
 matf x0;           // Initial state
 matf P0;           // Covariance for the initial state
 matf Q;            // Process noise covariance
 matf Q_sqrt;
 matf R;            // Measurement noise covariance
 matf R_inverse;
 matf R_sqrt;
 matf P0_sqrt;
};

struct z_struct
{
 matf xTrue;
 matf y;
};

struct gpf_struct
{
 matf xf;
 int info;
};

matf g;
// Inline function declaration for m.f
matf m_f(matf* x)
{
 g = zeros<matf>(x->n_rows, x->n_cols);
 for (int j = 0; j < b; j += 2)
 {
  g.row(j) = x->row(j) + 50 * (sin(t) - x->row(j))*dt;
  g.row(j + 1) = x->row(j + 1) + 50 * (3 * cos(t) - x->row(j + 1))*dt;
  t += dt;
 }
 return g;
}

// Inline function declaration for m.f for the particle filter one step forward
matf m_fTime(matf* x, matf t)
{
 g = zeros<matf>(x->n_rows, x->n_cols);
```

```
 for (int j = 0; j < b; j += 2)
 {
  g.row(j) = x->row(j) + 50 * (sin(t.row(j)) - x->row(j))*dt;
  g.row(j + 1) = x->row(j + 1) + 50 * (3 * cos(t.row(j + 1)) - x->row(j + 1))*dt;
 }
 return g;
}

matf h;

// Measurement equation
matf m_h(matf* x)
{
 h = zeros<matf>(x->n_rows, x->n_cols);
 for (int j = 0; j < b; j += 2)
 {
  h.row(j) = -0.5*x->row(j) + 0.1*x->row(j + 1);
  h.row(j + 1) = +0.1*x->row(j) + 0.5*x->row(j + 1);
 }
 return h;
}

// Particle filter function declaration
matf pf(const matf* y, const m_struct* m, int Npf);

// MC function declaration
void mc(matf& q, matf& i);

int main(int argc, char** argv)
{
 wall_clock timer;
 m_struct m;

 m.x0 = zeros<matf>(b, 1);        // Initial State
 m.nx = b;             // State dimension
 m.ny = b;              // Measurement dimension

 m.P0 = (1e-6)*eye<matf>(b, b);
 m.P0(0, 0) = 1;

 m.Q = 0.001*eye<matf>(b, b);
 m.R = 0.01*eye<matf>(b, b);
 m.R_inverse = inv<matf>(m.R);
 m.R_sqrt = sqrt(m.R);
 m.Q_sqrt = sqrt(m.Q);
 m.P0_sqrt = sqrt(m.P0);

 ///////////////////////////////
 //    Simulate the model     //
 ///////////////////////////////

 int Tfinal = 200;
 const int MC = 1;

 z_struct z[MC];

 timer.tic();

 for (int j = 0; j < MC; ++j)
 {
  matf x = zeros<matf>(m.nx, Tfinal + 1);
  matf y = zeros<matf>(m.ny, Tfinal + 1);

  x.col(0) = m.x0 + m.P0_sqrt*randn<matf>(m.nx, 1);

  for (int i = 0; i < Tfinal; i++)
  {
   matf tmp = x.col(i);
   x.col(i) = m_f(&tmp) + m.Q_sqrt*randn<matf>(m.nx, 1);
   y.col(i) = m_h(&tmp) + m.R_sqrt*randn<matf>(m.ny, 1);
   x.col(i + 1) = x.col(i);
   y.col(i + 1) = y.col(i);
```

```
 }
 z[j].xTrue = x.cols(0, Tfinal - 1);      // store the true states
 z[j].y = y.cols(0, Tfinal - 1);          // store the measurements
}


//////////////////////////////////////////////
//    Compute the estimates using the PF  //
/////////////////////////////////////////////

int Npf = 200;
gpf_struct gPF[MC];
for (int i = 0; i < MC; ++i)
{
 cout << "Monte Carlo iteration:" << i << endl;
 gPF[i].xf = pf(&z[i].y, &m, Npf);        // Run the particle filter


 }

 cout << endl << "Total Code Execution Time is:
\qquad \qquad \qquad " << timer.toc() << "  seconds" << endl;

 //Save csv file for plotting
 z[0].xTrue.save("z2.csv", csv_ascii);
 gPF[0].xf.save("gPF.csv", csv_ascii);


 // Compute RMSE values
 matf RMSE_PF = zeros<matf>(m.nx, 1);
 double eterm;
 matf estErrorPF = zeros<matf>(m.nx, Tfinal);

 for (int i = 0; i < MC; i++)
 {
  estErrorPF = gPF[i].xf - z[i].xTrue;

  RMSE_PF = RMSE_PF + sum<matf>(pow(estErrorPF, 2), 1);
 }
 eterm = MC*Tfinal;
 eterm = 1 / eterm;
 RMSE_PF = sqrt(eterm*RMSE_PF);

 cout << endl << "RMSE for particle filter:" << endl;
 RMSE_PF.print();

 cin.get();
 return 0;
}


matf pf(const matf* y, const m_struct* m, int Npf)
{
 int Tfinal = y->n_cols;
 matf q, index = zeros<matf>(1, Tfinal);

 matf q, index = zeros<matf>(1, Tfinal);
 matf e, yNow;

 matf xf = zeros<matf>(m->nx, Tfinal);
 matf x = repmat<matf>(m->x0, 1, Npf) + sqrt(m->P0)*randn<matf>(m->nx, Npf);
 matf tm = zeros<matf>(b, Npf);
 float t = 0.;

 for (int u = 0; u < Npf; u++)              // generating time matrix TM
  for (int v = 0; v < b; v += 2)
  {
   tm(v, u) = t;
   tm(v + 1, u) = t;
   t += dt;
  }

 for (int j = 0; j < Tfinal; j++)
```

```
{
 yNow = y->col(j);
 e = repmat<matf>(yNow, 1, Npf);
 e -= m_h(&x);
// Compute the importance weights
 q = exp(-0.12*(sum<matf>(e % (m->R_inverse*e))));

 if (accu<matf>(q)>1e-12)         // divergence check
 {
  q = q / accu(q);
  xf.col(j) = sum<matf>(repmat<matf>(q, m->nx, 1) % x, 1);
  mc(q, index);            // Resample using MC
  for (int i = 0; i < Npf; ++i)
  {
   x.col(i) = x.col(index(i));
  }
 }

 else
 {
  xf = zeros<matf>(m->nx, Tfinal);
  cout << "Weights close to zero at j= " << j << endl;
  break;
 }

 x = m_fTime(&x, tm);
// Predict the particles one step forward using normal random generates
 x = x + m->Q_sqrt*randn<matf>(m->nx, Npf);
}
return xf;
}


void mc(matf& q, matf& i)
{

 int M = q.n_cols;
 i = zeros<matf>(1, M);
 matf u, qc;
 qc = cumsum<matf>(q, 1);

// Random generates from uniform distribution (randu)
 u = cumsum<matf>(randu<matf>(1, M), 1) / M;
 int k = 1;

 for (int j = 0; j < M; ++j)
 {
  while (qc(k) < u(j))
  {
   ++k;
  }
  i(j) = k;
 } }
```

NOTES

[1]The acronym LQR means linear quadratic regulator.

[2]The acronym LQG means linear quadratic Gaussian.

[3]For simplicity, we are dealing with a square (n × n)-matrix of UGV configurations. The whole approach works also for a rectangular (n × m)-matrix, in which $\alpha = 1, \cdots, n$, $\beta = 1, \cdots, m$.

[4]c_{r-1} are the coefficients of the linear ODE of order r for the error function $e(t) = q_j(t) - q_j^{ref}(t)$:

$$e^{(r)} + c_{r-1}e^{(r-1)} + \cdots + c_1 e^{(1)} + c_0 e = 0.$$

[5]Half continuous-time, half discrete-time.

REFERENCES

1. Hui, K.-P. and Pourbeik, P. (2011) OPAL—A Survivability-Oriented Approach to Management of Tactical Military Networks. IEEE MILCOM, 7-10 November 2011.
2. Nicholas, P., et al. (2016) Measuring the Operational Impact of Military SATCOM Degradation. Winter Simulation Conference (WSC), 11-14 December 2016.
3. Jormakha, J., et al. (2011) UAV-Based Sensor Networks for Future Force Warriors. International Journal of Advances in Telecommunications, 4, 58-71.
4. Offensive Swarm-Enabled Tactics (OFFSET). https://www.grants.gov/web/grants/view-opportunity.html?oppId=291825
5. Vukobratovic, M. and Potkonjak, V. (1982) Scientific Fundamentals of Robotics, Vol. 1, Dynamics of Manipulation Robots: Theory and Application. Springer, Berlin.
6. Vukobratovic, M. and Stokic, D. (1982) Scientific Fundamentals of Robotics, Vol. 2, Control of Manipulation Robots: Theory and Application. Springer, Berlin.
7. Vukobratovic, M. and Kircanski, M. (1985) Scientific Fundamentals of Robotics, Vol. 3, Kinematics and Trajectories Synthesis of Manipulation Robots. Springer, Berlin.
8. Vukobratovic, M. and Kircanski, N. (1985) Scientific Fundamentals of Robotics, Vol. 4, Real-Time Dynamics of Manipulation Robots. Springer, Berlin.
9. Vukobratovic, M., Stokic, D. and Kircanski, N. (1985) Scientific Fundamentals of Robotics, Vol. 5, Non-Adaptive and Adaptive Control of Manipulation Robots. Springer, Berlin.
10. Vukobratovic, M. and Potkonjak, V. (1985) Scientific Fundamentals of Robotics, Vol. 6, Applied Dynamics and CAD of Manipulation Robots. Springer, Berlin.
11. Ivancevic, V. and Ivancevic, T. (2006) Human-Like Biomechanics. Springer, Dordrecht.
12. ASIMO (2016) The Honda Humanoid Robot. http://world.honda.com/ASIMO/
13. Ivancevic, V.G. (2010) Nonlinear Complexity of Human Biodynamics Engine, Nonlin. Dynamics, 61, 123-139.

14. Durrant-Whyte, H. and Bailey, T. (2006) Simultaneous Localization and Mapping (SLAM): Part I. IEEE Robotics & Automation Magazine, 13, 99-110. https://doi.org/10.1109/MRA.2006.1638022
15. Wikipedia (2016) Simultaneous Localization and Mapping.
16. Montemerlo, M., Thrun, S., Koller, D. and Wegbreit, B. (2002) FastSLAM: A Factored Solution to the Simultaneous Localization and Mapping Problem.
17. Montemerlo, M., Thrun, S., Koller, D. and Wegbreit, B. (2003) FastSLAM 2.0: An Improved Particle Filtering Algorithm for Simultaneous Localization and Mapping That Provably Converges. Proceedings of IJCAI, 1151-1156.
18. Thrun, S., Montemerlo, M., Koller, D., Wegbreit, B., Nieto, J. and Nebot, E. (2004) FastSLAM: An Efficient Solution to the Simultaneous Localization and Mapping Problem with Unknown Data Association. Journal of Machine Learning Research, 1-48.
19. OpenSLAM (2016). https://www.openslam.org/
20. Kalman, R.E. (1960) A New Approach to Linear Filtering and Prediction Problems. Journal of Basic Engineering, 82, 34-45. https://doi.org/10.1115/1.3662552
21. Kalman, R.E. and Bucy, R.S. (1961) New Results in Linear Filtering and Prediction Theory. Journal of Basic Engineering, 96, 95-108. https://doi.org/10.1115/1.3658902
22. Kalman, R.E., Falb, P. and Arbib, M.A. (1969) Topics in Mathematical System Theory. McGraw Hill, New York.
23. Ivancevic, V. and Yue, Y. (2016) Hamiltonian Dynamics and Control of a Joint Autonomous Land-Air Operation. Nonlinear Dynamics, 84, 1853-1865. https://doi.org/10.1007/s11071-016-2610-y
24. Ivancevic, V. and Ivancevic, T. (2006) Geometrical Dynamics of Complex Systems. Springer, Dordrecht. https://doi.org/10.1007/1-4020-4545-X
25. Ivancevic, V. and Ivancevic, T. (2007) Neuro-Fuzzy Associative Machinery for Comprehensive Brain and Cognition Modelling. Springer, Berlin. https://doi.org/10.1007/978-3-540-48396-0
26. Kosko, B. (1992) Neural Networks and Fuzzy Systems, A Dynamical Systems Approach to Machine Intelligence. Prentice-Hall, New York.
27. Arulampalam, S., Maskell, S., Gordon, N. and Clapp, T. (2002) A Tutorial on Particle Filters for Online Nonlinear/Non-Gaussian

Bayesian Tracking. IEEE Transactions on Signal Processing, 50, 174-188. https://doi.org/10.1109/78.978374

28. Ristic, B., Arulampalam, S. and Gordon, N. (2004) Beyond the Kalman Filter. Artech House.
29. Schön, T.B., Gustafsson, F. and Nordlund, P.-J. (2005) Marginalized Particle Filters for Mixed Linear/Nonlinear State-Space Models. IEEE Transactions on Signal Processing, 53, 2279-2289. https://doi.org/10.1109/TSP.2005.849151
30. Schön, T.B., Lindsten, F., Dahlin, J., et al. (2015) Sequential Monte Carlo Methods for System Identification. IFAC-PapersOnLine, 48, 775-786.
31. Gordon, N.J., Salmond, D.J. and Smith, A.F.M. (1993) Novel Approach to Nonlinear/Non-Gaussian Bayesian State Estimation. I Radar and Signal Processing, 140, 107-113.
32. Casella, G. and Robert, C.P. (1996) Rao-Blackwellisation of Sampling Schemes. Biometrika, 83, 81-94. https://doi.org/10.1093/biomet/83.1.81
33. Chen, R. and Liu, J.S. (2000) Mixture Kalman Filters. Journal of the Royal Statistical Society, 62, 493-508. https://doi.org/10.1111/1467-9868.00246
34. Doucet, A., Godsill, S.J. and Andrieu, C. (2000) On Sequential Monte Carlo Sampling Methods for Bayesian Filtering. Statistics and Computing, 10, 197-208. https://doi.org/10.1023/A:1008935410038
35. Doucet, A., Gordon, N. and Krishnamurthy, V. (2001) Particle Filters for State Estimation of Jump Markov Linear Systems. IEEE Transactions on Signal Processing, 49, 613-624. https://doi.org/10.1109/78.905890
36. Andrieu, C. and Doucet, A. (2002) Particle Filtering for Partially Observed Gaussian State Space Models. Journal of the Royal Statistical Society, 64, 827-836. https://doi.org/10.1111/1467-9868.00363
37. Schön, T., Gustafsson, F. and Nordlund, P.-J. (2003) Marginalized Particle Filters for Nonlinear State-Space Models. Tec. Report LiTH-ISY-R-2548, Linköping Univ.
38. Schön, T., Gustafsson, F. and Nordlund, P.-J. (2005) Marginalized Particle Filters for Mixed Linear/Nonlinear State-Space Models. IEEE Transactions on Signal Processing, 53, 2279-2289. https://doi.org/10.1109/TSP.2005.849151

39. Karlsson, R., Schön, T. and Gustafsson, F. (2005) Complexity Analysis of the Marginalized Particle Filter. IEEE Transactions on Signal Processing, 53, 4408-4411. https://doi.org/10.1109/TSP.2005.857061
40. Schön, T., Karlsson, R. and Gustafsson, F. (2006) The Marginalized Particle Filter in Practice. Aerospace Conference, 4-11 March 2006. https://doi.org/10.1109/AERO.2006.1655922
41. Wikipedia (2016) Armadillo (C++ Library).

SECTION 3
INTELLIGENT PROCESS CONTROL

CHAPTER 10

Automatic Intelligent Control System Based on Intelligent Control Algorithm

Zishan Huang

School of Electrical and Electronic Engineering, Hubei University of Technology, Wuhan 430068, China

ABSTRACT

In order to improve the effect of automatic intelligent control, this paper improves the intelligent vector control algorithm, analyzes the performance of the algorithm with common power pulses as an example, and obtains an improved intelligent vector control algorithm suitable for modern intelligent systems. By making two mutually adjacent vectors nonidentical combination in time, the two vectors have the difference between zero and nonzero, and the vector can be rotated into a certain area. In addition, with the support

Citation: Zishan Huang, "Automatic Intelligent Control System Based on Intelligent Control Algorithm", Journal of Electrical and Computer Engineering, vol. 2022, Article ID 3594256, 10 pages, 2022. https://doi.org/10.1155/2022/3594256.

of improved algorithms, this paper gets an automated intelligent control system based on intelligent control algorithms. Finally, this paper verifies the effect of the intelligent control system through experimental research. The analysis of the test results shows that the automated intelligent control system proposed in this paper has a better effect in intelligent control.

INTRODUCTION

Traditional control theory is the general term of classic control theory and modern control theory. Their main feature is model-based control. Traditional control uses inaccurate models and certain fixed control algorithms to place the entire control system under the model framework, which lacks flexibility and adaptability, so it is difficult to be competent for the control of complex systems. The core of intelligent control is control decision-making, which uses flexible decision-making methods to force the control system to approach the desired goal [1]. Traditional control is suitable for solving relatively simple control problems such as being linear and time-invariant. These problems can also be solved with intelligent methods. Intelligent control is the development of traditional control theory, and traditional control and intelligent control can be unified under the framework of intelligent control [2]. The generation of intelligent control comes from the high complexity and uncertainty of the controlled object and the increasingly high control performance that people demand. It is a type of automatic control that can independently drive intelligent machines to achieve their goals without human intervention. In addition, its creation and development require a high degree of integration and utilization of contemporary multiple frontier disciplines, multiple advanced technologies, and multiple scientific methods [3].

The intelligent behavior in the intelligent control system is essentially a mapping relationship from input to output. When the input of the system is not an example that has been learned, it can give a suitable output because of its adaptive function. Even when some parts of the system fail, the system can work normally. If the system has a higher degree of intelligence, it can also automatically find faults and even have the function of self-repair, thus reflecting stronger adaptability.

It has the function of self-organization and coordination for complex tasks and scattered sensor information. The intelligent controller can make decisions on its own and take actions proactively within the scope of the task requirements; and when there is a multitarget conflict, the controller has the right to make judgments on its own under certain restrictions. The intelligent control system can learn the inherent information of the unknown characteristics of a process or its environment and use the obtained experience for further estimation, classification, decision-making, or control, so as to improve the performance of the system. That is, it has the ability to learn and acquire new knowledge about the object and environment during operation and use the new knowledge to improve control behavior. The third feature of intelligent control is that it can comprehensively cross various technologies, making the design of intelligent control systems and intelligent controllers increasingly diversified, and the application scope of intelligent control technology is becoming more and more extensive. It is difficult for simple technology to achieve intelligent simulation, and the combination of multiple technologies is conducive to intelligent simulation; the simple control method makes the system lack intelligence, and the control of integrated adaptive, self-organization, (self)learning, and other technologies can improve the intelligence of the system control. The fourth characteristic of intelligent control is that it can describe the stability of the system, the controllability and observability of the system, and the optimal control of the system (that is, the description of the entropy function and the energy function) like the traditional control theory which analyzes the dynamic characteristics of the system, system complexity, etc. The fifth characteristic of intelligent control is that it is robust and of real time, is insensitive to environmental interference and uncertain factors, can respond to highly abstract discrete symbol commands, and has online real-time response capabilities.

This paper combines the intelligent control algorithm to study the automated intelligent control system, builds the intelligent control system model according to the current more advanced algorithm model, and verifies the system function through experimental research.

RELATED WORK

With the rapid development of artificial intelligence and big data technology, related multiagent technology has also been applied in many fields in recent years. Because multiagent technology has the characteristics of reasoning, cooperation, and self-healing, it creates the possibility to solve the problem of complex and dynamic scheduling system. Literature [4] conducted the research on agent-based distributed dynamic job shop scheduling and proposed a multiagent distributed dynamic job shop scheduling scheme based on contract network protocol bidding mechanism; literature [5] analyzed agent-based technology distributed job shop scheduling structure, proposes the concept of autonomous scheduling execution, and provides an autonomous scheduling execution algorithm based on disturbance analysis, so that the scheduling execution agent can autonomously make rapid and optimal decisions on system disturbances. Literature [6] proposes to use multiagent negotiation to solve the problem of intelligent transfer conflicts and finally designs a kind of information that can solve the problem of virtual enterprise transfer conflicts through rational and efficient use of the agent's scheduling, collaboration, and intelligence characteristics. Integrated system: Literature [7] studied the agile production line scheduling based on multiagent, combined with the latest theoretical results of many technologies in the field of computer and artificial intelligence, applied multiagent technology to the solution field of agile production line scheduling, and solved the actual production scheduling problem, providing useful exploration. Literature [8] studied the dynamic scheduling method of automatic surface treatment production line based on multiagent, established an architecture model based on multiagent, and abstracted the complex surface treatment automatic production line workshop scheduling problem into management, processing, and workpiece, transports four kinds of agents, focuses on the analysis of the communication and negotiation mechanism between these four agents, studies a tender-bid mechanism based on the contract network, and optimizes scheduling through comprehensive indicators. Literature [9] uses a combination of multiagent model and cellular automation model to construct a city dynamic model using geographic information system technology and takes the actual city as an empirical object, which proves that the model has a high degree of credibility. Literature [10] uses bionics to summarize the structural and functional similarities between the ship formation and the biological immune system and combines the multiagent theory with the biological immune system and proposes a new immune multiagent network (IMAN) model and applied it to the cooperative air defense system of the fleet. Literature [11] established a distributed and coordinated multiagent system and applied it to the coordination and optimization of the power system. The system includes multiple power supply agents and management coordination agents. Finally, the system is verified by simulation experiments. The feasibility: Literature [12] uses multiagent theory to establish a production scheduling system, simulates the production scheduling system according to actual production conditions, and obtains the optimal production scheduling plan, which reduces production costs and improves production efficiency. It is useful for the formulation of production scheduling plans. Certain reference value: Literature [13] studied the adaptive conditions of distributed structure, centralized structure, and decentralized structure under certain interactive equipment and data resources and further explored distributed collaborative solving and centralized collaborative solving problems. Literature [14] explored the control problem of the agent network with a fixed topology and finally proved through simulation experiments that the method based on the consensus protocol can be successfully applied to the multiagent formation control, and the effect is better. Literature [15] studied the dynamic scheduling strategy of multiagent production line based on MDP. Aiming at the dynamic task allocation problem of workshop production line, based on Markov decision process theory, a dynamic task scheduling mathematical model of multiagent production line was established. Literature [16] carried out modeling, optimization, and simulation research on the production line based on multiagent, built the model through Simulink/Stateflow simulation tool and conducted experimental research, and proposed related control strategies according to the actual production situation. The communication program is written by VC language, and the distributed simulation of the multiagent system is completed. Finally, through statistics and analysis of simulation data, the production line optimization strategy is verified.

VECTOR CONTROL BASED ON INTELLIGENT CONTROL ALGORITHM

When researching the multimodal control algorithm, it is necessary to use the MATLAB/Simulink platform to build the simulation model of this variable structure AC servo pressurization system. If we want to build a physical model of an AC motor, we should understand its characteristics. One is a nonlinear system, the other is a strongly coupled system, and the third is a high-order system, all of which are multivariable.

These characteristics indicate that it should be expressed by nonlinear formulas, which is very inconvenient when controlling such systems. The main reason for the complexity of the AC motor model is the coupling between the magnetic flux and the magnetic flux. If the motor model is decomposed into a model controlled separately by torque and flux linkage, then it can be simulated as a DC motor control model to control a non-DC motor. Vector control can achieve the abovementioned control effect.

The principle of vector control is to measure the stator current vector of the motor and to decompose the stator current through the basic principle of the magnetic field, so as to realize the control of the motor. The specific method is to decompose the current vector into the torque current and the excitation current and then control them.

The control here refers to controlling the amplitude and phase of the current component of the magnetic field and the torque current component. This control method is a vector control method. Therefore, the vector control strategy is introduced to the simulation of the motor when doing the optimization algorithm part this time.

Coordinate transformation is a very critical link in vector control technology. Only by transforming its coordinates can the motor torque and flux linkage be decoupled.

Coordinate transformation is the core step of AC motor vector control and an important mathematical method to simplify the complex physical model of the motor. The purpose of coordinate transformation is to transform the physical model of an AC motor into a model similar to a DC motor. The AC motor model after coordinate transformation is more convenient for analysis and control.

The essence of coordinate transformation is to replace the original variables in the formula with a set of new variables obtained after transformation. It is based on the principle of generating the same rotating magnetomotive force. *e variable in the three-phase stator coordinate system is equivalent to the variable αβ0 in the two-phase stationary coordinate system through Clarke transformation, which is equivalent to the alternating currents i_α and i_β.

Through the Park transformation, the variable αβ0 in the two-phase stationary coordinate system can be equivalent to the variable dq0 in the two-phase rotating coordinate system, which is equivalent to the DC currents i_d and i_q. The variable ABC in the three-phase stator coordinate system, the variable αβ0 in the two-phase static coordinate system, and the variable dq0 in the two-phase rotating coordinate system are three commonly used coordinate systems in the vector control of AC motors. According to the above equivalent relationship, the structure diagram shown in Figure 1 can be obtained. It can be seen from Figure 1 that the vector control input is three-phase power A, B, and C. To obtain a DC motor, which is input by i_d and i_q and output by ω, it is necessary to perform Park conversion and Clarke conversion according to the oriented rotor flux [17].

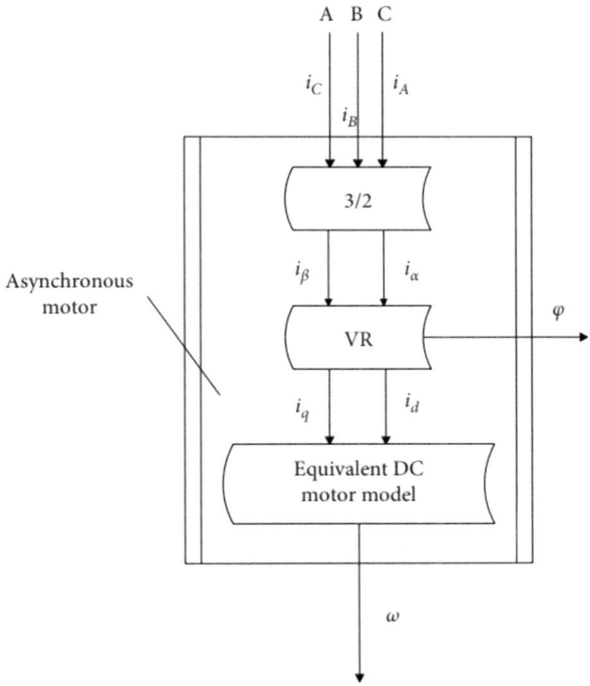

Figure 1. Coordinate transformation structure diagram of motor.

Clarke Transformation (3S/2S Transformation)

There are two types of Clarke transformation: equal-amplitude vector transformation and equal-power vector transformation. If equal-amplitude vector transformation is used, compensation needs to be added in the later Park transformation. Therefore, the Clarke transformation used this time is an equal-power transformation method. The following two points must be done in the constant power coordinate transformation:

a. The effect of the magnetic field generated by the current before and after the coordinate transformation remains unchanged.

b. The power of the motor before and after the coordinate transformation remains at the original value.

The vector u and current i in the original coordinates are transformed into u' and current i' in the new coordinates. Since we want them to have the same transformation matrix C, then [18]

$$u = Cu',$$
$$i = Ci'. \quad (1)$$

In order to realize the inverse transformation, the transformation matrix C must have an inverse matrix C^{-1}, so the transformation matrix C must be a square matrix, and the value of its determinant must not be equal to zero. Because u = zi, z is the impedance matrix,

$$\begin{aligned} u' &= C^{-1}u \\ &= C^{-1}zi \\ &= C^{-1}zCi' \\ &= z'i'. \end{aligned} \quad (2)$$

In the above formula, the changed impedance matrix is represented by z', which can be expressed as

$$z' = C^{-1}zC. \quad (3)$$

In order to follow the principle of constant power consumption, the electrical power consumption $i^T u = u_1 i_1 + u_2 i_2 + \cdots u_n i_n$ under one coordinate is equal to the electrical power consumption $i^{T'} u' = u_1' i_1' + u_2' i_2' + \cdots u_n' i_n'$ under the other coordinate, namely [19]:

$$i^T u = i^T u', \quad (4)$$

$$i^T u = (Ci')^T Cu' = i^{T'} C^T Cu'. \quad (5)$$

In order to make formula (4) the same as formula (5), the transformation matrix *c* should be an orthogonal matrix.

In the above formula, C^{-1} is the inverse matrix of *C*; i^T is the transposed matrix of i; $i^{T'}$ is the transposed matrix of i'; C^T is the transposed matrix of C; I is the identity matrix; Z and Z' are impedance matrices, respectively; u, u', i, i' are the vector and current column and row matrix, respectively. At the same time, according to the matrix algorithm, there are $C^{-1}C = I$; $(Ci')^T = i^{T'}$ CT; $(KC)^T = KC^T$; u = Cu', and $u' = C^{-1}u$.

Figure 2 is the relative positional relationship of the magnetic potential vector. The magnetic potential vectors are the magnetic potential vectors of the nonrotor three-phase motor windings *A*, *B*, and *C* and the magnetic potential vectors of the two-phase motor windings α and β, respectively. The *A* axis coincides with the α axis. It can be known from the vector coordinate transformation principle that the two are completely equal magnetic fields; that is, the resultant magnetic potential vector has equal components on the two position and space axes.

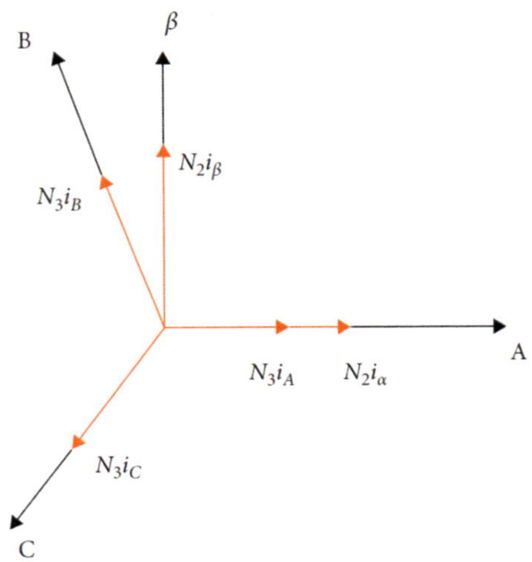

Figure 2. Vector coordinate system.

Therefore, there are

$$N_2 = N_3 i_A + N_3 i_B \cos 120^\circ + N_3 i_C \cos(-120^\circ),$$
$$N_2 i_\beta = 0 + N_3 i_B \sin 120^\circ + N_3 i_C \sin(-120^\circ). \quad (6)$$

That is:

$$i_\alpha = \frac{N_3}{N_2}\left[i_A - \frac{1}{2} i_B - \frac{1}{2} i_C\right], \quad (7)$$

$$i_\beta = \frac{N_3}{N_2}\left[0 + \frac{\sqrt{3}}{2} i_B - \frac{\sqrt{3}}{2} i_C\right]. \quad (8)$$

In the formula, N_2 and N_3, respectively, represent the effective number of turns of each phase of the stator of a three-phase motor and a two-phase motor. Formulas (7) and (8) are expressed by matrix, namely:

$$\begin{bmatrix} i_\alpha \\ i_\beta \end{bmatrix} = \frac{N_3}{N_2} \begin{bmatrix} 1 & -\frac{1}{2} & -\frac{1}{2} \\ 0 & \frac{\sqrt{3}}{2} & -\frac{\sqrt{3}}{2} \end{bmatrix} \begin{bmatrix} i_\alpha \\ i_\beta \\ i_C \end{bmatrix}. \tag{9}$$

The transformation matrix $\begin{bmatrix} 1 & -1/2 & -1/2 \\ 0 & \sqrt{3}/2 & -\sqrt{3}/2 \end{bmatrix}$ cannot be changed inversely, because it is a nonsquare matrix. To this end, it is necessary to introduce the zero-axis current, that is, an unrestricted nonoriginal variable i_0 of i_α and i_β. In the formed $\alpha - \beta - 0$-axis coordinate system, the zero axis is perpendicular to the α and β axes at the same time. We define

$$N_2 i_0 = k(N_3 i_A + N_3 i_B + N_3 i_C), \tag{10}$$

$$i_0 = \frac{N_3}{N_2} k(i_A + i_B + i_C). \tag{11}$$

In the formula, k is the undetermined coefficient. Therefore, formula (10) is rewritten as

$$\begin{bmatrix} i_\alpha \\ i_\beta \\ i_0 \end{bmatrix} = \frac{N_3}{N_2} \begin{bmatrix} 1 & -\frac{1}{2} & -\frac{1}{2} \\ 0 & \frac{\sqrt{3}}{2} & -\frac{\sqrt{3}}{2} \\ k & k & k \end{bmatrix} \begin{bmatrix} i_A \\ i_B \\ i_C \end{bmatrix}. \tag{12}$$

In the formula, the matrix C is defined as

$$C = \frac{N_3}{N_2} \begin{bmatrix} 1 & -\frac{1}{2} & -\frac{1}{2} \\ 0 & \frac{\sqrt{3}}{2} & -\frac{\sqrt{3}}{2} \\ k & k & k \end{bmatrix}. \tag{13}$$

The transpose matrix C^T of C is

$$C^T = \frac{N_3}{N_2}\begin{bmatrix} 1 & 0 & k \\ -\frac{1}{2} & \frac{\sqrt{3}}{2} & k \\ -\frac{1}{2} & -\frac{\sqrt{3}}{2} & k \end{bmatrix}. \qquad (14)$$

The inverse matrix C^{-1} of C is

$$C^{-1} = \frac{2N_2}{3N_3}\begin{bmatrix} 1 & 0 & \frac{1}{2k} \\ -\frac{1}{2} & \frac{\sqrt{3}}{2} & \frac{1}{2k} \\ -\frac{1}{2} & -\frac{\sqrt{3}}{2} & \frac{1}{2k} \end{bmatrix}. \qquad (15)$$

In order to comply with the principle of constant power conversion, $C^T = C^{-1}$ is known. That is, the above formula (14) is equal to the formula (15); then:

$$\frac{2N_2}{3N_3} = \frac{N_3}{N_2}; \frac{1}{2k} = k. \qquad (16)$$

We can obtain separately

$$\frac{N_3}{N_2} = \sqrt{\frac{2}{3}},$$

$$k = \sqrt{\frac{1}{2}}. \qquad (17)$$

By substituting formula (17) into formula (13) and formula (15), then

$$C = \begin{bmatrix} 1 & -\frac{1}{2} & -\frac{1}{2} \\ 0 & \frac{\sqrt{3}}{2} & -\frac{\sqrt{3}}{2} \\ \frac{1}{\sqrt{2}} & \frac{1}{\sqrt{2}} & \frac{1}{\sqrt{2}} \end{bmatrix},$$

$$C^{-1} = \sqrt{\frac{2}{3}}\begin{bmatrix} 1 & 0 & \frac{1}{\sqrt{2}} \\ -\frac{1}{2} & \frac{\sqrt{3}}{2} & \frac{1}{\sqrt{2}} \\ -\frac{1}{2} & -\frac{\sqrt{3}}{2} & \frac{1}{\sqrt{2}} \end{bmatrix}. \qquad (18)$$

Therefore, the Clarke transformation (or 2/3 transformation) formula is

$$\begin{bmatrix} i_\alpha \\ i_\beta \\ i_0 \end{bmatrix} = C \begin{bmatrix} i_A \\ i_B \\ i_C \end{bmatrix} = \sqrt{\frac{2}{3}} \begin{bmatrix} 1 & -\frac{1}{2} & -\frac{1}{2} \\ 0 & \frac{\sqrt{3}}{2} & -\frac{\sqrt{3}}{2} \\ \frac{1}{\sqrt{2}} & \frac{1}{\sqrt{2}} & \frac{1}{\sqrt{2}} \end{bmatrix} \begin{bmatrix} i_A \\ i_B \\ i_C \end{bmatrix}. \tag{19}$$

Park Transformation (2s/2r Transformation)

Generally speaking, the 2s/2r conversion refers to the conversion between the two-phase nondynamic coordinate system and the two-phase rotating coordinate system. To obtain the rotating magnetomotive force, two-phase nondynamic winding is required, and two-phase balanced nondirect current is added to it. If the two-phase windings whose rotational angular velocity is equal to the rotational angular velocity of the synthesized magnetomotive force are turned, then the space-rotating magnetomotive force is generated in the two-phase winding with nonalternating current.

Figure 3 shows the Park transformation relationship diagram, and the Park transformation is derived from Clark transformation as

$$\begin{cases} u_d \cos\alpha + u_q \sin\alpha = u_\alpha, \\ u_d \sin\alpha - u_q \cos\alpha = u_\beta, \end{cases} \tag{20}$$

$$\begin{pmatrix} u_\alpha \\ u_\beta \end{pmatrix} = \begin{pmatrix} \cos\alpha & \sin\alpha \\ \sin\alpha & -\cos\alpha \end{pmatrix} \begin{pmatrix} u_d \\ u_q \end{pmatrix}. \tag{21}$$

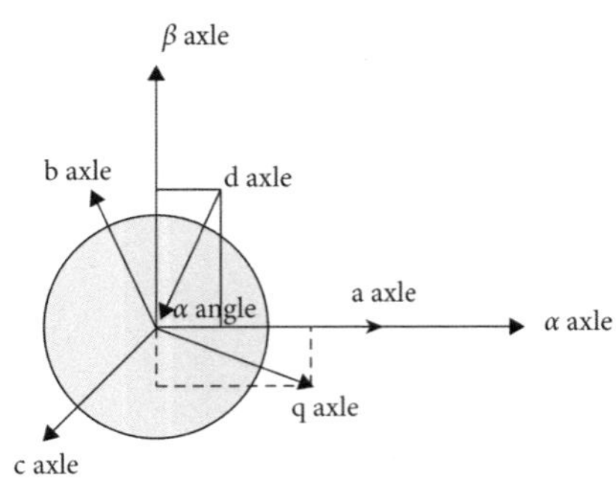

Figure 3. Park transformation relationship diagram.

From formula (20), we can get

$$\begin{cases} u_d\cos^2\alpha + u_q\sin\alpha\cos\alpha = u_\alpha\cos\alpha, \\ u_d\sin^2\alpha + u_q\sin\alpha\cos\alpha = u_\beta\sin\alpha. \end{cases} \tag{22}$$

By adding the two types, there are

$$u_d = u_\alpha\cos\alpha + u_\beta\sin\alpha,$$

$$\begin{cases} u_d\cos\alpha\sin\alpha + u_q\sin^2\alpha = u_\alpha\sin\alpha, \\ u_d\cos\alpha\sin\alpha - u_q\cos^2\alpha = u_\beta\cos\alpha. \end{cases} \tag{23}$$

By adding the two formulas, we get

$$u_q = u_\beta\cos\alpha - u_\alpha\sin\alpha, \tag{24}$$

and we get

$$\begin{pmatrix} u_d \\ u_q \end{pmatrix} = \begin{pmatrix} \cos\alpha & \sin\alpha \\ -\sin\alpha & \cos\alpha \end{pmatrix}\begin{pmatrix} u_\alpha \\ u_\beta \end{pmatrix},$$

$$\begin{pmatrix} i_{sm} \\ i_{st} \end{pmatrix} = \begin{pmatrix} \cos\alpha & \sin\alpha \\ -\sin\alpha & \cos\alpha \end{pmatrix}\begin{pmatrix} i_{sa} \\ i_{sb} \end{pmatrix}. \tag{25}$$

Park is inversely transformed into the inverse matrix of matrix (29):

$$C_{\text{park}}^{-1} = \begin{pmatrix} \cos\alpha & -\sin\alpha \\ \sin\alpha & \cos\alpha \end{pmatrix}. \tag{26}$$

The function of space vector pulse width modulation (SVPWM) is to make the output waveform closer to a sine waveform. The specific method is as follows: a three-phase power inverter is composed of six power switching elements, and the switching mode is appropriately switched to control the generation of a pulse width modulated wave. Compared with the traditional pulse width modulation technology, it has the advantages of reducing the motor torque pulsation to a certain extent, and the rotation trajectory is closer to the circular flux trajectory.

The space vector pulse width modulation technology uses the principle of average equivalent utility as the basis of theoretical analysis; that is, by combining the basic vector vectors in a single open and closed cycle, the average vector is equal to the given vector. At a certain moment, by making two mutually adjacent vectors carry out a nonidentical combination in time, the two vectors have a difference between zero and nonzero, and the vector can be rotated into a certain area. In order to obtain the PWM waveform, the action time of the two vectors can be added multiple times, and the timing of the addition is within a complete sampling time period. In this way, the action time of each vector can be controlled, and a space vector rotating according to a circular trajectory can be obtained. The actual magnetic flux close to the ideal magnetic flux circle is generated by changing the open and closed state of the inverter, and then the switching state of the inverter is determined by the comparison result of the two. Figure 4 shows the inverter circuit.

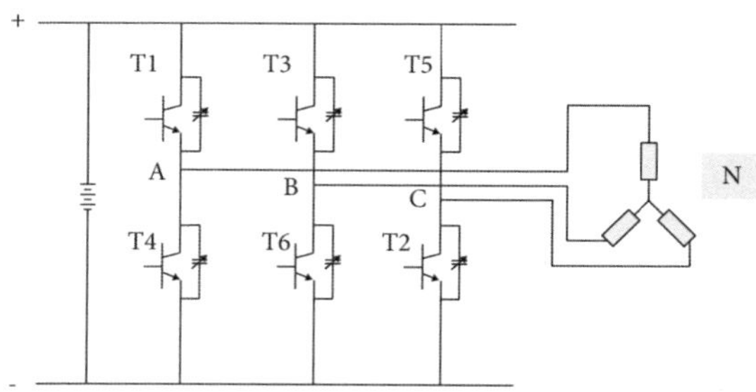

Figure 4. Inverter circuit.

U_{dc} is the non-AC bus side vector, and the three-phase vectors output by the inverter are U_A, U_B, and U_C, respectively, which form an arrangement of 120° on each other in space, and the coordinate system is a three-phase plane nondynamic coordinate system. Three nonplanar vectors $U_A(t)$, $U_B(t)$, and *n* $U_C(t)$ can be defined. The directions of the three are always on the reference line of each phase, and the magnitude is changed with time according to the sine law, and the time phase differs by 120. If it is assumed that the effective value of the phase vector is U_m and the power supply frequency is *f*, then the formula can be obtained:

$$\begin{cases} U_A(t) = U_m \cos(\theta), \\ U_B(t) = U_m \cos\left(\dfrac{\theta - 2\pi}{3}\right), \\ U_C(t) = U_m \cos\left(\dfrac{\theta - 2\pi}{3}\right). \end{cases} \quad (27)$$

In the formula, $\theta = 2\pi ft$; then the resultant space vector U(t) added by the three-phase vector space vector can be expressed as

$$U(t) = U_A(t) + U_B(t)e^{j2\pi/3} + U_C(t)e^{j4\pi/3} = \frac{3}{2}U_m e^{j\theta}. \tag{28}$$

According to the above, U(t) is a rotating space vector whose amplitude is 1.5 times the peak value of the phase vector, and U_m is the peak value of the phase vector, and the space vector rotates at an angular frequency $\omega = 2\pi f$ in the counterclockwise direction at a uniform speed. The projection of the space vector U(t) on the three-phase coordinate axis (*a*, *b*, *c*) is the symmetrical three-phase sine. Since the inverter's three-phase bridge arm is composed of 6 opening and closing tubes, in order to explore the space vector output by the inverter when the upper and lower arms of each phase are opened and closed in different combinations, the opening and closing expression $S_\chi(\chi = a, b, c)$ is specifically described as

$$S_\chi = \begin{cases} 1, & \text{Upper bridge arm conduction,} \\ 0, & \text{Lower bridge arm conduction.} \end{cases} \tag{29}$$

There are eight possible combinations of (S_a, S_b, S_c), including 6 nonzero vectors $U_1(001)$, $U_2(010)$, $U_3(011)$, $U_4(100)$, $U_5(101)$, $U_6(110)$ and two zero vectors $U_0(000)$ and $U_7(111)$.

The magnitude modulus of a nonzero vector is $2U_{dc}/3$, the distance between two adjacent vectors is 60°, and the magnitude of the two zero vectors at the center of the sector is zero. The components of any vector in each sector are taken from the two vectors and the zero vector that are adjacent to each other in each sector. The resultant vector is

$$\int_0^T U_{\text{ref}}\text{d}t = \int_0^{T_x} U_x\text{d}t + \int_{T_x}^{T_x+T_y} U_y\text{d}t + \int_{T_x+T_y}^{T} U_0^*\text{d}t. \tag{30}$$

Among them, U_{ref} is the expected vector; T is the sampling time; T_x, T_y, and T_0, respectively, correspond to the action time of the two nonzero vectors U_x, U_y and the zero vector U_0 in one cycle. Among them, U_0 includes two zero vectors, U_0 and U_7. The meaning of formula (30) is that the sum of integral effect produced by vector U_{ref} in time *T* is the same as the sum of integral effects produced by values U_x, U_y, and U_0 in times T_x, T_y, and T_0, respectively.

The equivalent rotation vector is usually a combination of a three-phase sine wave vector in a vector space vector. The input power angle frequency is its rotation speed, and its trajectory is shown in Figure 5.

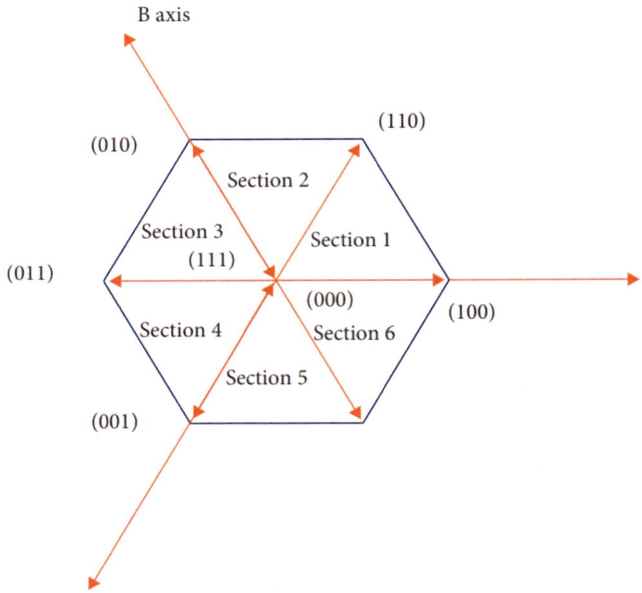

Figure 5. Vector pie chart.

Using the abovementioned vector synthesis technique, a three-phase sine wave vector can be obtained. The acquisition method is to use the set vector to start from position $U_4(100)$ on the vector space vector and use the two adjacent nonzero vectors and zero vector in the area to synthesize the vector as a small increase. The tentative vector obtained by a small increase each time is equivalent to a vector space vector, and the vector space vector plane can be used for smooth rotation of the vector to achieve the purpose of pulse width modulation of the vector.

AUTOMATIC INTELLIGENT CONTROL SYSTEM BASED ON INTELLIGENT CONTROL ALGORITHM

The overall architecture design is the first step to complete the design and realization of the intelligent control system. It points out the working principle and control flow of the entire system. The software and hardware platforms implemented by the system determine the functional characteristics of the system, and the design of the communication protocol specifies all the data formats in the system. The system is mainly composed of an embedded intelligent terminal and a wireless sensor network. The controlled object is placed within the coverage of the wireless sensor network, and the embedded intelligent terminal communicates with the wireless sensor network through a serial port. The overall system architecture is shown in Figure 6.

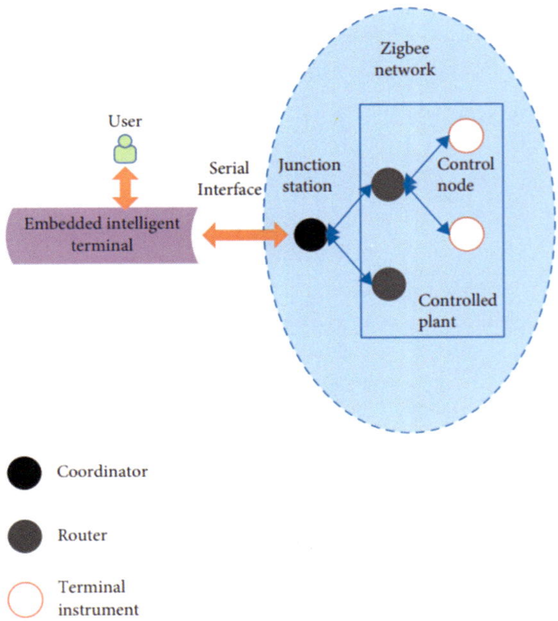

Figure 6. The overall architecture of the intelligent control system.

The overall control flowchart of the intelligent control system in this paper is shown in Figure 7. The specific control process is as follows: (1) The embedded intelligent terminal performs initialization operations and starts core services. (2) The embedded intelligent terminal is connected to the convergence point through the serial port, so that the coordinator can access the network. After the coordinator accesses the network, it accesses the control node, so that the control node accesses the network. (3) The sensor collects the data information of the controlled object, and after relevant processing such as A/D conversion of the control node, it actively periodically reports it to the embedded intelligent terminal according to the ZigBee private communication protocol format. (4) After the embedded intelligent terminal receives the data from the ZigBee network, it is handed over to the parsing module for processing, and the network data is distinguished from the data reported actively according to the function code. In the network access data, it extracts ZigBee network equipment information and stores it in the net equipment table of the database. The sensor data information and collected data information are extracted from the actively reported data and stored in the sensor table and collection data table, respectively, in the database. (5) It extracts the specific data information of the controlled object from the collected data and obtains the control vector corresponding to the current data through the offline look-up table method of the fuzzy intelligent control algorithm. (6) It encapsulates the obtained control vector into a control command in the ZigBee private communication protocol format and sends it to the corresponding device in the underlying sensor network. (7) The corresponding device at the bottom layer receives the control instruction, and after D/A conversion and other processing, the corresponding analog vector is output from the AO output interface of the control node and added to the control unit to realize real-time control of the controlled object.

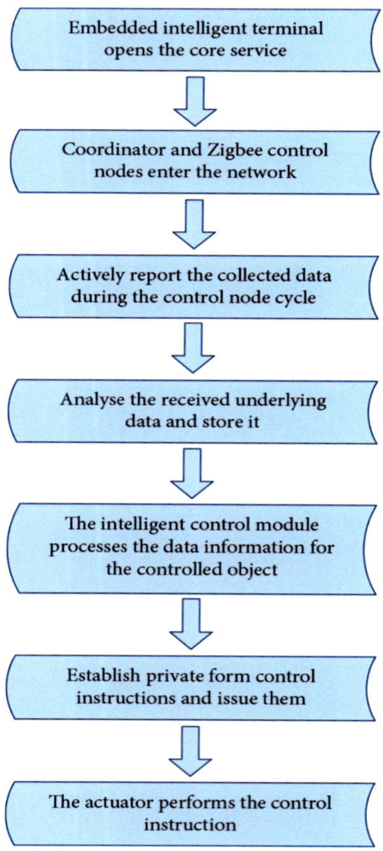

Figure 7. The control flow of the intelligent control system.

Embedded intelligent terminals play a central role in the entire intelligent control system. The specific content can be seen in Figure 8. The interface between it and the coordinator is a serial port, which realizes the realization of the intelligent control function of the system by collecting information from the control node, intelligently analyzing the information, and intelligently issuing control commands.

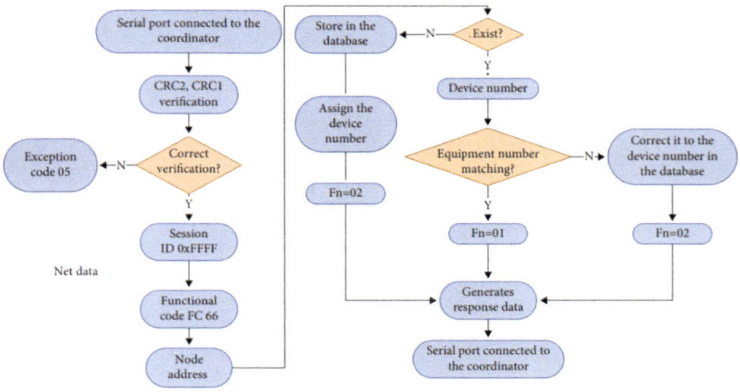

Figure 8. The flowchart of terminal processing of network access data.

If the node exists and matches, the actively reported data will be handed over to the two processing database subclasses for processing. Figure 9 shows the process of the smart terminal processing the actively reported data.

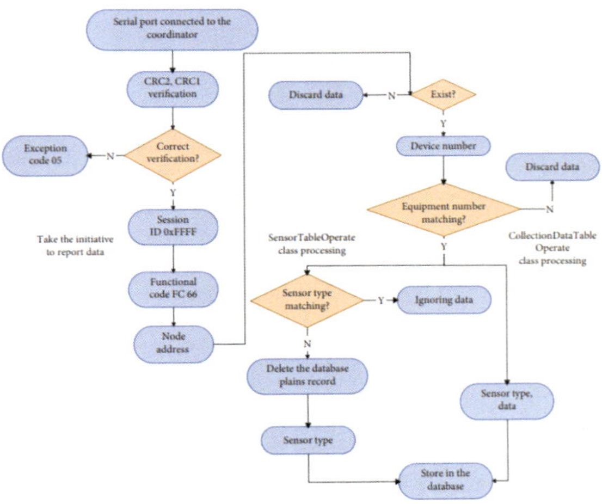

Figure 9. The flowchart of the terminal processing the actively reported data.

Through the above research, the performance verification of the intelligent control system in this paper is carried out. First, the algorithm effect evaluation of the intelligent vector control algorithm proposed in this paper is carried out. The statistical test results are shown in Figure 10.

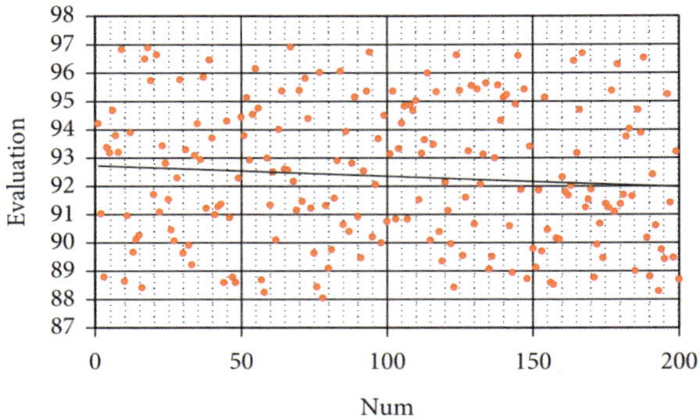

Figure 10. Evaluation of the intelligent effect of intelligent vector control algorithm.

Through the above research, it can be seen that the intelligent vector control algorithm proposed in this paper has a good intelligent effect, and then the evaluation of intelligent control effect of the automatic intelligent control system is carried out, and the result shown in Figure 11 is obtained.

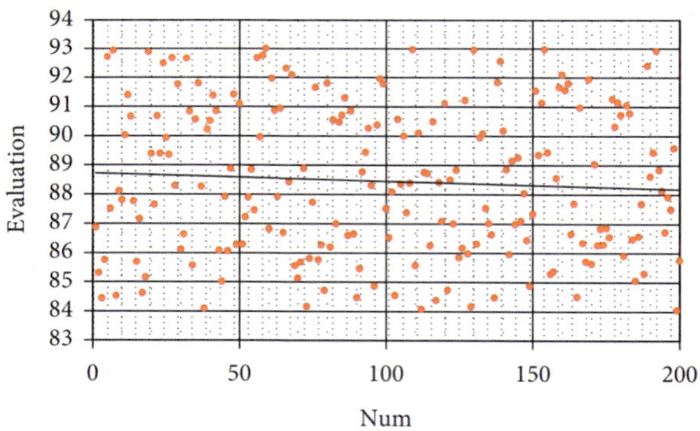

Figure 11. Evaluation of intelligent control effect of automated intelligent control system.

Through the above experimental analysis, it can be seen that the automated intelligent control system proposed in this paper has a better effect in intelligent control, and the embedded intelligent system of this paper can be applied to multiple systems with intelligent control requirements.

CONCLUSION

One of the characteristics of intelligent control is that it is suitable for uncertain or difficult-to-define process control, complex nonlinear controlled object control, and time-varying process control. As a complex model, the description can be quantitative (number or parameter) or qualitative (causal relationship) description. The second characteristic of intelligent control is that it uses self-adaptation, self-organization, self-learning, and other methods to improve the automation and intelligent control of the system. This enables it to process various uncertain, qualitative information and data structures, as well as unstructured information and data, process and utilize knowledge of different natures, identify changes in the structure or composition of the main control system, and modify or reconstruct its own parameters or structure according to changes in the main control system or environment. In addition, it is also fault-tolerant, capable of self-diagnosing, shielding, and self-recovering various faults. This paper combines the intelligent control algorithm to study the automated intelligent control system, builds the intelligent control system model according to the current more advanced algorithm model, and verifies the system function through experimental research.

ACKNOWLEDGMENTS

This study was sponsored by Hubei University of Technology.

REFERENCES

1. V. Syrotiuk, S. Syrotyuk, V. Ptashnyk et al., "A hybrid system with intelligent control for the processes of resource and energy supply of a greenhouse complex with application of energy renewable sources," *Przeglad Elektrotechniczny*, vol. 96, no. 7, pp. 149–152, 2020.
2. L. Vladareanu, V. Vladareanu, H. Yu, D. Mitroi, and A.-C. Ciocîrlan, "Intelligent control interfaces using e multidimensional theory applied on VIPRO platforms for developing the IT INDUSTRY 4.0 concept," *IFAC-PapersOnLine*, vol. 52, no. 13, pp. 922–927, 2019.
3. P. Zheng, J. Zhang, H. Liu, J. Bao, Q. Xie, and X. Teng, "A wireless intelligent thermal control and management system for piglet in large-scale pig farms," *Information Processing in Agriculture*, vol. 8, no. 2, pp. 341–349, 2021.
4. M. Nadafzadeh and S. Abdanan Mehdizadeh, "Design and fabrication of an intelligent control system for determination of watering time for turfgrass plant using computer vision system and artificial neural network," *Precision Agriculture*, vol. 20, no. 5, pp. 857–879, 2019.
5. A. P. Araújo Neto, G. W. Farias Neto, T. G. Neves, W. Ramos, and K. Brito, "Changing product specification in extractive distillation process using intelligent control system," *Neural Computing & Applications*, vol. 32, no. 17, pp. 13255–13266, 2020.
6. Y. Arya, "Automatic generation control of two-area electrical power systems via optimal fuzzy classical controller," *Journal of the Franklin Institute*, vol. 355, no. 5, pp. 2662–2688, 2018.
7. N. Harrabi, M. Souissi, A. Aitouche, and M. Chaabane, "Intelligent control of grid-connected AC-DC-AC converters for a WECS based on T-S fuzzy interconnected systems modelling," *IET Power Electronics*, vol. 11, no. 9, pp. 1507–1518, 2018.
8. H. Yang, A. Alphones, Z. Xiong, D. Niyato, J. Zhao, and K. Wu, "Artificial-intelligence-enabled intelligent 6G networks," *IEEE Network*, vol. 34, no. 6, pp. 272–280, 2020.
9. M. Al-Khafajiy, S. Otoum, T. Baker et al., "Intelligent control and security of fog resources in healthcare systems via a cognitive fog model," *ACM Transactions on Internet Technology*, vol. 21, no. 3, pp. 1–23, 2021.
10. L. Haibo, L. Zhenglin, and L. Guoliang, "Neural network prediction model to achieve intelligent control of unbalanced drilling's

underpressure value," *Multimedia Tools and Applications*, vol. 78, no. 21, pp. 29823–29851, 2019.

11. L. Peng and Z. Jiang, "Intelligent automatic pig feeding system based on PLC," *Revista Científica de la Facultad de Ciencias Veterinarias*, vol. 30, no. 5, pp. 2479–2490, 2020.
12. S. Anila, B. Saranya, G. Kiruthikamani, and P. Devi, "Intelligent system for automatic railway gate controlling and obstacle detection," *International Journal Of Current Engineering And Scientific Research (IJCESR)*, vol. 4, no. 8, pp. 2394–0697, 2017.
13. L. B. Prasad, H. O. Gupta, and B. Tyagi, "Intelligent control of nonlinear inverted pendulum system using Mamdani and TSK fuzzy inference systems: a performance analysis without and with disturbance input[J]," *International Journal of Intelligent Systems Design and Computing*, vol. 2, no. 3-4, pp. 313–334, 2018.
14. J. Yang, P. Wang, W. Yuan, Y. Ju, W. Han, and J. Zhao, "Automatic generation of optimal road trajectory for the rescue vehicle in case of emergency on mountain freeway using reinforcement learning approach," *IET Intelligent Transport Systems*, vol. 15, no. 9, pp. 1142–1152, 2021.
15. X. Yao, J. H. Park, H. Dong, L. Guo, and X. Lin, "Robust adaptive nonsingular terminal sliding mode control for automatic train operation," *IEEE Transactions on Systems, Man, and Cybernetics: Systems*, vol. 49, no. 12, pp. 2406–2415, 2018.
16. M. J. Mahmoodabadi and M. Taherkhorsandi, "Intelligent control of biped robots: optimal fuzzy tracking control via multi-objective particle swarm optimization and genetic algorithms," *AUT Journal of Mechanical Engineering*, vol. 4, no. 2, pp. 183–192, 2020.
17. W. He, Z. Li, and C. L. P. Chen, "A survey of human-centered intelligent robots: issues and challenges," *IEEE/CAA Journal of Automatica Sinica*, vol. 4, no. 4, pp. 602–609, 2017.
18. Y. Huang, S. Liu, C. Zhang, X. You, and H. Wu, "True-data testbed for 5G/B5G intelligent network," *Intelligent and Converged Networks*, vol. 2, no. 2, pp. 133–149, 2021.
19. H. Dong, C. Roberts, Z. Lin, and F.-Y. Wang, "Guest editorial introduction to the special issue on intelligent rail transportation," *IEEE Transactions on Intelligent Transportation Systems*, vol. 20, no. 7, pp. 2677–2680, 2019.

CHAPTER 11

Skeleton of Implementing Voice Control for Building Automation Systems

Waleed Dweik[1], Musa Abdalla[2], Yazan AlHroob[1], Anas AlMajali[3], Suad Alhaj Mustafa[2], and Mohammad Abdel-Majeed[1]

[1]Computer Engineering Department, University of Jordan, Amman, Jordan

[2]Mechanical Engineering Department, University of Jordan, Amman, Jordan

[3]Department of Computer Engineering, Faculty of Engineering, The Hashemite University, Zarqa, Jordan

ABSTRACT

Building automation is one of the most popular trends in the 21st century, as it provides buildings with modern and luxurious touches. Furthermore, voice-controlled automation systems are very convenient for the current fast-paced life style as people need to quickly, efficiently, and safely control

Citation: W. Dweik, M. Abdalla, Y. AlHroob, A. AlMajali, S.A. Mustafa, M.A.Majeed, "Skeleton of Implementing Voice Control for Building Automation Systems", Scientific Programming, vol. 2022, Article ID 6886086, 15 pages, 2022. https://doi.org/10.1155/2022/6886086.

everything around them. The goal of this article is to provide a skeleton for implementing voice control to a building automation system. The article also presents and evaluates a use case of the proposed skeleton integrated with a KNX-based office automation system. Through the integrated system, lights and blinds are controlled and sensors are polled using voice commands from an authorized personal. This article serves as an educational reference for engineering students interested in applying authorized voice control capability to automation systems.

INTRODUCTION

With the emerging Internet of things (IoT), smart building automation is gaining massive popularity [1–4]. A building automation system (BAS) is normally installed to control and monitor services such as heating, cooling, ventilation, lighting, shading, and security. Nowadays, automation systems are deployed in commercial buildings (e.g., offices, malls, factories, hotels, and medical institutions) as well as residential buildings (e.g., houses and apartments). In the commercial domain, the main goals of a BAS are to reduce energy consumption and facilitate building monitoring, maintenance, and operation. In the residential domain, a BAS helps people achieve many of their daily routines efficiently while maintaining their comfort, which is especially needed for elderly and individuals with disabilities. To achieve the aforementioned goals, a BAS allows users to control devices/appliances/services either according to the readings of a network of specialized sensors or according to their personal preferences. For example, a home automation system might allow a resident to control the electrical light intensity according to his/her mood or according to the measured natural light intensity in the room in order to save electricity.

Traditional building automation systems interact with the user using a graphical user interface [5–8]. On the other hand, recent automation systems (e.g., Alexa [9], Apple [10]) integrate voice control for extra convenience and flexibility. There are two main approaches to implement voice recognition module for a BAS: first, an offline approach that relies on trained machine learning algorithms capable of recognizing uttered control commands. While this approach provides better availability as it does not require Internet connection, it requires substantial amount of training data in order to achieve high recognition accuracy. The second approach is to leverage one of the voice recognition services that are available through cloud APIs such that the actual voice processing is performed in the cloud and the BAS is only responsible for processing the returned textual data and translating it into actions.

Integrating voice recognition with a BAS is especially important for people who suffer from paralysis. Every year, around 18,000 new spinal core injuries, which are the main cause of paralysis, are registered in the United States [11]. The main causes of these new cases are accidents, diseases, and acts of violence. The most severe types of physical paralysis are monoplegia, hemiplegia, paraplegia, and quadriplegia [12]. People with monoplegia suffer from paralysis in one upper or lower limb (i.e., one arm or one leg), while people with hemiplegia suffer from paralysis in one arm and one leg on the same side of the body. More severely, people with paraplegia cannot move both lower limbs, while people who suffer from quadriplegia lose the ability to voluntarily move any of their upper or lower limbs. Although paralyzed people can be intellectually capable, they normally cannot perform their daily activities like ordinary people and they need continuous support and monitoring. By integrating voice recognition with home/office automation systems, paralyzed people can control all devices and appliances around them; hence, live a normal life and be productive members of the society with minimal supervision.

The main contributions of this article are as follows:

i. Deriving a skeleton for integrating voice control with any building automation system. The proposed voice control skeleton supports online and offline modes. During online mode (i.e., Internet connection is available), the skeleton relies on Google Speech Recognition API to recognize user commands and to allow other smart features (e.g., web search and run applications). During offline mode, the skeleton utilizes an adapted version of the CMUSphinx open-source speech recognizer [13]. In addition, the proposed skeleton uses a wake-up-word (WUW) model with speaker verification, so that the control commands are only initiated by the authorized user.

ii. Implementing the proposed skeleton and integrating it with an office automation system as a proof of concept use case. The office is equipped with Konnex (KNX) components (e.g., lightning, blinds/shutters, heating, and air-conditioning) [14]. The integrated system allows the user to issue voice commands in order to control the KNX components (e.g., "turn on all lights").

KNX AUTOMATION SYSTEMS

Building automation or smart building is the process of transforming traditional buildings into ones that can be automatically controlled using electronic devices (e.g., computers). Most modern buildings are smart with many smart devices [15]. One of the most widely used building automation standards is Konnex (KNX) [16]. KNX is an open standard that is based on three previous standards: the European Home Systems Protocol (EHS), BatiBUS, and the European Installation Bus (EIB or Instabus). KNX is a standardized technology, which controls the automation of integral functions of any residential, commercial, or industrial building.

KNX uses a wired bus system for building automation using a central software provided by KNX association, which is called Engineering Tool Software (ETS). ETS provides creating projects, configuring KNX devices, and a huge products' catalog and database of KNX devices from various manufacturers such as Siemens and ABB. KNX enables various areas of a building to be controlled, link them together, and adjust them to each other. This includes, for example, controlling, managing, and monitoring of lighting, roller shutters, security, energy management, heating, ventilation, air-conditioning (HVAC) systems, and many other applications.

KNX achieves its functionality by utilizing dedicated controllers that are connected with abundant and diverse sensors and actuators using an open worldwide standard with over 300 different manufacturers producing products that all interwork and operate together seamlessly. KNX operates by ensuring all components, devices, features, and functions of any building (or outdoor space) communicate via one common language (i.e., KNX standard) instantly and remotely as depicted by Figure 1. Moreover, KNX offers interfaces to many other technologies, such as Ethernet (LAN) and lighting controls with DALI and BACnet network, making it easy to exchange information and data via the KNX network. In particular, the KNXnet/IP supports connection to building control (e.g., OPC, PROFINET, SIMATIC S7).

The basic unit of the KNX communication protocol is the Telegram, which is similar to the TCP/IP data packet. A telegram contains several fields such as control field, source address, and receiver address [17]. Figure 2 shows the telegram structure. Each KNX device deployed as part of an automation system is configured with an individual address of 16 bit.

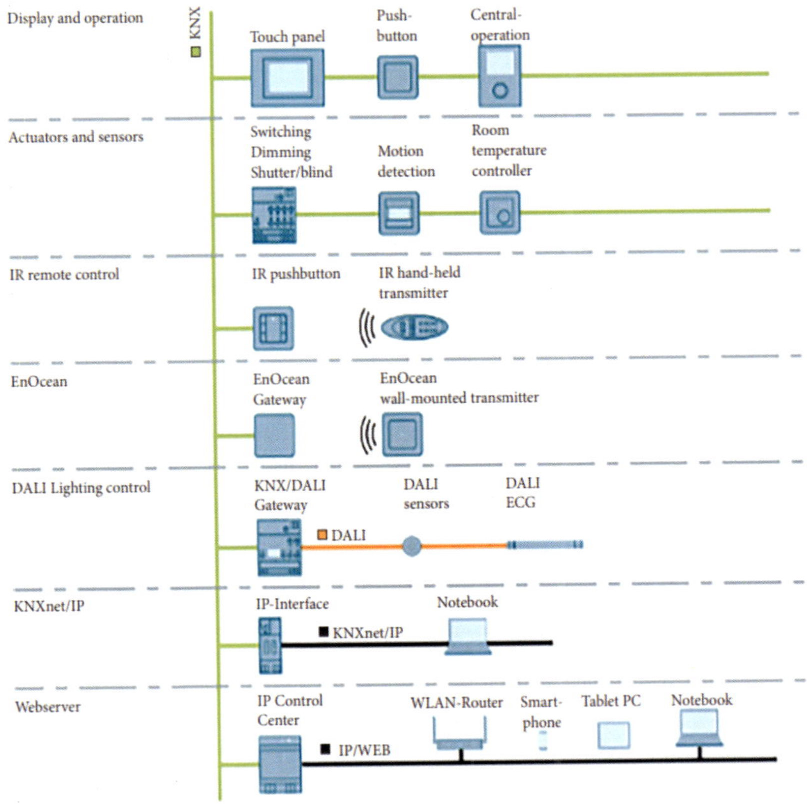

Figure 1. KNX devices and network capabilities illustration.

The individual address is divided into three fields—4-bits for area, 4-bits for line, and 1-byte for device as shown in Figure 3. The source address field in the telegram represents the individual address of the sender.

<table>
<tr><td>Control Field <8 bit></td><td>Source Address <16 bit></td><td>Receiver Address <16 bit></td><td colspan="3">N-PDU</td><td>Check Field <8 bit></td></tr>
<tr><td></td><td></td><td></td><td><8 bit></td><td colspan="2">T-PDU</td><td></td></tr>
<tr><td></td><td></td><td></td><td></td><td><6 bit></td><td>A-PDU</td><td></td></tr>
</table>

Figure 2. KNX telegram structure.

Area <4 bit>	Line <4 bit>	Device <8 bit>

Figure 3. Individual address structure.

In addition, each functionality or channel of KNX devices has a group address. The group address can be shared between two or more functionalities/channels inside the same device or different devices. Figure 4 shows how the group address can be formatted in two levels (i.e., main group and subgroup) or three levels (i.e., main group, middle group, subgroup) [17]. The group address requires only 15-bits (i.e., bit D15 is reserved). The receiver address field in the telegram can be either an individual address in the case of point-to-point connection or a group address in the case of multicast/broadcast connection.

<table>
<tr><td>D15</td><td>D14</td><td>D13</td><td>D12</td><td>D11</td><td>D10</td><td>D9</td><td>D8</td><td>D7</td><td>D6</td><td>D5</td><td>D4</td><td>D3</td><td>D2</td><td>D1</td><td>D0</td></tr>
<tr><td rowspan="2"></td><td colspan="4">Main Group</td><td colspan="11">Sub Group</td></tr>
<tr><td colspan="4">Main Group</td><td colspan="3">Middle Group</td><td colspan="8">Sub Group</td></tr>
</table>

Figure 4. Group address structure.

The N-PDU field in the telegram is destined/generated by the network layer, and it consists of two parts—8-bit value and T-PDU as shown in Figure 2. The most significant bit in the 8-bit field determines whether the receiver address is an individual or group address. The next 3-bits provide routing information, and the following 4-bits provide the length of the payload. The T-PDU part is destined/generated by the transport layer, and it consists of two parts: 6-bit value and A-PDU. The most significant 2-bits in the 6-bit field determine the type of the transport layer communication, and the next 4-bits represent the sequence number for numbered communication types. The A-PDU is destined/generated by the application layer, and it represents the actual useful information (i.e., payload). More information about the telegram structure can be found in Ref. [17].

RELATED WORK

In this section, we first discuss previous work related to building automation systems in general. Second, we present the most recent related work in the area of voice-controlled building automation systems. The second part of the related work is divided into three subsections according to the type of the deployed speech recognition mechanism. The author in Ref. [18] provided the basic principles and technologies for building automation. The work discusses four main issues related to building automation: network communication protocols and standards, integration and interfacing of building automation subsystems and multiple building systems, control strategies, configuration, and technologies for automation control.

Sauter et al. discussed the historical evolution of factory and building automation systems [19]. The article lists typical services provided by building automation systems: climate control, visual comfort, safety, security, transportation, one-way audio, energy management, supply and disposal, communication, and information exchange. Regarding building automation, the authors concluded that the main challenge lies in the applications due to the large amount of data collected from an increasing number of data points. To overcome this challenge, smart ways of data processing are required. In Ref. [20], the authors provided a more detailed overview by discussing the fundamental concepts, features, standards, and requirements of building automation systems. Different building automation technologies (e.g., BACnet, KNX, and ZigBee) were analyzed and compared according to the coverage of the functional aspects (i.e., grouping, event notification, alarm notification, historical data access, scheduling, and scenarios). The authors concluded that no specific technology is capable of providing all required functionalities of building automation.

An important domain of building automation is its efficiency in reducing energy and improving occupants' quality of life. The authors in Ref. [21] discussed the importance of building automation and control systems (BACSs) and technical building management (TBM) systems in optimizing building performance and gave a general framework for the analysis of the performance for a wide range of BACS. The framework focuses on the need for building performance simulation, data analytics, performance tracking, and the need for sensors and actuators to measure the quality of building environment.

The research work in Ref. [22, 23] focused on the interaction opportunities between occupants and building automation systems. In Ref. [22], the authors focused on scaling the demand management programs from the level of a single building to the level of a district microgrids. Although modern building automation systems integrate thermostatic control with improved sensors and data analytics to reduce energy consumption by regulating the thermostat set point, the demand management actions are either predetermined by a field engineer or set by the user with static rule-based options. As a result, the efficacy of these systems is not fully exploited as aspects such as thermal inertia, real-time user activity, and weather conditions are ignored. To overcome this shortage, the authors proposed a distributed demand management system with three main features: first, using feedback mechanisms to scale up the states of the different buildings (i.e., temperatures, occupancy schedule, availability of renewable energy) to district-level microgrids; second, integrating occupant behavior using multiobjective optimization; And third, leveraging feedback-based "personalized" corrections for each building to provide consistent performance against weather conditions and user behavior at the district level. The simulation results of the proposed system show improvement in the energy cost and reduction in the percentage of people dissatisfied.

In Ref. [23], Baldi et al. tackled the inefficiency associated with using fixed rules to manage thermostatic loads of buildings. These rules represent a barrier against dynamic load regulation according to rooms geometry, orientation, and human occupancy (i.e., smart zoning). The authors proposed to formulate the problem as multiple-mode feedback-based optimal control problem and solve it by specifying the parameters of the possible control strategies and adaptively finding the optimal strategy. The proposed solution reduces energy consumption by 15% and increases thermal comfort by 25% when compared to strategies with predetermined rules. On the other hand, the proposed solution reduces energy consumption by 4% and increases thermal comfort by 8% when compared to optimized strategies without smart zoning.

The main idea in Ref. [24–26] is to improve the efficiency of building energy management systems (BEMS). The authors in Ref. [24] discussed the potential of using Internet of Energy (IoE) to solve the efficiency problems of BEMS such as energy data loss and energy overloading. Analogous to data generation, storage, and transmission in IoT and smart grid, IoE facilitates the generation, storage, and transmission of energy using different types of nodes such as renewable sources, batteries, and transformers, respectively. The demand and supply information from IoE can help to reduce buildings power consumption. In Ref. [25], the authors proposed an optimization control system for indoor thermal comfort and building energy consumption using predictive mean vote and simplified thermal model. The proposed system can efficiently regulate heating, ventilation, and air-conditioning systems of buildings. Hettiarachchi et al. proposed an IoT-based energy management system to reduce energy wastage caused by shortage of monitoring and automation [26]. The proposed system utilizes a trained XGBoost machine learning model hosted on the cloud to predict consumption based on device and sensor data (e.g., electricity load, weather data, and temperature) collected and stored on the cloud as well. The predictions can be used either to perform manual adjustment of device operation by the user or to perform automatic adjustment with priority given to devices tagged as critical.

Hardware-Based Speech Recognition for Voice Control

In this subsection, we focus on the related work, which leverages specialized hardware components for the speech recognition of control commands. In Ref. [27], Guo et al. described the implementation of a voice-controlled home automation system using the embedded sound-controller, SUNPLUS SPCE061A, combined with Microchip 2.4 GHz ZigBee modules. A multistage algorithm and the speech recognition library provided by SUNPLUS are used to determine the target room, appliance, and operation. Examples of tested commands are (living room, light, ON) and (bedroom, TV, volume up). No recognition accuracy results are presented. Similarly, the authors in Ref. [28] added multiple LD3320 voice recognition modules to ZigBee-based home automation network. The LD3320 is based on speaker-independent automatic speech recognition (SI-ASR) technology with a built-in library. Although the recognition accuracy of the LD3320 module is high (i.e., 95%), only 50 keywords of specific languages can be recognized.

The authors in Ref. [29] proposed a smart system to control distributed home appliances using either voice or gesture commands. The main purpose of the proposed system is to improve the quality of life for senior citizens and people with disabilities. The system mainly consists of a transmitter module and a receiver module. In the transmitter module, a microcontroller receives input commands from an audio-capturing device and a tilt sensor. After processing the input command, an RF module is used to transmit the signals to the appliance of interest. In the receiver module, the received signals are forwarded to a microcontroller, which control the relay of the device of interest.

In Ref. [30], the authors proposed a multifunctional smart home automation system. The system leverages a dedicated hardware called Voice Recognition Module (V3) [31], which can be adapted to a particular speaker's voice. The module supports up to 80 voice commands and requires two setup steps: training and loading. The authors tested the system using 10 speakers (5 males and 5 females) to train the voice recognition module and then measure the best distance between the speaker's mouth and the microphone and the average number of attempts for the command to be correctly recognized. The experimental results showed that the best distance is between 1 cm and 15 cm and the average number of attempts is between 1 and 1.3 for various commands.

Off-the-Shelf Software-Based Speech Recognition for Voice Control

In this subsection, we focus on the related work, which leverages off-the-shelf free software packages for the speech recognition of control commands. The Wolfram Alpha, a cloud-based API, is used in Ref. [32] to recognize the spoken command and send it to the desired home appliance through a Raspberry Pi board. Unfortunately, no analysis of the recognition accuracy is provided for the system. In Ref. [33], the authors presented the design and implementation details of the wireless home automation system. The system consists of three main components: a microphone module to capture voice commands, a centralized controller to receive and process (i.e., filter and sample) commands before sending them to a personal computer for recognition, and appliances' control modules to receive the recognized commands and control the appliances' operations accordingly. The speech recognition part was accomplished using Microsoft Speech AI library. Using 1225 commands from 35 male and female subjects, the recognition accuracy was 79.8%.

A mobile application is proposed in [34] to provide voice and touch control of home automation. The mobile application leverages free applications on Google Play Store (i.e., AMR-Voice and Arduino Bluetooth Controller) to receive voice commands and provide a graphical user interface of a remote control. The Arduino Uno board acts as the main controller and is connected to the application via Bluetooth. However, the authors did not provide any recognition accuracy results. In Ref. [35], the authors proposed to use Kinect V2 to receive voice commands and then send them to a trained computer system with Microsoft Speech Recognition Engine for identification. The reported accuracy is 95% when the distance between the user and the Kinect is less than four meters and when the noise level is less than 53 dB.

The authors in [36] discussed and evaluated three approaches to implement voice control for smart home automation. The first approach relies on a cloud-based voice recognition service and a cloud-based home automation system (e.g., Google Speech API, Amazon Alexa voice services, and Bing speech API). The second presented architecture relies on a local home automation system with a gateway to a cloud-based voice recognition service (e.g., Google Speech API and Bing speech API). The third approach leverages an offline voice recognition engine integrated with the local home automation system.

In [37], the authors proposed a smart home automation system with voice control using BitVoicer software [38]. The system is built using an Arduino board, set of specialized sensors, RF receiver/transmitter, and Ethernet shield. The system has two operational modes: automatic mode, which controls the appliances according to sensor readings without user intervention, and manual mode, which controls the appliances according to user voice commands. The system was tested using light emitting diodes (LEDs) instead of actual appliances, and no accuracy results are provided.

The authors in Ref. [39] developed a voice control home security and automation system. The security aspect of the project is simply represented by an SMS message sent to the user whenever an ultrasonic sensor detects any motion. Voice recognition is achieved by capturing the voice commands and converting them to text using an Android mobile application (i.e., AMR-Voice), and then, the text is sent to the main microcontroller using Bluetooth for analysis. Only ON/OFF commands are used for testing, and LEDs are used as outputs instead of real appliances.

Putthapipat et al. implemented a speech recognition gateway [40] on top of the previously published home automation framework, PiFrame [41]. PiFrame runs full-feature Linux OS to shorten the development cycle and reduce system's limitation. Google Speech API is used to convert voice commands into text and send it back to the PiFrame, which controls the home appliances accordingly. Most recently, Vishwakarma et al. [42] presented a smart home automation system using IoT. User commands can be received as voice commands through Google Assistant or through a web-based application in cases of noisy backgrounds.

Bespoke Software-Based Speech Recognition for Voice Control

In this subsection, we focus on the related work, which implements customized software speech recognition engines to identify control commands. In Ref. [43], the authors presented a voice-controlled home automation system using natural language processing (NLP) and Internet of Things (IoT). According to the authors, the system provides better control of the home appliances than the traditional mimicking of ON/OFF switching. The central component of the system is a mobile application, which is responsible for processing the user voice commands using artificial intelligence (AI) and NLP algorithms. In addition, the system contains MKR1000 Arduino boards connected to home appliances. The MKR1000 boards receive the commands from the mobile application and collect responses and send them back. The authors neither provided any elaborations on the supported commands nor presented any voice commands recognition accuracy results.

A phoneme-based voice recognition engine is proposed in Ref. [44] and is utilized to control a home automation system. The proposed engine consists of four stages: endpoint detection, feature extraction using Mel frequency cepstral coefficients (MFCCs), phoneme recognition using support vector machine (SVM), and word recognition using SVM. The reported average recognition accuracy using a sample of 7 words and 10 utterances per word (i.e., 70 utterances) is 90.1% with an average recognition time of 0.91 second. In Ref. [45], the authors implemented a voice command interface (VCI) for smart home automation systems. The proposed VCI defines the patterns of commands that can be mapped to the functionalities and locations of home devices. The VCI is compatible with the existing voice recognition services (e.g., Alexa from Amazon) and customized ones. The authors only confirmed that the VCI is successfully operational without any recognition accuracy results.

Compared to previous literature, this article describes a general approach that can be used to integrate voice control with wake-up word and speaker verification to any existing building automation system. The complete design details of the proposed approach are presented. In addition, the proposed approach is implemented and tested using an office automation system with KNX equipment. Finally, recognition accuracy for three speakers is measured and reported.

SKELETON DESIGN DETAILS

In this section, we start by discussing the flow diagram of the proposed voice control system shown in Figure 5. Upon initializing the system, a wake-up word (WUW) engine is activated. The WUW engine recognizes and records human voice, and then, it checks if the recorded audio file contains a predetermined wake-up word (e.g., Hey Computer). Once the wake-up word is detected, the recorded audio file is used to authenticate the user via a speaker verification module. After the speaker is verified, the Internet connectivity is checked to select the most appropriate speech recognition scheme (i.e., online or offline recognition). For online speech recognition, the Google Speech recognition API and a natural language processor are used to identify the desired functionality and whether there are any defined responses to be generated and processed. For offline speech recognition, the proposed system leverages an adapted version of the famous CMUSphinx toolkit to recognize predefined automation commands. Next, we discuss the details of each stage separately.

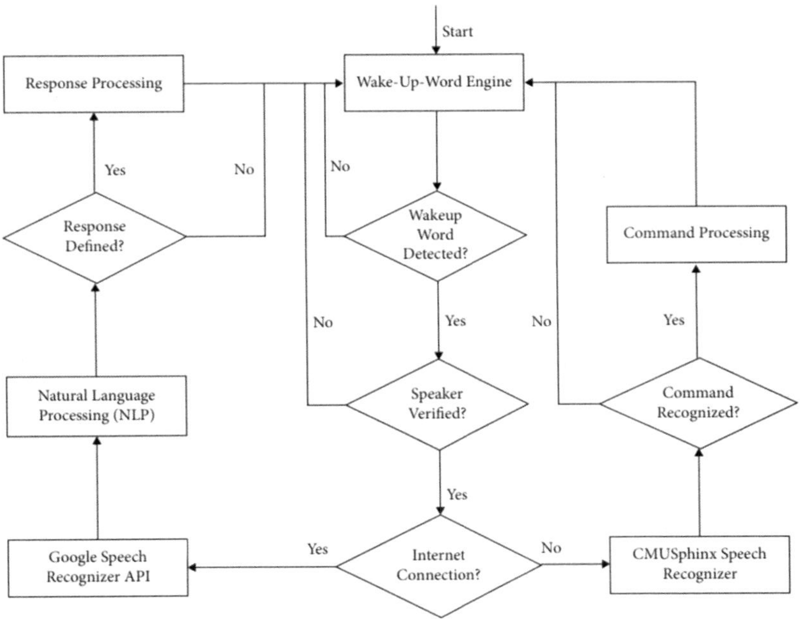

Figure 5. General flow diagram.

Wake-Up Word Engine

Wake-up word (WUW) detection, also called automatic voice trigger, is one of the most prominent applications related to speech recognition. In order to attract system's attention, the user should utter a predefined WUW that the system is trained to recognize. WUW detection can be achieved using algorithms that are based either on pattern matching with predefined templates or on the automatic speech recognition (ASR) framework. The latter type is more robust and generalized.

The steps associated with a typical ASR-based WUW detection process are discussed in Ref. [46] and reproduced in Figure 6. After speech enhancement algorithm, a voice activity detection (VAD) module continuously detects human speech and acoustic events. Afterwards, the features of the detected speech are extracted and inputted to a decoder, which cooperates with an acoustic model to complete the delicate alignment of the speech phoneme boundaries. For each speech unit that is output from the decoder, a variety of confidence metrics are calculated from multiple perspectives and a classifier is used to determine whether the processed speech is the predetermined WUW or not.

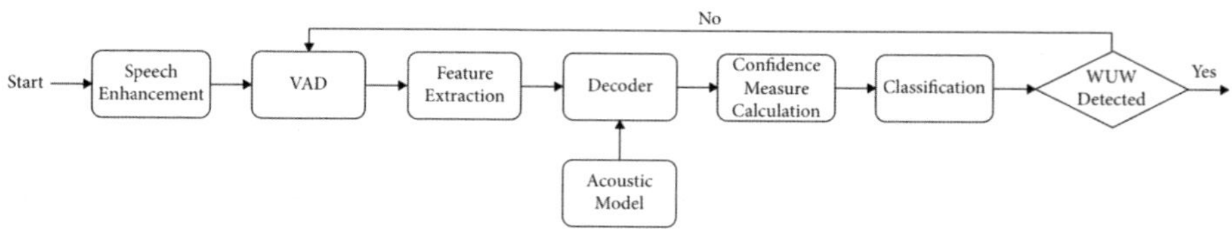

Figure 6. WUW engine diagram.

There are multiple readily implemented WUW engines that can be utilized to add voice control for automation systems such as the Snowboy hotword detection engine [47] and the Porcupine wake word detection engine [48]. Both engines are lightweight (i.e., have low memory footprint), highly customizable (i.e., user can define and train his/her own single or multiple wake-up words and determine the detection sensitivity), and portable. However, when comparing the two engines in terms of detection accuracy and CPU usage, Porcupine is 5.7 times more accurate than Snowboy while requiring 6.5 times less CPU usage [49]. In addition to the licensed business editions, these engines have open-source personal editions.

Speaker Verification

Speaker recognition (SR) is the process of recognizing the speaking person based on the voice bio-metrics extracted from the speech signal. SR includes speaker identification (SI), speaker diarization (SD), and speaker verification (SV). SI is the process of determining which preregistered speaker is speaking. In other words, SI answers the question (who is speaking?) [50]. On the other hand, SD receives an audio file with multiple speakers as an input and the output is a separate audio signal for each speaker. In other words, SD answers the question (who spoke what and when?) [51]. SV is the process of authenticating the identity of a speaker.

In building automation, SV is the most relevant as it is used to determine whether or not a specific speaker has permissions to use the system. SV can be classified either as text-dependent (i.e., verification is done for a predetermined phrase) or text-independent (i.e., verification is done for any spoken phrase) [52]. In this work and as shown in Figure 5, a recorded audio file with the correctly detected predefined WUW is provided by the WUW engine to the SV engine in order to authenticate the speaker. Next, we present two possible techniques to implement the more general text-independent SV despite that the proposed skeleton always validates the speaker for the same phrase (i.e., the predetermined WUW).

The first technique is based on a Gaussian mixture model (GMM) fitted to the voice features of the user with permissions to control the automation system of concern. A Gaussian mixture is a function consisting of several Gaussians, each of which is determined by k $(1, \ldots, K)$, where K is the number of clusters in the dataset. Figure 7 shows the process of generating a GMM file for a specific authorized user. The features vectors of N audio files from the targeted user are extracted and used to train the GMM and generate its output file.

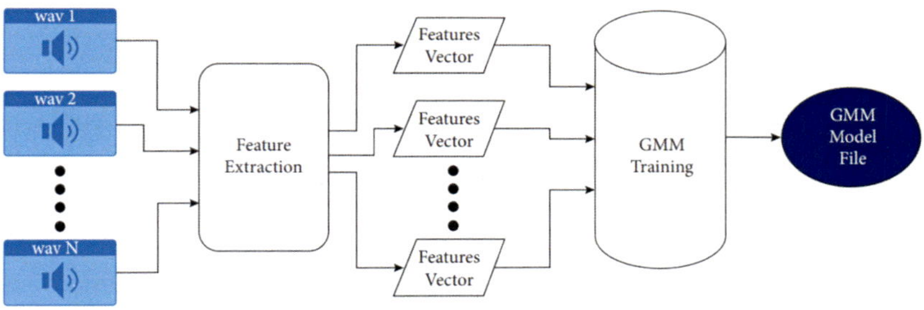

Figure 7. GMM for speaker verification.

The feature extraction in SV is normally based on MFCCs, filterbank energies, and delta functions, which can be applied using python_speech_features library. SciKit-learn Python machine learning can be used to define the GMM. The input audio files might need to be processed for noise reduction, sample rate changing, and format conversion using libraries such as Librosa and pysndfx with sound exchange (SoX) package. With every access trial to the automation system, the features vector extracted from the audio file, which contains the correctly detected WUW is inputted to the created GMM file. The GMM uses the score method to compute the cosine similarities between the input vector and the dataset used for training. The output of the score method is compared to a predetermined similarity threshold (e.g., 95%) to authenticate the user.

The second technique is based on residual neural network (ResNet), an artificial neural network (ANN) that utilizes skip connections or shortcuts to jump over some network layers in order to mimic the pyramidal cells in the cerebral cortex. Normally, ResNet models are implemented using two- or three-layer skips with in-between rectified liner units (ReLUs) and batch normalization. Similar to the first technique, feature extraction for the dataset can also be achieved using the python_speech_features library.

In order for the ResNet model to verify a specific speaker, the features extracted from the audio files of this speaker need to be included in the training set of the model. During training, enrollment embeddings are generated from the audio features of each speaker and saved. With every access trial to the automation system, the feature extracted from the audio file, which contains the correctly detected WUW, is inputted to the trained ResNet to generate its embeddings and compare them to the enrolled embeddings. The comparison score determines whether speaker is verified or not according to predetermined threshold.

Automatic Speech Recognition

Automatic speech recognition (ASR) is the field in which algorithms and procedures are developed to translate the spoken languages into text. To achieve this task, the ASR process converts the spoken sentences into audio signals and apply statistical models to extract the audio statistical properties such as Mel Frequency Cepstrum Coefficients (MFCCs), Linear Predictive Cepstral Coefficients (LPCCs), and discrete Fourier transform (DFT). Based on the collected features, the phoneme and word identification is done using pattern recognition probabilistic models such as hidden Markov model (HMM) and Gaussian mixture model (GMM). Recent ASR techniques leverage the advances in deep learning and big data fields. The latter is clear by the wide industrial adoption of deep learning techniques to perform ASR tasks. Siri, Alexa, Google Assistant, and Cortana are examples of voice assistants that use state-of-the-art speech recognition technologies to translate and understand the spoken sentences and take actions such as schedule a meeting and place purchase orders.

One way to perform ASR is by handing off the task to cloud servers on which the recognition models and data are installed. We refer to this approach as online speech recognition. On the other hand, offline speech recognition does not require Internet connection because the models are installed locally on the device. Accuracy, speed, security, and maintenance are aspects that should be taken into consideration when deciding which ASR approach to use (i.e., online vs offline). The skeleton proposed in this article deploys an online ASR model that is used whenever an Internet connection is present to benefit from the high recognition accuracy. At the same time, the skeleton deploys an offline model that is only used when no Internet connection is present to provide availability.

In a typical online speech recognition engine, a voice signal is captured in the local system through a microphone and sent over the Internet to designated servers in order to be processed by machine learning techniques and converted to text (i.e., speech to text). The text is then returned to the local system and inputted to a natural language processor for recognition. There are multiple popular cloud-based speech-to-text APIs that one can leverage such as Google Cloud Speech API, IBM Watson Speech to Text, Microsoft Azure Bing Speech API, and Amazon Transcribe.

Natural language processing (NLP) is a subfield of AI that allows computers to understand and process human spoken languages. In this skeleton, the purpose of NLP is to extract a meaningful command (e.g., light_one_on) from a spoken sentence (e.g., switch on light number one). A typical natural language processor performs the following steps [53]:

- Sentence segmentation: each paragraph is divided into separate sentences.
- Word tokenization: each sentence is divided into separate tokens (i.e., words).
- Predicting parts of speech: each token is identified as a noun, verb, adjective, etc.
- Lemmatization: each token is converted to its base or dictionary form; that is, all inflections are removed (e.g., "switching" becomes "switch").
- Identifying stop words: each token that represents a stop word (e.g., "and," "a," "the") is filtered out.
- Dependency parsing: the relationships between tokens are identified.
- Named entity recognition: each noun token is labeled with real-word concept that represent it. For example, "John" is labeled as "People's name."
- Conference resolution: each token that represents a pronoun (e.g., "it," "he," "she") is associated with the token to which it refers.

Depending on the nature of the targeted speech, a subset of the above steps can be used to a perform the required NLP. The proposed skeleton targets identifying user commands for automation systems, which are normally short sentences with specific desired actions; hence, word tokenization, lemmatization, and identifying stop words can be sufficient to recognize the spoken command.

In offline speech recognition, the entire recognition process is completed locally using dedicated software. The process consists of receiving an acoustic waveform, dividing it in order to isolate sounds from noise and silence, and analyzing the sounds to determine the best matching set of words. There are multiple offline speech recognition platforms that are free, open source, and customizable such as CMUSphinx and Mozilla DeepSpeech. CMUSphinx accumulates 20 years of research at Carnegie Mellon University and is continuously updated to utilize most recent technologies. DeepSpeech engine is based on open-source code algorithms and Google TensorFlow machine learning toolkit.

Security Aspects

Any automation system with authorized access might be at risk of attacks that could bypass the authentication step. For systems with speaker verification, examples of such attacks include deepfake audio attacks and replay attacks. Deepfake audio attacks are still no the rise as there are many reports of deepfake attacks that tricked human beings but there are no reported cases where machines were exploited. On the other hand, there are many reports of successful replay attacks. For example, the authors in Ref. [54] claim that when a recorded voice of an Amazon Echo device owner was replayed asking: “Alexa, who am I?”, the system replied with the owner’s name. Similar vulnerability was reported in “Google Home” system because it only verifies the wake-up word, which is “OK Google”; whatever the user says after that is not verified. Similarly, the proposed skeleton only authenticates the WUW, which makes it vulnerable to replay attacks. So, if another user gives a command after the authenticated owner utters the WUW, the command will be executed. The question here becomes what are the risks of such attack? To answer that, the probability of the attack and its impact need to be analyzed. The impact of reply attacks heavily depends on the criticality of the system controlled by the skeleton. For example, the use case of the office automation system has low criticality; hence, no high risks are posed (e.g., high electricity bill for keeping lights ON for long hours).

However, if highly critical systems are controlled (e.g., automation systems in hospitals), then the probability of such attacks needs to be reduced by deploying techniques such as higher-order spectral analysis (HOSA)-based features to capture traces of replay attacks [54]. The detailed analysis of the probability and impacts of the security aspects will be addressed in future work.

EXPERIMENTAL SETUP AND RESULTS

As a test case, the proposed skeleton is integrated into a KNX-based office automation system, which is described in the following subsection. Afterwards, the implementation details of the skeleton and the experimental results are presented.

KNX-Based Office Automation System

The under-study office automation system is built using the KNX devices and components described below. The drivers of devices used in the system are downloaded from ETS devices catalog or from the SIEMENS official website. The office monitored physical parameters include presence, humidity, temperature, and carbon dioxide (CO_2) levels, while the control parameters were selected to be three rows of lights and a blind shutter. These parameters were selected to demonstrate the ability to voice control the main aspects of a BAS that is energy management, control, and monitoring. Figure 8 illustrates a reduced KNX diagram of the full schematics of the tested system.

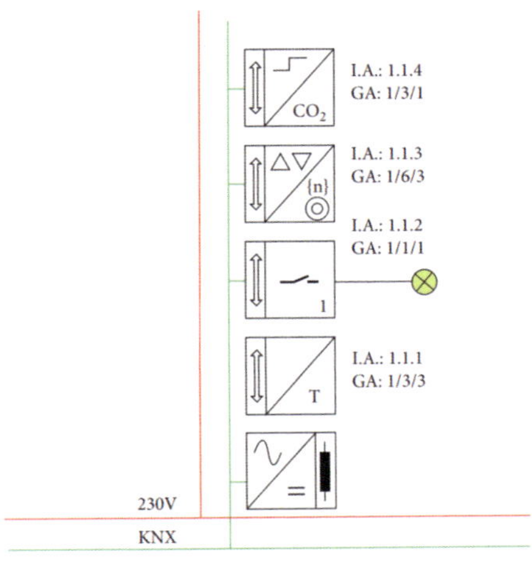

Figure 8. Office basic KNX schematics.

The programming/configuration of the KNX devices in our office is done using the ETS. Each KNX device has a specific button that must be pressed such that it can be programmed/configured. Moreover, each device and functionality must be configured with a group address and a specific datapoint type (DPT) to determine the value set or sent to that device/functionality. Table 1 lists the KNX devices installed in our experimental office along with their configuration parameters.

Table 1. KNX device configuration.

Device/functionality	Group address	KNX Datapoint id and name	Datapoint size
Switch channel A (light 1)	1/1/1	1.001 DPT_switch	1 bit
Switch channel B (light 2)	1/1/2	1.001 DPT_switch	1 bit
Switch channel C (light 3)	1/1/3	1.001 DPT_switch	1 bit
Presence detector with brightness sensor	1/1/150	9.004 DPT_value_lux	2 byte (float)
CO_2 sensor	1/3/1	9.008 DPT_value_air-quality	2 byte (float)
Humidity sensor	1/3/2	9.007 DPT_value_humidity	2 byte (float)
Temperature sensor	1/3/3	9.001 DPT_value_temp	2 byte (float)
Blind shutter actuator	1/6/3	1.008 DPT_updown	1 bit

KNX Switching/Dimming Actuator

This device provides three channels (i.e., A, B, and C) of switching and dimming functionalities for lights and three sensor channels. In our office, three rows of lights are connected to the three switching/dimming channels in order to control each light independently. The installed lights do not support dimming functionality; hence, it is sufficient to only use a 1-bit KNX DPT as shown in Table 1 in order to switch each row of lights ON/OFF. If the dimming functionality was available, a more complex structured DPT would be required (i.e., 3.007 DPT_Control_Dimming). In addition to lighting, a presence detector with integrated brightness sensor is also connected to this actuator. At the current stage of our implementation, only the brightness sensor is leveraged to report the office brightness level through a 2-byte float DPT in the illuminance unit (i.e., Lux). The measured brightness can be used to control the movement of the shutter blinds through the KNX shutter actuator described below.

KNX CO_2, Humidity, and Temperature Sensor

This sensor has three functionalities: first, it measures the CO_2 concentration level in the air (i.e., air quality measurement) with 2-byte float DPT in parts per million (ppm) unit. Typical CO_2 concentrations in occupied indoor spaces with good air exchange are in the range of 350-to-1000 ppm. Second, the sensor measures the air humidity percentage in the room using a 2-byte float DPT. Third, the room temperature is measured in Celsius degree unit using a 2-byte float DPT. The three readings can be used to control heating, ventilation, and air-conditioning (HVAC) in the office.

KNX Shutter Actuator

The shutter actuator has a switch functionality that is used to move the office blinds up or down. The configuration in Table 1 shows that 1-bit DPT (i.e., 1.008 DPT_UpDown) is used to control the movement of the office blinds (i.e., 0 for up and 1 for down). This indicates that the blinds in our testing environment will be either all the way up or all the way down. However, one could configure an extra 1-bit DPT to start/stop blinds movement (i.e., 1.010 DPT_Start) or a more complex structured DPT (i.e., 3.008 DPT_Control_Blinds) given that it is supported by the actuator.

KNX IP Router

The KNX IP router is the interface between the KNX bus and the computer system, which runs the ETS for configuration and the proposed voice command application for controlling the office automation system. Using Ethernet connections, the router forwards telegrams with required functionalities to the targeted devices. The telegrams are forwarded between different KNX lines through a LAN as a fast backbone (i.e., KNXnet/IP Routing connection).

Voice Control Application

The proposed skeleton is implemented as a voice control desktop application using python programming language version 3.6. The application imports the asynchronous python library for KNX (i.e., XKNX) that supports a huge number of devices' DPTs, which are defined and ready to use. In addition, the XKNX library leverages the asynchronous start () and stop () functions along with the async/await syntax of the Asynchronous I/O python library (i.e., asyncio) to provide gateway scanning and tunneling to the KNX IP router.

For WUW detection engine, the application utilizes the Porcupine wake word detection engine version V1.8, which was chosen for the reasons mentioned in above section and because it supports Windows OS. Since the proposed voice control application is implemented on a desktop computer, we chose "Hey Computer" as the WUW in our experimental work. The optimizer tool within the Porcupine engine is used to train the wake word model on the chosen WUW. Notice that there is a dedicated optimizer tool for each platform supported by the Porcupine engine. Since the Porcupine engine is leveraged as is without any modification, its accuracy is the same as the one reported in [49] (i.e., miss rate of 5.9% at one false alarm per 10 hours). Hence, no testing is required for the WUW engine.

For speaker verification, our application utilizes the ResNet model developed by the Korea Advance Institute of Science and Technology (KAIST) [55]. The model is trained on a dataset that contains 240 speakers and 100 utterances per speaker, all in Korean language. In order to be able to verify a user who is authorized to control the experimental office automation system, the extracted audio features of that user need to be enrolled in the model using the process. Similar to Ref. [55], in our work, feature extraction is also achieved using the python_speech_features library.

After verifying the speaker using the audio file that contains the correctly detected WUW, the next step in our proposed skeleton is to receive voice commands from the user and identify them using ASR. The implemented control application tests Internet connection to choose between offline and online speech recognition. For online speech recognition, our application leverages the online Google Speech API supported by the Python Speech Recognition 3.8.1 library. This API supports multiple languages, and in our application, the US English language is chosen. The captured voice signals are transmitted over the Internet to Google servers where machine learning techniques are used to process and convert them to text. In order to reduce the effect of ambient noise, the microphone's energy threshold is automatically calibrated so that the voice control application can reliably recognize the speech.

Consequently, the generated text (e.g., switch on light 1) is returned to the local host where a natural language processor is used to extract a meaningful command (e.g., light_one_on). The natural language processor in our application is built using Python Natural Language Toolkit (NLTK) library and TensorFlow AI library. NLTK is used to extract features from the received text using the bag-of-words (BoW) approach. The extracted features in a form of a vector are inputted to a trained machine learning model developed using TensorFlow in order to output the expected desired control command.

The dataset used to train, validate, and test the machine learning model of the natural language processor is represented by the supported commands (e.g., light_one_on, shutter_up, temperature, open_youtube). Each command is represented by a tag, multiple patterns, and a response stored in JSON format. For example, the command "light_one_on" is represented by the tag "light_one_on," seven patterns (i.e., "can you turn on light one," "please turn on light one," "turn on light one," "switch light one on," "switch on light one," "light one on," "on light one"), and the response "light_one_on." Patterns represent the speech commands that the user might say and must lead to the same response.

The preprocessing of the dataset includes the following steps: first, extract the words from all patterns using word tokenization from NLTK. Second, perform stemming using NLTK LancasterStemmer to reduce each word to its root. Third, implement BoW approach using the unique words found in the second step. For each pattern in the dataset, a vector of bits is generated such that each bit is associated with one unique word. When the word exists in the pattern, its associated bit is set to one; otherwise, it is set to zero. For example, if the unique words are ["light," "on," "off," "one," "two," "three"], then the vector for pattern "light one on" is [1, 1, 0, 1, 0, 0]. The fourth step is to map each pattern vector to an output vector that indicates which tag (i.e., command or response) is associated with the respective pattern. Every bit in the output vector is associated with exactly one tag; hence, only one bit in an output vector is set to one, while the remaining bits are set to zero. Lastly, store the list of pattern vectors (inputs) and their respective output vectors into variables and convert them to Numpy arrays in order to be used for training, validation, and testing.

The machine learning model of the natural language processor in our application is a four-layer neural network (i.e., one input layer, two hidden layers, and one output layer). The dataset described above is divided into three parts—training, validation, and testing sets, and the results. After the model is trained, validated, and tested, it is deployed in our voice control application. Once the text is returned from the online Google API, it is processed using NLTK to generate its associated vector based on the words in the text. The vector is used as an input to the neural network, which generates an output vector that indicates the expected command.

On the other hand, the offline speech recognition in our application is implemented using CMUSphinx system. The recognition accuracy results of using the CMUSphinx system as is were very low. To improve recognition accuracy, CMUSphinx provides Sphinxbase library and Sphinxtrain tools, which can be used to train a new acoustic model. However, this approach requires substantial audio for training and testing the new model. Another way to leverage the CMUSphinx system is adapting the default acoustic model, which is the approach used in our application. This process requires preparing a ".fileids" file and a ".transcription" file; the first contains the names of the adaptation ".wav" files, and the second contains the sentences uttered in those files.

In order to improve the recognition accuracy of the adapted model, a special dictionary is adopted instead of the full one. The special dictionary only contains the command words that are used to control our experimental office automation system and their phonemes. In addition, a new grammar that is specified for our control commands is built with the Java Speech Grammar Format (JSGF) and used. The integration of the adapted CMUSphinx recognizer into our application is straightforward because the Python speech recognition library contains the PocketSphinx system with general language model for US English language. It is worth noting that higher accuracy is achieved when using the CMUSphinx recognizer with noise reduction techniques (e.g., ambient noise reduction) and noise cancelling microphones.

Once the desired command is identified either through online or offline speech recognition, the application sends a telegram to the respective KNX device/functionality requesting the command to be executed. For example, the application declares an object of class "Light" for every installed light line, gives it the group address programmed to it, and uses set_on () and set_off () functions to turn the light ON and OFF when "light on" and "light off" commands are received, respectively. For sensors, it is the same process, but the function used is sync (), which is used to synchronize the measured value.

The KNX switch, shutter actuator, and sensors are connected to the PC on which the ETS and the voice control application are running through the KNX IP router. After configuring the KNX devices, the communications between the PC and the KNX devices were successfully tested through direct commands from the PC. Next, the functionalities of the voice control application were tested, and the results are presented in the following three subsections.

Speaker Verification Results

The speaker verification ResNet model was tested with three users. For each user, 16 English language sentences were recorded in ".wav" format. For each user, one ".wav" file is used for enrollment, while the remaining 15 files and the files of other speakers are used for testing. The threshold of comparison score is set to 95%. Table 2 lists the verification results of each user after being enrolled. For example, when User1 is enrolled, 14 out of his 15 ".wav" file are correctly verified and all User2 and User3 files are correctly not verified (i.e., verification accuracy=44/45=97.8%). For all users, the verification accuracy is higher than 95%.

Table 2. Speaker verification results.

Enrolled speaker	Tested speaker	Test files	Accepted files	Rejected files	Verification accuracy	False positive rate	False negative rate
User1	User1	15	14	1	97.8%	0%	2.2%
	User2 and User3	30	0	30			
User2	User2	15	15	0	100%	0%	0%
	User1 and User3	30	0	30			
User3	User3	15	15	0	95.5%	4.5%	0%
	User1 and User2	30	2	28			

Online Speech Recognition Results

The dataset used to evaluate the accuracy of the implemented NLP consists of 167 patterns associated with 29 commands (i.e., on average six patterns per command). Table 3 lists a sample of the dataset that consists of three commands and their respective patterns. The patterns represent the text generated by the online Google Cloud Speech API after converting the recorded audio files. Once the dataset is preprocessed, it is divided into training, validation, and testing sets according to the percentages: 70%, 15%, and 15%, respectively.

Table 3. Sample of NLP dataset.

Command	Patterns
Light_two_three_on	"Please turn on light two and three," "Turn on light two and three," "Switch light two and three on," "Switch on light three and two"
Shutter_down	"Move down the shutter," "Move shutter down," "Move down the blinds," "Close the blinds," "Blinds down," "Shutter down"
Get_tempera-ture	"What is the temperature," "How cold is it," "How hot is it," "Tell me the temperature," "How is the weather," "Read the temperature sensor"

Figure 9 shows the training and validation accuracies of the proposed NLP. The training accuracy saturates at 100% after 500 steps and the validation accuracy saturates at 97.5% after 450 steps. The testing set consists of 25 patterns (i.e., 0.15 x 167), four of which were incorrectly predicted (i.e., testing accuracy=21/25=84%). For example, the pattern "Close the blinds" was wrongly predicted as the command "Shutter_up" because the word "Close" was neither included in the training set nor in the validation set. The accuracy of the implemented online speech recognition is expected to increase and the size of the dataset increases.

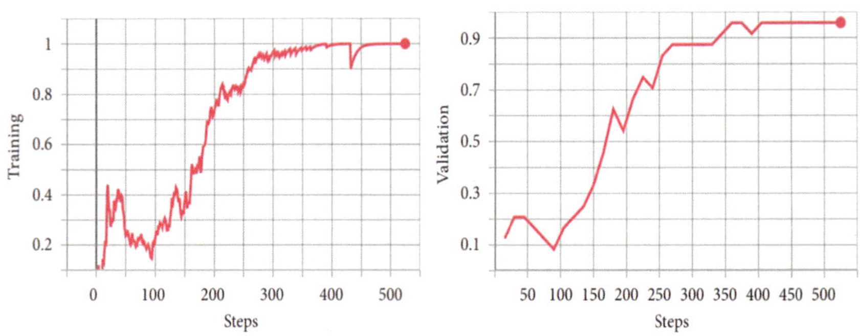

Figure 9. Training and validation accuracies of the NLP.

Offline Speech Recognition Results

The implemented CMUsphinx recognizer is tested using 300 ".wav" audio files recorded from the same three users in the speaker verification testing (i.e., 100 files from each user). The files contain commands that are used to control our experimental office automation system such as turn on all lights, turn off all lights, please turn off all lights, can you turn on light 1 and 2, what is the temperature now, please tell me what is the temperature, move shutter up, and please move down the blind. The test results are shown in Table 4. For User2, all 100 commands were accurately recognized and executed. On the other hand, 98 and 96 commands were accurately recognized and executed for User1 and User3, respectively. The inaccurate recognition of the two and four commands for User1 and User3 is mainly attributed to the high noise in the recorded commands.

Table 4. Offline speech recognition results.

Speaker	Recognition accuracy (%)
User1	98
User2	100
User3	96

CONCLUSIONS

Advancements in building automation are evolving rapidly with the development of highly intelligent control systems through computers and personal mobile devices. Moreover, the emergence of IoT has revolutionized building automation systems by making them proactive, intuitive, and easily managed by occupants. One of the most important features in modern building automation systems is reliable voice control. In this article, we propose a generic skeleton of implementing voice control for building automation systems. The skeleton consists of three main stages: WUW detection, speaker verification, and dual-mode speech recognition (i.e., online and offline). The proposed skeleton is implemented and integrated into a KNX-based office automation system, which allows the authorized personal to successfully control lights and shutters and poll sensors' readings using voice commands.

REFERENCES

1. R. K. Kodali, V. Jain, S. Bose, and L. Boppana, "IoT based smart security and home automation system," in *Proceedings of the 2016 International Conference on Computing Communication and Automation*, pp. 1286–1289, Greater Noida, India, April 2016.
2. P. Gupta and J. Chhabra, "IoT based Smart Home design using power and security management," in *Proceedings of the 2016 International Conference on Innovation and Challenges in Cyber Security*, pp. 6–10, Greater Noida, India, Feburary 2016.
3. D. Pavithra and R. Balakrishnan, "IoT based monitoring and control system for home automation," in *Proceedings of the 2015 Global Conference on Communication Technologies*, pp. 169–173, Thuckalay, India, April 2015.
4. S. Tanwar, P. Patel, K. Patel, S. Tyagi, N. Kumar, and M. S. Obaidat, "An advanced internet of thing based security alert system for smart home," in *Proceedings of the 2017 International Conference on Computer, Information and Telecommunication Systems*, pp. 25–29, Dalian, China, July 2017.
5. P. Priyadarshni and S. Sharma, "GUI-MATLAB based home/industrial automation using MCU89S52," *International Journal of Science and Research*, vol. 3, pp. 158–161, 2014.
6. A. Telepatil, A. N. Hambar, and P. Terwadkar, "Home automation with MATLAB and ARDUINO interface," *International Journal of Innovative Research in Computer and Communication Engineering*, vol. 5, pp. 5501–5507, 2017.
7. Archiexpo, "Commercial home automation system," 2021, https://www.archiexpo.com/architecture-design-manufacturer/commercial-home-automation-system-57152.html.
8. C. S. Homes and LIVE LIFE BRILLIANTLY, 2021, https://www.control4.com/.
9. A. A. team, "Alexa voice service (AVS) overview," 2021, https://developer.amazon.com/en-US/docs/alexa/alexa-voice-service/get-started-with-alexa-voice-service.html.
10. Apple, "Your home at your command," 2021, https://www.apple.com/ios/home/.

11. N. B. Jain, G. D. Ayers, E. N. Peterson et al., "Traumatic spinal cord injury in the United States, 1993-2012," *JAMA*, vol. 313, no. 22, pp. 2236–2243, 2015 06.
12. A. J. Spittle, C. Morgan, J. E. Olsen, I. Novak, and J. L. Y. Cheong, "Early diagnosis and treatment of cerebral palsy in children with a history of preterm birth," *Clinics in Perinatology*, vol. 45, no. 3, pp. 409–420, 2018, https://www.sciencedirect.com/science/article/pii/S0095510818313708.
13. CMUSphinx, "OPEN. SOURCE," *SPEECH RECOGNITION TOOLKIT*, 2020, https://cmusphinx.github.io/.
14. KNX, "Turn every idea into a top project," 2021, https://www.knx.org/knx-en/for-offices/smart-home-control/.
15. M. Intelligence, "Open Source S. P. E. E. C. H. Recognition T. O. O. L. K. I. T," 2021, https://www.mordorintelligence.com/industry-reports/smart-building-market.
16. Knx, "Smart Building Market - Growth, T. R. E. N. D. S," *COVID-19 IMPACT, AND FORECASTS*, 10 pages, 2021, https://www.knx.org/knx-en/for-professionals/newsroom/en/press/KNX-is-the-leading-communication-protocol-in-the-European-Chinese-smart-homes-light-commercial-market/.
17. K. Association, "Knx basic course," 2021, http://knx.com.ua/attachments/article/132/KNX-basic_course_full.pdf.
18. S. Wang, *Intelligent Buildings and Building Automation*, Spon Press, London, United Kingdom, 2010, https://books.google.jo/books?id=BA2LuAAACAAJ.
19. T. Sauter, S. Soucek, W. Kastner, and D. Dietrich, "The evolution of factory and building automation," *IEEE Industrial Electronics Magazine*, vol. 5, no. 3, pp. 35–48, 2011.
20. P. Domingues, P. Carreira, R. Vieira, and W. Kastner, "Building automation systems: concepts and technology review," *Computer Standards & Interfaces*, vol. 45, pp. 1–12, 2016, https://www.sciencedirect.com/science/article/pii/S0920548915001361.
21. N. Aste, M. Manfren, and G. Marenzi, "Building Automation and Control Systems and performance optimization: a framework for analysis," *Renewable and Sustainable Energy Reviews*, vol. 75, pp. 313–330, 2017, https://www.sciencedirect.com/science/article/pii/S1364032116307365.

22. C. D. Korkas, S. Baldi, and E. B. Kosmatopoulos, "9 - grid-connected microgrids: demand management via distributed control and human-in-the-loop optimization," *Advances in Renewable Energies and Power Technologies Elsevier*, pp. 315–344, 2018, https://www.sciencedirect.com/science/article/pii/B9780128131855000255.
23. S. Baldi, C. D. Korkas, M. Lv, and E. B. Kosmatopoulos, "Automating occupant-building interaction via smart zoning of thermostatic loads: a switched self-tuning approach," *Applied Energy*, vol. 231, pp. 1246–1258, 2018, https://www.sciencedirect.com/science/article/pii/S0306261918315095.
24. V. T. Nguyen, T. Luan Vu, N. T. Le, and Y. Min Jang, "An overview of internet of energy (IoE) based building energy management system," in *Proceedings of the 2018 International Conference on Information and Communication Technology Convergence*, pp. 852–855, ICTC, Jeju, Korea (South), November2018.
25. R. Carli, G. Cavone, M. Dotoli, N. Epicoco, and P. Scarabaggio, "Model Predictive Control for thermal comfort Optimization in Building Energy Management Systems," in *Proceedings of the 2019 IEEE International Conference on Systems, Man and Cybernetics (SMC)*, pp. 2608–2613, Bari, Italy, November2019.
26. D. G. Hettiarachchi, G. M. A. Jaward, V. P. V. Tharaka, J. M. D. S. Jeewandara, and K. T. M. U. Hemapala, "IoT based building energy management system," in *Proceedings of the 2021 3rd International Conference on Electrical Engineering*, pp. 69–73, Colombo, Sri Lanka, November 2021.
27. J. K. Guo, C. L. Lu, J. Y. Chang et al., "Interactive voice-controller applied to home automation," in *Proceedings of the 2009 Fifth International Conference on Intelligent Information Hiding and Multimedia Signal Processing*, pp. 828–831, Kyoto, Japan, November 2009.
28. J. Zhu, X. Gao, Y. Yang, H. Li, Z. Ai, and X. Cui, "Developing a Voice Control System for ZigBee-Based home Automation Networks," in *Procedings of the 2010 2nd IEEE InternationalConference on Network Infrastructure and Digital Content*, pp. 737–741, Beijing, China, September 2010.
29. V. Jeet, H. S. Dhillon, and S. Bhatia, "Radio frequency home appliance control based on head tracking and voice control for disabled person," in *Proceedings of the 2015 Fifth International Conference on Communication Systems and Network Technologies*, pp. 559–563, Gwalior, India, April 2015.

30. S. Ajgaonkar, M. Chauhan, H. Humane, K. Jain, and A. Malhotra, "Voice controlled multi-functional smart device," *International Research Journal of Engineering and Technology (IRJET)*, vol. 7, pp. 7634–7638, 2020.
31. Y. F. Kung, S. W. Liou, G. Z. Qiu, B. C. Zu, Z. H. Wang, and G. J. Jong, "Home monitoring system based internet of things," in *Proceedings of the 2018 IEEE International Conference on Applied System Invention*, pp. 325–327, ICASI), Taiwan, China, April 2018.
32. S. Hidayat and S. F. Firmanda, "Scheduler and voice recognition on home automation control system," in *Proceedings of the 2015 3rd International Conference on Information and Communication Technology*, pp. 150–155, ICoICT, Nusa Dua, Bali, Indonesia, September 2015.
33. A. Paul, M. Panja, M. Bagchi, N. Das, R. M. Mazumder, and S. Ghosh, "Voice recognition based wireless room automation system," in *Proceedings of the 2016 International Conference on Intelligent Control Power and Instrumentation*, pp. 84–88, ICICPI, Kolkata, October 2016.
34. S. Kumar and S. S. Solanki, "Voice and touch control home automation," in *Proceedings of the 2016 3rd International Conference on Recent Advances in Information Technology*, pp. 495–498, RAIT, Dhanbad, India, March 2016.
35. K. A. S. V. Rathnayake, S. I. A. P. Diddeniya, W. K. I. L. Wanniarachchi, W. H. K. P. Nanayakkara, and H. N. Gunasinghe, "Voice operated home automation system based on Kinect sensor," in *Proceedings of the 2016 IEEE International Conference on Information and Automation for Sustainability*, pp. 1–5, ICIAfS, Galle, Sri Lanka, December 2016.
36. T. Erić, S. Ivanović, S. Milivojša, M. Matić, and N. Smiljković, "Voice control for smart home automation: evaluation of approaches and possible architectures," in *Proceedings of the 2017 IEEE 7th International Conference on Consumer Electronics - Berlin*, pp. 140–142, ICCE-Berlin, Sepetmber 2017, Berlin, Germany.
37. I. Krishna and K. Lavanya, "Intelligent home automation system using BitVoicer," in *Proceedings pf the 2017 11th International Conference on Intelligent Systems and Control*, pp. 14–20, ISCO, Dhaka, Bangladesh, December 2017.
38. B. Tecnologia and B. V. Server, "Speech recognition and synthesis," 2020, http://www.bitsophia.com/en-US/BitVoicer/Overview.aspx.

39. M. Ebrahim Abidi, A. L. Asnawi, N. F. Azmin et al., "Development of voice control and home security for smart home automation," in *Proceedings of the 2018 7th International Conference on Computer and Communication Engineering*, pp. 1–6, ICCCE, Kuala Lumpur, Malaysia, September 2018.
40. P. Putthapipat, C. Woralert, and P. Sirinimnuankul, "Speech recognition gateway for home automation on open platform," in *Proceedings of the 2018 International Conference on Electronics, Information, and Communication*, pp. 1–4, ICEIC, Honolulu, HI, USA, January 2018.
41. P. Putthapipat and K. Techakittiroj, "PiFrame: a framework for home automation platform on the full feature OS," in *Proceedings of the n2016 International Conference on Electronics, Information, and Communications*, pp. 1–4, ICEIC, Danang, Vietnam, September 2016.
42. S. K. Vishwakarma, P. Upadhyaya, B. Kumari, and A. K. Mishra, "Smart energy efficient home automation system using IoT," in *Procedings of the 2019 4th International Conference on Internet of Things: Smart Innovation and Usages*, pp. 1–4, IoT-SIU, Ghaziabad, India, April 2019.
43. P. J. Rani, J. Bakthakumar, B. P. Kumaar, U. P. Kumaar, and S. Kumar, "Voice controlled home automation system using natural language processing (NLP) and internet of things (IoT)," in *Proceedings of the 2017 Third International Conference on Science Technology Engineering Management*, pp. 368–373, ICONSTEM, Chennai, India, March2017.
44. G. B. Karan, D. Kumar, K. Pai, and J. Manikandan, "Design of a phoneme based voice controlled home automation system," in *Proceedings of the 2017 IEEE International Conference on Consumer Electronics-Asia*, pp. 31–35, ICCE-Asia, Bengaluru, India, October 2017.
45. S. Milivojša, S. Ivanović, T. Erić, M. Antić, and N. Smiljković, "Implementation of voice control interface for smart home automation system," in *Proceedings of the 2017 IEEE 7th International Conference on Consumer Electronics - Berlin*, pp. 263-264, ICCE-Berlin, Berlin, Germany, September 2017.
46. F. Ge and Y. Yan, "Deep neural network based wake-up-word speech recognition with two-stage detection," in *Proceedings of the 2017 IEEE International Conference on Acoustics, Speech and Signal Processing*, pp. 2761–2765, ICASSP, New Orleans, LA, USA, June 2017.

47. K. Ai, "Snowboy hotword detection," 2020, https://github.com/Kitt-AI/snowboy.
48. P. ai, "Porcupine," 2020, https://github.com/Picovoice/Porcupine.
49. P. ai, "Porcupine™ wake word engine - V1.8 feature tour," 2021, https://picovoice.ai/blog/porcupine-wake-word-engine-v1-8-feature-tour/.
50. L. Sun, T. Gu, K. Xie, and J. Chen, "Text-independent speaker identification based on deep Gaussian correlation supervector," *International Journal of Speech Technology*, vol. 22, pp. 449–457, 2019.
51. R. Milner and T. Hain, "DNN-based Speaker Clustering for Speaker Diarisation," INTERSPEECH, 2016.
52. E. Variani, X. Lei, E. McDermott, I. L. Moreno, and J. Gonzalez-Dominguez, "Deep neural networks for small footprint text-dependent speaker verification," in *Proceedings of the 2014 IEEE International Conference on Acoustics, Speech and Signal Processing*, pp. 4052–4056, ICASSP, May 2014.
53. A. Geitgey, "natural language processing is fun," 2025 pages, 2020, https://medium.com/@ageitgey/natural-language-processing-is-fun-9a0bff37854e.
54. K. M. Malik, H. Malik, and R. Baumann, "Towards vulnerability analysis of voice-driven interfaces and countermeasures for replay attacks," in *Proceedings of the 2019 IEEE Conference on Multimedia Information Processing and Retrieval*, pp. 523–528, MIPR, San Jose, CA, USA, April 2019.
55. Y. Jung, Y. Kim, H. Lim, Y. Choi, and H. Kim, "Spatial pyramid encoding with convex length normalization for text-independent speaker verification," *Interspeech*, 2019.

CHAPTER 12

Data-Driven Model Identification and Control of the Inertial Systems

Irina Cojuhari

Department of Software Engineering and Automatics, Technical University of Moldova, Chisinau, Republic of Moldova.

ABSTRACT

In the synthesis of the control algorithm for complex systems, we are often faced with imprecise or unknown mathematical models of the dynamical systems, or even with problems in finding a mathematical model of the system in the open loop. To tackle these difficulties, an approach of data-driven model identification and control algorithm design based on the maximum stability degree criterion is proposed in this paper. The data-driven model identification procedure supposes the finding of the mathematical model of the system based on the undamped transient response of the closed-

Citation: Cojuhari, I. (2023), "Data-Driven Model Identification and Control of the Inertial Systems. Intelligent Control and Automation", 14, 1-18. doi: 10.4236/ica.2023.141001.

loop system. The system is approximated with the inertial model, where the coefficients are calculated based on the values of the critical transfer coefficient, oscillation amplitude and period of the underdamped response of the closed-loop system. The data driven control design supposes that the tuning parameters of the controller are calculated based on the parameters obtained from the previous step of system identification and there are presented the expressions for the calculation of the tuning parameters. The obtained results of data-driven model identification and algorithm for synthesis the controller were verified by computer simulation.

Keywords: Data-Driven Model Identification, Controller Tuning, Undamped Transient Response, Closed-Loop System Identification, PID Controller

INTRODUCTION

Almost all electrical drivers, power converters, technological processes in industrial applications are controlled based on the proportional-integral-derivative (PID) control algorithm [1] [2] . This algorithm is widely used in different industrial applications, due to its simplicity, feasibility and the advantages that offer to automatic control systems, such as good performance and robustness in case of uncertainties and disturbances.

One of the difficulties in automatic control systems is the problem of synthesis of the PID control algorithm, which supposes the procedure of calculation of the tuning parameters according to the dynamics of the control object [2] [3] . The incorrect tuning of the PID controller can lead to bad performance and in the worst case can lead to the instability of the system. In this case, ensuring the closed loop system stability is one of the most important aspects in the synthesis of the control algorithm [4] [5] .

For the last decades have been developed a big variety of PID tuning techniques in continuous—time, discrete time and frequency domain [6] [7] . In general, these methods can be grouped into the following categories as: experimental tuning methods, analytical tuning methods and optimization tuning methods, which are based on the algorithms from artificial intelligence domain [6] [8] .

The experimental methods do not require to be known preliminary the mathematical model that describes the dynamics of the control object. These methods are based on some simple assumptions for calculation of the tuning parameters. One of the most known and used experimental methods for tuning the typical controllers is Ziegler-Nichols method, which supposes the calculation of the tuning parameters based on the undamped transient response of the closed loop system and it does not require to be known the mathematical model of the control object [2] [9] [10] [11] . Due to its simplicity, it is used as self-tuning method of P, PI or PID controller, though this method is characterized by the drawbacks as:

- It is recommended for slow processes;
- It does not offer the procedure for synthesis of the controller with imposed performance;
- It provides the oscillating step response of the system;
- It does not take into account the requirements related to the system stability reserve;
- This method does not offer the procedure for optimization of the tuning parameters.

The analytical methods require to be known the mathematical model of the control object and based on some algorithms for calculation of the tuning parameters and graphical representation in the time, or frequency domain, it can offer the designed closed loop system satisfactory performance and good robustness [12] [13] [14] . The disadvantages of these methods are the necessity of using a big volume of calculations and the mathematical model of the control object should be given, or should be obtained based on the identification procedure.

Nowadays, there are developed many identification methods and procedures of the mathematical model, allowing with high accuracy to identify different structures of object models with or without time delay, with high or low order inertia. Most of these methods offer procedures for approximation of the dynamics of control object with transfer function with first and second-order inertia and these methods in most cases are applied in the open loop systems [2] [15] [16] [17] . The existence of the procedure, which will permit to obtain the mathematical model of the control object in the closed loop system essentially will simplify the procedure of tuning the typical controllers [18] .

Another category of tuning techniques is related to optimization methods and artificial intelligence approaches such as genetic algorithms, fuzzy logic control, neural networks [18] . These approaches are widely used in the case of optimization problems of tuning parameters, especially for the case, when it is needed to control the nonlinear process and offer good performance and robustness of the system. However, these methods usually involve operator implication in formation the initial data set and hardly can be realized as self-tuning methods of the controller.

This paper provides a new method for experimental identification, inspired by the Ziegler and Nichols tuning method when the closed loop control system is marginally stable. The new procedure of tuning the PID controller ensures to the system critically damped step response.

BASIC PRINCIPLES OF THE PID CONTROL ALGORITHM AND TUNING METHODS

PID Control Algorithm

The PID control algorithm is widely used in different industrial applications and it can be easily implemented for various control problems. The PID controller as input receives the error signal $e(t)$ and provides command signal—$u(t)$, where the typical structure of the PID controller is given by the transfer function [8] :

$$H_{\text{PID}}(s) = k_p + \frac{k_i}{s} + k_d s, \tag{1}$$

where k_p—is the proportional tuning parameter, k_i—integral tuning parameter, k_d—derivative tuning parameter of the PID controller.

In Figure 1, it is presented the structural scheme of the automatic control system.

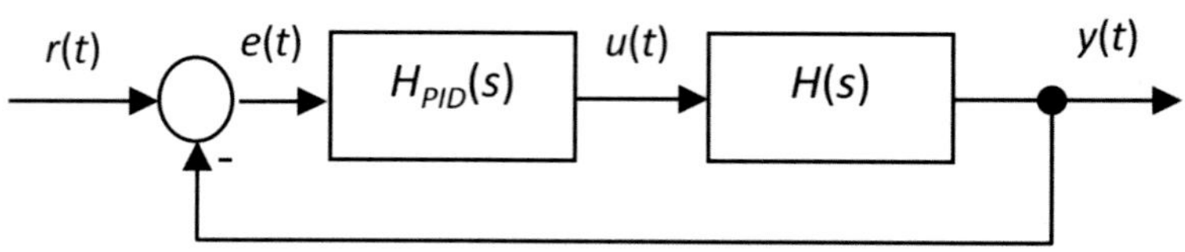

Figure 1. Structural scheme of the automatic control system.

In Figure 1, $H_{PID}(s)$ is the transfer function of the PID controller, described by the transfer function (1) and H(s) is the transfer function of the control object:

$$H(s)=\frac{k}{(T_1s+1)(T_2s+1)\cdots(T_ns+1)}$$
$$=\frac{k}{a_0s^n+a_1s^{n-1}+\cdots+a_{n-1}s+a_n}=\frac{B(s)}{A(s)},$$
(2)

where T_0, T_1, $\cdots$, T_{n-1}, T_n are time constants; a_0, a_1, $\cdots$, a_{n-1}, a_n are the parameters of the characteristic equation and depend on the internal proprieties of the control object; k is the static gain; n is the order of the characteristic equation $A(s)$.

The static gain of the system is calculated [2] :

$$k=\lim_{t\to\infty}\frac{\Delta y}{\Delta u}=\lim_{t\to\infty}\frac{y_{st}-y_{\text{initial}}}{u-u_{\text{initial}}},$$
(3)

where y_{st} is the steady-state output, y_{initial} is the initial value of the output, $u(t)$—input signal, u_{inital} is the initial value of the input signal.

In the Figure 1, the $r(t)$ is the reference signal, $y(t)$—output signal, $u(t)$—command signal and $e(t)$—error signal:

$$e(t)=r(t)-y(t).$$
(4)

According to the Equation (1) and Equation (2) the characteristic equation of the closed loop system with PID controller can be presented by:

$$A(s)=\frac{1}{k}\left(a_0s^{n+1}+a_1s^n+\cdots+a_{n-1}s^2+a_ns\right)+k_ds^2+k_ps+k_i.$$
(5)

Tuning Methods

Next, there are analyzed two methods of tuning the PID controller:

- The experimental method of tuning the PID controller—Ziegler—Nichols method;
- Graph-analytical method—maximum stability degree method with iterations.

Ziegler-Nichols Method

The Ziegler-Nichols method for tuning typical controllers from 1942, when it was proposed for the first time, has a big impact on linear feedback control systems and in implementation of self-tuning methods of typical controllers in different industrial applications [10] . The procedure for tuning the PID controller according to the Ziegler-Nichols method, for now, is accepted as a standard in control system practices. Ziegler and Nichols proposed two empirical methods for tuning the typical controllers. The first method is supposed, that the process is aperiodic and it can be obtained the transient response in the open loop and that is approximated with transfer function with inertia first order and time delay. According to this approximation, there are given the analytical expressions for the calculation of the tuning parameters of the P, PI and PID controllers [2] [9] [10] [11] .

The second method doesn't require to be known the mathematical model of the control object and it offers the rules for determination in experimental way the tuning parameters. This method of tuning is performed for the closed loop systems with PID controller (Figure 1), when the integral tuning parameter—k_i and derivative tuning parameter—k_d, are settled to the zero value. The proportional tuning parameter—k_p is slowly increased until the system reaches the limit of stability and it determinates the period of the un-amortized oscillations T_{cr} of the system. The value of the proportional tuning parameter, when the system achieves the limit of stability is the critical transfer coefficient—k_{cr}. Based on these two parameters T_{cr} and k_{cr}, there are given the expressions for calculations of the tuning parameters of the P, PI and PID controllers (Table 1).

Maximal Stability Degree Method with Iterations

One of the analytical methods, which permit tuning the P, PI and PID controllers and it is taken into accounts the stability degree of the system is the maximum stability degree method with iterations [17] .

Affirmation 1. If the coefficients of the characteristic equation are known and constant, then the stability degree of the system J obtains the maximum possible value by the [19] :

$$J = \frac{a_1}{(n+1)a_0}, \tag{6}$$

and in this case the real parts of all characteristic equation roots are equal with each other:

$$\alpha_1 = \alpha_2 = \cdots = \alpha_n = J, \tag{7}$$

where $\alpha_1, \alpha_2, \cdots, \alpha_n$ are real parts of characteristic equation roots $p_i = -\alpha_i \pm j\omega_k$, $i,k = \overline{1,n+1}$, $k = \overline{1,m}$.

For the case when the number of the tuning parameters is equal or less then the characteristic equation order, the value of maximum stability degree value can be chosen as arbitrary value and the 1st Affirmation will not be valid.

In concordance with the 1st Affirmation, it is possible to be found the tuning parameters of the PID controller so as to be satisfied the maximum stability degree of the system:

$$J = \eta_{\max} = \max \eta\left(k_p, k_i, k_d\right). \tag{8}$$

Table 1. Expressions for calculation the tuning parameters of the P, PI and PID controllers [2] .

Type of controller	k_p	k_i	k_d
P	$0.5\ k_{cr}$	-	-
PI	$0.45\ k_{cr}$	$1/(0.8T_{cr})$	-
PID	$0.75\ k_{cr}$	$1/(0.6T_{cr})$	$0.1T_{cr}$

According to the maximum stability degree method with iterations, in the characteristic Equation (5) of the closed loop system with PID controller is introduced the substitution s = −J, and the characteristic equation becomes [19] :

$$A(-J) = \frac{1}{k}\left(a_0\left(-J\right)^{n+1} + a_1\left(-J\right)^n + \cdots + a_{n-1}J^2 - a_nJ\right) + k_dJ^2 - k_pJ + k_i. \tag{9}$$

The Equation (9) derives two times and there are obtained the analytical expressions for calculating the tuning parameters of the PID controller:

$$k_p = \frac{1}{k}\left(\left(-1\right)^{n+1}\left(n+1\right)a_0J^n + \left(-1\right)^n na_1J^{n-1} + \cdots + 2a_{n-1}J - a_n\right) + 2k_dJ; \tag{10}$$

$$k_i = \frac{1}{k}\left(\left(-1\right)^n a_0J^{n+1} - \left(-1\right)^n a_1J^n + \cdots - a_{n-1}J^2 + a_nJ\right) - k_dJ^2 + k_pJ; \tag{11}$$

$$k_d = \frac{1}{2k}\left(\left(-1\right)^n n\left(n+1\right)a_0J^{n-1} - \left(-1\right)^n n\left(n-1\right)a_1J^{n-2} + \cdots - 2a_{n-1}\right). \tag{12}$$

The analytical expressions (10)-(12) permit to calculate the tuning parameters of the PID controller and they represent the dependences of the known parameters of the control object k, a_0, a_1, ···, a_n and an unknown parameter J—stability degree of the system: $k_p = f(J)$, $k_i = f(J)$, $k_d = f(J)$. According to the 1st Affirmation, if the degree of the system is higher than

2nd order, the maximum stability degree value can be calculated based on the Equation (6) [19] .

Otherwise, according to the maximum stability degree method with iterations, the value of the stability degree can be varied $J \geq 0$ and based on the dependencies $k_p = f(J)$, $k_i = f(J)$, $k_d = f(J)$, there are obtained different values of the tuning parameters of the PID controller, that offer the different performance for the automatic control system.

SYNTHESIS OF THE PID CONTROL ALGORITHM BASED ON THE UNDAMPED TRANSIENT RESPONSE OF THE CLOSED LOOP SYSTEM

The procedure for synthesis of the PID controller supposes to be known the mathematical model that approximates the dynamics of the control object, this implies to be used the procedure of experimental identification. The new tuning algorithm of the PID controller is presented in Figure 2.

Data-Driven System Identification

In order to solve the problem of mathematical identification in the closed loop system, it is proposed the procedure for experimental identification, based on the undamped transient response of the closed loop control system. According to this procedure, the dynamics of the control object is proposed to be approximated with transfer function with inertia third order:

$$H(s)=\frac{k}{(T_1s+1)(T_2s+1)(T_3s+1)}=\frac{k}{a_0s^3+a_1s^2+a_2s+a_3}=\frac{B(s)}{A(s)}, \tag{13}$$

where T_1, T_2, T_3 are time constants; k is transfer coefficient, that is calculated according to the Equation (3); $a_0 = T_1T_2T_3$, $a_1 = T_1T_2 + T_1T_3 + T_2T_3$, $a_2 = T_1 + T_2 + T_3$, $a_3 = 1$.

The characteristic equation of the transfer function (13) is following:

$$A(s)=a_0s^3+a_1s^2+a_2s+a_3. \tag{14}$$

The coefficients a_0, a_1, a_2, a_3 are proposed to be calculated based on the values of the k_{cr}—critical transfer coefficient, A_{cr}—amplitude of unamortized oscillations and T_{cr}—period of the unamortized oscillations, that are obtained for the case, when automatic control system gains the limit of stability.

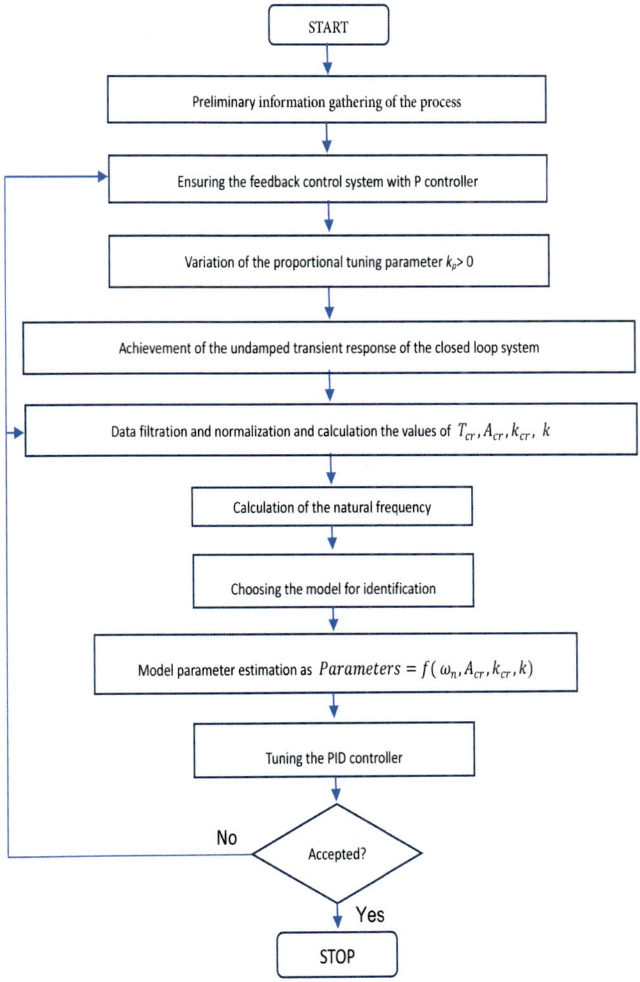

Figure 2. The algorithm of data-driven model identification and control.

To achieve this state of the system as in the Ziegler-Nichols method, it is given the closed loop system with PID controller (Figure 1), the coefficients k_i, k_d are settled to the zero, and the proportional tuning parameter $k_p > 0$ is varied, until the system achieves the limit of stability.

If it is known the value of period oscillations, it can be calculated the natural frequency by the [2] :

$$\omega_n = \frac{2\pi}{T_{cr}}. \tag{15}$$

In this way, it is proposed to find the dependencies for the calculation of the parameter of the control object in an experimental way, which depends on the natural frequency, oscillation amplitude and critical transfer coefficient:

$$a_0, a_1, a_2, a_3 = f\left(\omega_n, A_{cr}, k_{cr}\right). \tag{16}$$

The characteristic equation of the closed loop system (13) with critical transfer coefficient is the following:

$$A(s) = \frac{1}{k}\left(a_0 s^3 + a_1 s^2 + a_2 s + a_3\right) + k_{cr}. \tag{17}$$

In case, when the control system achieves the limit of stability, it has the poles placed on the imaginary axe, with real parts equal to zero and it is proposed to make the following substitution in the characteristic Equation (17):

$$s = j\omega_n. \tag{18}$$

Based on the substitution (18), the characteristic Equation (17) will become:

$$\begin{aligned} A(j\omega) &= \frac{1}{k}\left(a_0\left(j\omega_n\right)^3 + a_1\left(j\omega_n\right)^2 + a_2\left(j\omega_n\right) + a_3\right) + k_{cr} \\ &= \left(-a_1\omega_n^2 + a_3 + k_{cr}k\right) + j\left(-a_0\omega_n^3 + a_2\omega_n\right) \\ &= P(\omega) + jQ(\omega). \end{aligned} \tag{19}$$

Next, it is proposed to equal the real and imaginary parts with zero—$P(\omega) = 0$ and $Q(\omega) = 0$, and based on these equalling there are obtained the following expressions:

$$\begin{cases} \omega_n = \sqrt{\dfrac{a_2}{a_0}}; \\ a_1 = \dfrac{k_{cr}k + 1}{\omega_n^2}. \end{cases} \tag{20}$$

The transfer function of the control object (13) can be rewritten in the following form [2] :

$$H(s)=\frac{\alpha\omega_n^2}{(s+\alpha)(s^2+2\xi\omega_n s+\omega_n^2)}=\frac{Y(s)}{X(s)}, \tag{21}$$

where ξ is a damping ratio.

In this case, the unit—step response of the system is given by:

$$Y(s)=\frac{1}{s}H(s), \tag{22}$$

$$y(t)=1+C_1\mathrm{e}^{-\alpha t}+C_2\mathrm{e}^{-\xi\omega_n t}\sin(\omega_d-\theta), \tag{23}$$

where

$$C_1=-\frac{\omega_n^2}{\omega_n^2-2\zeta\omega_n\alpha+\alpha^2}, \tag{24}$$

$$C_2=-\frac{\alpha}{\sqrt{(\omega_n^2-2\zeta\omega_n\alpha+\alpha^2)(1-\xi^2)}}, \tag{25}$$

$$\theta=\tan^{-1}\frac{\sqrt{1-\xi^2}}{-\xi}+\tan^{-1}\frac{\omega_n\sqrt{1-\xi^2}}{\alpha-\xi\omega_n}, \tag{26}$$

$$\omega_d=\omega_n\sqrt{1-\xi^2}. \tag{27}$$

In case, when system achieves the limit of stability there is $\xi = 0$ and the transfer function (21) becomes:

$$H(s)=\frac{\alpha\omega_n^2}{(s+\alpha)(s^2+\omega_n^2)}=\frac{B(s)}{A(s)}. \tag{28}$$

The characteristic equation of the transfer function (28):

$$A(s)=(s+\alpha)(s^2+\omega_n^2)$$

has three poles— $s_1=-\alpha$, $s_2=j\omega_n$, $s_3=-j\omega_n$.

According to the Vieta's theorem for calculation the roots of the cubic equation, it is obtained the following dependence:

$$\alpha=\frac{a_1}{a_0}. \tag{29}$$

where a_0, a_1 —are parameters of the characteristic Equation (14).

The value of the amplitude of the unamortized oscillations from the undamped transient response can be calculated according to the step response of the system (23). There is obtained the following expression for calculation the value of oscillations amplitude:

$$A_{cr} = \frac{\alpha}{\sqrt{\alpha^2 + \omega_n^2}}. \tag{30}$$

Thus, then it is known the value of the oscillation amplitude, it is possible to calculate the value of α:

$$\alpha = \frac{A_{cr}\omega_n}{\sqrt{1 - A_{cr}^2}}. \tag{31}$$

In concordance with Equations (20), (29) and (31) the control object's parameters are calculated according to the following expressions:

$$\begin{cases} a_0 = \dfrac{a_1}{\alpha}; \\ a_1 = \dfrac{k_{cr}k + 1}{\omega_n^2}; \\ a_2 = a_0\omega_n^2; \\ a_3 = 1. \end{cases} \rightarrow \begin{cases} a_0 = \dfrac{(k_{cr}k + 1)\sqrt{1 - A_{cr}^2}}{A_{cr}\omega_n^3}; \\ a_1 = \dfrac{k_{cr}k + 1}{\omega_n^2}; \\ a_2 = \dfrac{(k_{cr}k + 1)\sqrt{1 - A_{cr}^2}}{A_{cr}\omega_n}; \\ a_3 = 1. \end{cases} \tag{32}$$

Based on Equations (32), the parameters of the control object can be calculated in dependency of the natural frequency, oscillation amplitude and critical transfer coefficient, that are obtained from the undamped transient response of the closed loop system. In this way, the procedure of identification of the mathematical model can be performed in the closed-loop system.

Data-Driven Control Algorithm Synthesis

It is considered, that control process is described by the transfer function with inertia third order—Equation (13) and the characteristic equation is the following:

$$A(s) = a_0 s^3 + a_1 s^2 + a_2 s + a_3. \quad (33)$$

In this case, the system is characterized by the three poles—s_1, s_2, s_3, that can be calculated based on the Vieta's theorem for the calculation the roots of the cubic equation:

$$\begin{cases} s_1 + s_2 + s_3 = -\dfrac{a_1}{a_0}; \\ s_1 s_2 + s_1 s_3 + s_2 s_3 = \dfrac{a_2}{a_0}; \\ s_1 s_2 s_3 = -\dfrac{a_3}{a_0}. \end{cases} \quad (34)$$

The dominant poles are the poles, that are situated more closely to the imaginary axe, and according to the [14] , the sum of dominant poles from characteristic Equation (33) is:

$$\Sigma_{\text{dom\_pol}} = s_1 + s_2 \approx -\frac{\dfrac{a_1}{a_0} - \dfrac{a_3}{a_0}}{2}, \quad (35)$$

where s_1 and s_2 are the dominant poles.

The characteristic Equation (33) has three poles and if it is known the sum of dominant poles (35), based on the Vieta's theorem, it is possible from the system (34) to calculate the production of the dominant poles, that will be equal with:

$$\Pi_{\text{dom\_pol}} = s_1 s_2 \approx -\frac{2a_3}{a_1 + a_3}. \quad (36)$$

The transfer function of the PID controller is given by Equation (1) and in conformity with Vieta's theorem for the second order equation, it is imposed that the zeros of the transfer function of the PID controller to be equal with dominant poles of the control object:

$$\begin{cases} s_1 + s_2 = -\dfrac{k_p}{k_d}; \\ s_1 s_2 = -\dfrac{k_i}{k_d}. \end{cases} \tag{37}$$

Based on Equations (35) and (36) the system of Equations (37) can be rewritten as:

$$\begin{cases} k_p = \dfrac{a_1 - a_3}{2a_0} k_d; \\ k_i = \dfrac{2a_3}{a_1 + a_3} k_d. \end{cases} \tag{38}$$

From (38), it can be observed that tuning parameters for P and I components depend on the parameter of the control process (13), which are known and derivative component that is unknown:

$$k_p, k_i = f\left(k_d, a_0, a_1, a_2, a_3\right). \tag{39}$$

It is proposed to use the maximum stability degree method with iterations [20] , in order to obtain the analytical expression for the calculation of the derivative tuning parameter.

Next, based on the maximum stability degree method with iterations, there is obtained the analytical expression for calculation the k_d tuning parameter of the PID controller:

$$k_d = \frac{1}{2k}\left(-12a_0 J^2 + 6a_1 J - 2a_2\right), \tag{40}$$

where J is the stability degree.

The value of the stability degree is calculated according to the expression [20] :

$$J = \frac{a_1}{4a_0}. \tag{41}$$

Based on Equations (40) and (41), the expression for calculation the k_d tuning parameter is:

$$k_d = \frac{3a_1^2 - 8a_0a_2}{8ka_0}. \tag{42}$$

In this way, the tuning parameters of the PID controller for the case then control process is described by the transfer function with inertia third order are the following:

$$\begin{cases} k_p = \Sigma_{\text{dom\_pol}} \cdot k_d; \\ k_i = \Pi_{\text{dom\_pol}} \cdot k_d; \\ k_d = \dfrac{3a_1^2 - 8a_0a_2}{8ka_0}. \end{cases} \rightarrow \begin{cases} k_p = \dfrac{a_1 - a_3}{2a_0} k_d; \\ k_i = \dfrac{2a_3}{a_1 + a_3} k_d; \\ k_d = \dfrac{3a_1^2 - 8a_0a_2}{8ka_0}. \end{cases} \rightarrow \begin{cases} k_p = \dfrac{(a_1 - a_3)(3a_1^2 - 8a_0a_2)}{16ka_0^2}; \\ k_i = \dfrac{3a_1^2a_3 - 8a_0a_2a_3}{4ka_0(a_1 + a_3)}; \\ k_d = \dfrac{3a_1^2 - 8a_0a_2}{8ka_0}, \end{cases} \tag{43}$$

where the parameters a_0, a_1, a_2, are known and there are calculated according to the equations (32).

STUDY CASES

Identification of the Mathematical Model of the DC Motor

In automatic control systems, the DC motor is frequently used as an actuator element.

Next, it is done the experimental identification of the mathematical model of the DC motor and in this study it was chosen the 2342L012 series Coreless encoder motor with technical parameters [20] :

- J rotor inertia, equal to 5.7 gcm^2;
- k_e back EMF constant, equal to 1.4 mV/min;
- k_m torque constant, equal to 13.4 mNm/A;
- R terminal resistance, equal to 1.9 Ω;
- k_n speed constant, equal to 713 min^{-1}/V;
- L rotor inductance, equal to 65 μH.

The input and output values of the DC motor are:

- Input value U_a: supply voltage;
- Output value $\dot{\theta} = \omega$: rotor shaft speed;

- Output value θ: rotor shaft position.

The transfer function in the open loop, in which the control value is the rotational speed of the rotor shaft and the input is a voltage applied to the armature windings U_a is:

$$\frac{\dot{\theta}(s)}{U_a(s)}=\frac{k_m}{(Ls+R)(Js+b)+k_e k_m}.$$

(44)

According to the technical parameters, the analytical model of the DC motor is:

$$H(s)=\frac{\dot{\theta}(s)}{U_a(s)}=\frac{641.885}{0.00000177s^2+0.052s+1}.$$

(45)

Based on the proposed procedure of identification in the closed loop, the automatic control system was brought to the critically regime and the experimental curve is presented in Figure 3.

The presented data were normalised and based on the undamped transient response of the closed loop system, there are obtained the following parameters:

$k_{cr}=5,\ A_{cr}=0.1185,\ T_{cr}=0.0011\,\text{s},\ \omega_n=5712.$

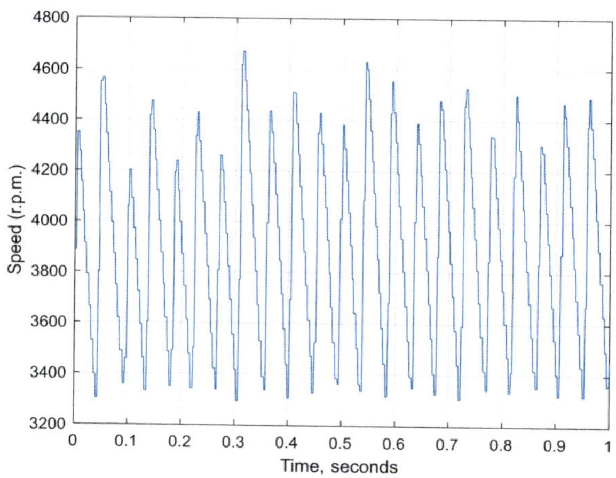

Figure 3. Undamped transient response of the system.

Based on Equation (32) there are calculated the parameters of the control object:

$a_0 = 2.698\text{e}-10,\ a_1 = 1.839\text{e}-07,\ a_2 = 0.008802,\ a_3 = 1.$

The identified transfer function is the following:

$$H(s) = \frac{1}{2.698\text{e}-10s^3 + 1.839\text{e}-07s^2 + 0.008802s + 1} = \frac{B(s)}{A(s)}. \tag{46}$$

Next in Figure 4, it is presented the comparison between the step response of the model of object analytical identified (transfer function (45))—curve 1; step response of the model of object identified by the proposed method (transfer function (46))—curve 2; the experimental curve of speed motor variation at the 5800 rpm—curve 3.

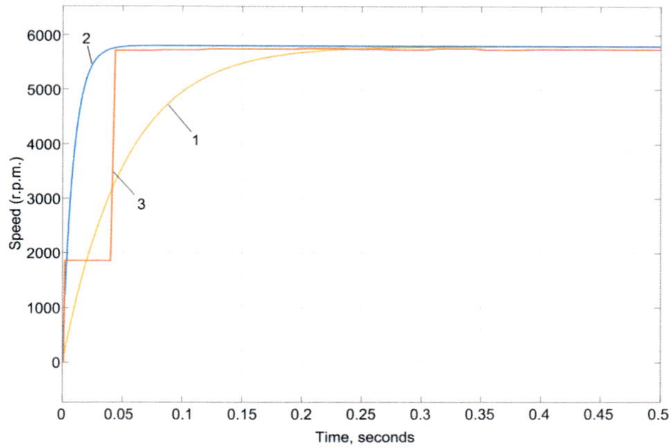

Figure 4. Comparison of the DC motor output: 1—step response of the system with transfer (45); 2—step response of the system with transfer function (46); 3—experimental curve of DC motor speed variation.

PID Control Algorithm Synthesis

It is supposed that control process is described by the transfer function:

$$H(s) = \frac{1}{(2s+1)(5s+1)(10s+1)} = \frac{1}{100s^3 + 80s^2 + 17s + 1} = \frac{B(s)}{A(s)}, \tag{47}$$

The automatic control system with P controller was simulated and it was obtained the undamped transient response of the closed-loop system and according to the identification algorithm there are obtained the following parameters of the model:

$$k_{cr} = 12.5993,\ T_{cr} = 15.2394\ \text{s},\ A_{cr} = 0.874,\ \omega_n = 0.4123.$$

According to the obtained values—the natural frequency, oscillation amplitude, critical transfer coefficient and based on the Equations (32), there are calculated the parameters of the model:

$$\begin{cases} a_0 = \dfrac{(k_{cr}k+1)\sqrt{1-A^2}}{A\omega_n^3} = 108.7; \\ a_1 = \dfrac{k_{cr}k+1}{\omega_n^2} = 80.43; \\ a_2 = \dfrac{(k_{cr}k+1)\sqrt{1-A^2}}{A\omega_n} = 18.39; \\ a_3 = 1. \end{cases} \tag{48}$$

Thus, the data-driven identified model is the following:

$$H(s) = \frac{1}{108.7s^3 + 80.43s^2 + 18.39s + 1} = \frac{B(s)}{A(s)}. \tag{49}$$

Next, it is done the comparison between original step response of the control object and step response of the identified model, that are presented in Figure 5.

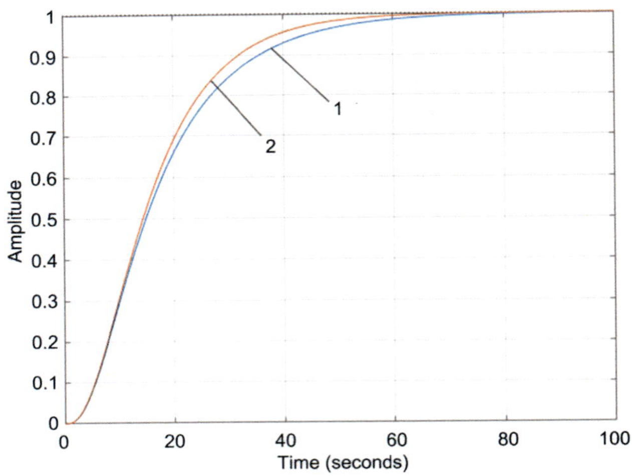

Figure 5. Comparison of the step responses of the control object: 1—step response of the system with transfer (47); 2—step response of the identified model (49).

Next, to the model of object (49) is tuned the PID controller based on Equations (43):

$$\begin{cases} k_p = \dfrac{a_1 - a_3}{2a_0} k_d = 3.9193; \\ k_i = \dfrac{2a_3}{a_1 + a_3} k_d = 0.0963; \\ k_d = \dfrac{3a_1^2 - 8a_0 a_2}{8ka_0} = 1.4313. \end{cases}$$

The obtained results of tuning the PID controller were compared with maximum stability degree method with iterations (MSMI), where the tuning parameters were calculated according to the relations (10)-(12). For comparison, it was used the Ziegler-Nichols, where the tuning parameters were calculated according to Table 1 and it was used the parametrical optimization, based on the PID Tuner Toolbox from MATLAB. The obtained tuning parameters and obtained performances of the automatic control system (t_r—rise time, t_s—settling time, —overshoot) are presented in Table 2.

Table 2. Tuning parameters and the performance of the automatic control system.

No.	Tuning method	k_p	k_i	k_d	t_s	t_r	σ, %
1	MSD method with dominant poles allocation	1.43	0.096	3.91	30.5	30.5	0.0
2	MSMI	1.74	0.12	3.91	37.57	18.72	3.33
3	Ziegler-Nichols	9.44	0.109	1.52	171.79	6.55	56.75
4	Parametrical optimization	3.29	0.245	9.64	10.98	27.5023	6.39

In Figure 6, there is presented the computer simulation of the automatic control system with PID controller tuned by: the proposed method in this paper—curve 1, maximum stability degree method with iterations—curve 2, Ziegler—Nichols method—curve 3, parametrical optimization—curve 4.

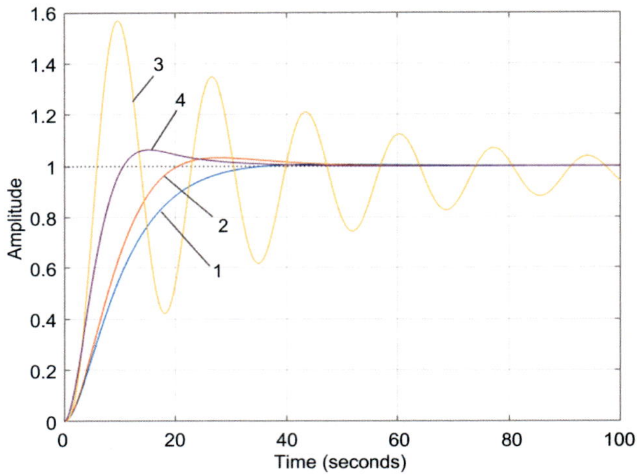

Figure 6. Transient responses of the automatic control system: 1—PID controller tuned by MSD method with dominant poles allocation; 2—PID controller tuned by MSMI; 3—PID controller tuned by Ziegler-Nichols method; 4—PID controller tuned by parametrical optimization.

From Figure 6, it was observed that in case of using the proposed method for tuning the PID controller, it is obtained the critically damped step response without overshoot. In the case of using the Ziegler-Nichols method, it is obtained oscillated transient response. In the case of using MSMI and parametrical optimization method, there are obtained the underdamped transient responses of the automatic control system.

CONCLUSIONS

Synthesis of the control algorithm is one of the main problems that concerned engineers during the design of automatic control systems. The solution to this problem supposes to be known some key parameters of the control process, or the mathematical model, that approximates the dynamics of the control process. Frequently the mathematical model of the control process is obtained in the experimental, or analytical way in the open loop. This implies being involved with the operator for system identification in the open loop. The Ziegler-Nichols method is become so used, due to the fact that does not require to be known the mathematical model of the control process and the procedure of tuning the controller supposes the extraction of some key parameters from step response of the closed-loop control system, when system is marginally stable.

In this paper, it is presented the procedure for data-driven model identification and control of the inertial systems. According to this procedure the first step involves being done the system identification, based on the undamped step response of the closed-loop system. The algorithm for mathematical identification supposes to be achieved the limit of stability of the closed-loop system and based on the parameters, that are extracted from the undamped step response, there are presented some simple expressions for calculation the parameters of the model.

In the second step, there are presented the expressions for calculation of the tuning parameters of the PID controller. These expressions were obtained based on the value of the maximum stability degree criterion and the tuning procedure of the PID controller ensures the system the critically damped step response.

The procedure of the data driven identification and control was verified by computer simulation and gave good performances in the case of tuning the PID controller, and as further research, the proposed tuning method can be extended to the frequency domain.

ACKNOWLEDGEMENTS

This research was funded by the postdoc project "Models and algorithms for control and behaviour approach of the interoperable systems", No. 22.00208.5007.12/ PD I, running at the Technical University of Moldova.

REFERENCES

1. Wang, L., Chai, S., Yoo, D., Gan, L. and Ng, K. (2015) PID and Predictive Control of Electrical Drivers and Power Converters Using Matlab®/ Simulink®. Wiley, Hoboken. https://doi.org/10.1002/9781118339459
2. Dumitrache, I. (2005) Ingineria Reglarii Automate [Automatic Control Systems]. Politehnica Press, Bucureşti.
3. Hornsey, S. (2012) A Review of Relay Auto-Tuning Methods for the Tuning of PID-Type Controllers. An International Journal of Undergraduate Research, 5, 2.
4. Basilio, J. and Matos, S.R. (2002). Design of PI and PID Controllers with Transient Performance Specification. IEEE Transactions on Education, 45, 364-370. https://doi.org/10.1109/TE.2002.804399
5. Bucz, Š. and Kozáková, A. (2018) Advanced Methods of PID Controller Tuning for Specified Performance. In: Shamsuzzoha, M., Ed., PID Control for Industrial Processes, IntechOpen, London, 74-119. https://doi.org/10.5772/intechopen.76069
6. O'Dwyer, A. (2009) Handbook of PI and PID Controller Tuning Rules. 3rd Edition, Imperial College Press, London.
7. Astrom, K.J. and Hagglund, T. (1995) PID Controllers: Theory, Design and Tuning. 2nd Edition, Instrument Society of America, Research Triangle Park.
8. Lazăr, C., Vrabie, D. and Carari, S. (2004) Sisteme automate cu regulatoare PID [Automatic systems with PID controllers]. Matrix Rom, Bucuresti.
9. Fišer, J. and Zítek, P. (2019) PID Controller Tuning via Dominant Pole Placement in Comparison with Ziegler-Nichols Tuning. IFAC-PapersOnLine, 52, 43-48. https://doi.org/10.1016/j.ifacol.2019.12.204
10. Ziegler, J.G. and Nichols, N.B. (1993) Optimum Settings for Automatic Controllers. Journal of Dynamic Systems, Measurement, and Control, 115, 220-222. https://doi.org/10.1115/1.2899060
11. Scherlozer, A., Orsini, M. and Patole, S. (2016) Simulation and Numerical Analysis and Comparative Study of a PID Controller Based on Ziegler-Nichols and Auto Turning Method. Proceedings of the 12th IEEE International Conference on Control and Automation, Kathmandu, 1-3 June 2016, 1-16.
12. Opelit, V. (1960) Osnovi Avtomaticeskogo Upravlenia [Bases of the Automatic Control]. Gosenergoizdat, Moskva.

13. Preitl, Ş. and Precup, R.E. (2001) Introducere în Ingineria Reglării Automate [Introduction to the Automatic Control System]. Editura Politehnica, Timişoara.
14. Lukas, V.A. (1990) Teoria Avtomaticheskogo Upravlenia [Automatic Control Theory]. 2nd Edition, Nedra, Moskva.
15. Dincel, E. and Söylemez, M.T. (2020) Robust PID Controller Design via Dominant Pole Assignment for Systems with Parametric Uncertainties. Asian Journal of Control, 24, 834-844. https://doi.org/10.1002/asjc.2484
16. Schubladze, A.M. (1980) Sposobi Sinteza System Upravlenia Maksimalinoi Stepeni Ustoichivosti [Methods of Synthesis of Control Systems with Maximum Stability Degree]. Avtomatika i Telemehanika, 1, 28-37.
17. Izvoreanu, B. (2010) The Iterative Algorithm of Tuning Controllers to the Model Object with Advanse Delay and Inertia Second Order. Proceedings of the 10th International Conference on Development and Application Systems DAS-2010, Suceava, 27-29 May 2010, 111-115.
18. Salem, A., Moustafa, Hassan, M.A. and Ammar, M.E. (2014) Tuning PID Controllers Using Artificial Intelligence Techniques Applied to DC-Motor and AVR. System Asian Journal of Engineering and Technology, 2, 129-138.
19. Fiodorov, I. (2013) Synthesis Algorithms of Controllers for Automatic Control Systems with Maximum Stability Degree. Annals of the University of Craiova, Electrical Engineering Series, 37, 139-143.
20. Cojuhari, I., Fiodorov, I., Izvoreanu, B. and Moraru, D. (2022) Synthesis of PID Algorithm for Speed Control of the DC Motor. Proceedings of the 2022 International Conference on Development and Application Systems (DAS), Suceava, 26-28 May 2022, 1-5. https://doi.org/10.1109/DAS54948.2022.9786125

CHAPTER 13

An Intelligent System for Real-Time Condition Monitoring of Tower Cranes

Aaron K. Adik[1] and Wilson Wang[2]

[1]Department of Electrical and Computer Engineering, Lakehead University, Thunder Bay, Canada.

[2]Department of Mechanical Engineering, Lakehead University, Thunder Bay, Canada.

ABSTRACT

Reliability and safety are major issues in tower crane applications. A new adaptive neurofuzzy system is developed in this work for real-time health condition monitoring of tower cranes, especially for hoist gearboxes. Vibration signals are measured using a wireless smart sensor system. Fault detection is performed gear-by-gear in the gearbox. A new diagnostic classifier is proposed to integrate strengths of several signal processing techniques for

Citation: Adik, A. and Wang, W. (2019), "An Intelligent System for Real-Time Condition Monitoring of Tower Cranes". Intelligent Control and Automation, 10, 155-167. doi: 10.4236/ica.2019.104011.

fault detection. A hybrid machine learning method is proposed to facilitate implementation and improve training convergence. The effectiveness of the developed monitoring system is verified by experimental tests.

Keywords: Adaptive Neuro-Fuzzy Systems, Machine Learning, Diagnostics, Pattern Classification, Tower Cranes, Smart Sensors

INTRODUCTION

Tower cranes are used extensively at construction sites to lift and move materials. They often rise hundreds of feet into the air, and can reach out just as far. As illustrated in Figure 1, a tower crane consists of several units: 1) the mast is the supporting tower of the crane; 2) slewing unit sits at the top of the mast, which allows the crane to rotate; 3) the operator cabin usually sits just above the slewing unit; 4) the jib is the operating arm, along which a trolley runs to move the load; 5) counter jib is the shorter horizontal arm, which carries the hoist unit, control electronics, and the concrete counter weights.

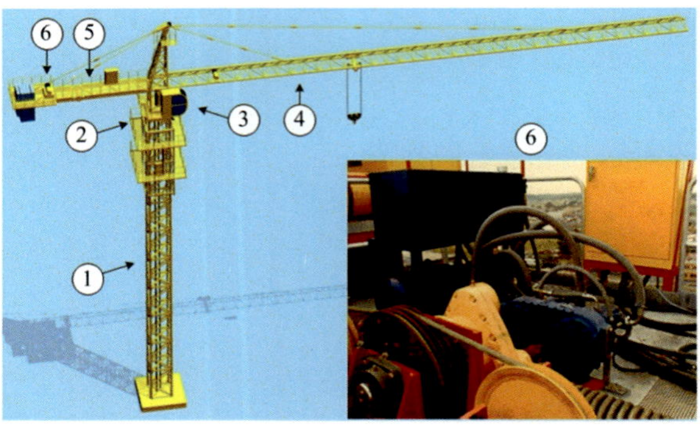

Figure 1. Illustration of a tower crane structure: (1) mast, (2) slewing unit, (3) operator cabin, (4) jib, (5) counter jib, (6) hoist unit.

Reliability and safety are the main issues for tower crane applications. Machinery failures in tower cranes could result in catastrophic damage to life and properties [1]. Several techniques and tools have been proposed in literature for tower crane condition monitoring such as based on the maximum load, maximum moment, maximum lifting height, trolley position on the jib, arm positions to prevent collision, and structure strength [2]. However, unexpected failures always occur in tower crane applications. As a matter of fact, machinery failure of tower cranes has become the top hazard at construction sites [3]. As a result, more strict regulations and clauses have been imposed by the Canadian government [4]; and the manufacturers and contractors have to develop and implement more efficient safety monitoring technologies for tower cranes [5].

This work will focus on the condition monitoring of hoist units in tower cranes. The hoist unit is located in the counter jib as shown in Figure 1. The most dangerous failure in a hoist drive is related gear tooth breakage. The gear materials are subjected to dynamic loading and impacts. If a tooth is broken partly or completely due to fatigue, the following teeth in that gear will be damaged quickly due to extra impacts. The gearbox could become "free rotation", and the lifted load would drop to the ground freely, which could cause serious danger to the people underneath it [6]. New technologies and tools are critically needed for more accurate health condition monitoring of the hoist gearboxes of tower cranes.

Condition monitoring is an act of fault diagnosis by means of appropriate information carriers such as temperature, acoustic signal, or vibration [7]. Vibration-based monitoring is the most commonly used approach because it is ease of measurement and has relatively high signal-to-noise ratio [8], which also will be used in this study.

Fault diagnosis is a process of classifying the vibration features into different categories corresponding to different equipment health conditions. Automatic fault diagnosis can be conducted by either model-based or data-driven techniques [9]. Although model-based approaches are relatively easy to develop and implement, accurate analytical models are difficult to derive especially for complex machinery systems [10]. Data-driven paradigms conduct fault diagnosis by using some intelligent tools such as neural networks, fuzzy logic, or synergetic schemes [11] [12]. The author's research team has also proposed neurofuzzy (NF) schemes for machinery fault diagnostics [13] [14], in which the system parameters are trained online/offline. Although these techniques have been applied effectively in many applications [15], these NF schemes could not be used effectively for applications using a standalone computing device with limited RAM capability; in addition they do not have sufficient adaptive capability to accommodate time-varying dynamic effects as in tower crane monitoring applications.

In summary, when a monitoring system is used in real-world applications, unreasonably missed alarms (i.e., the monitor cannot pick up existing faults) and false alarms (i.e., the monitor triggers alarms because of noise instead of real faults) will seriously mitigate its validity. In addition, machinery dynamics may also change suddenly, such as just after repair or regular maintenance. To tackle these aforementioned challenges, the objective of this work is to propose a new adaptive NF (i.e., ANF in short), system for real-time health condition monitoring of hoist gearbox in tower cranes. It is new in the following aspects: 1) A new adaptive NF classifier is proposed for more accurate fault diagnosis in gear systems in order to provide a more reliable real-time condition monitoring tool for tower cranes. 2) A new hybrid training technique is proposed to facilitate training operation and improve training convergence. The effectiveness of the developed classifier is verified by experimental tests.

THE DEVELOPED INTELLIGENT MONITORING SYSTEM

Overview

The developed intelligent monitor consists of several modules as illustrated in Figure 2. The hardware includes smart sensor-based data acquisition (DAQ) system and a computing device (e.g., a small laptop). The software module consists of functions such as signal processing, database management, diagnostic classifier, and training, which will be discussed in this section.

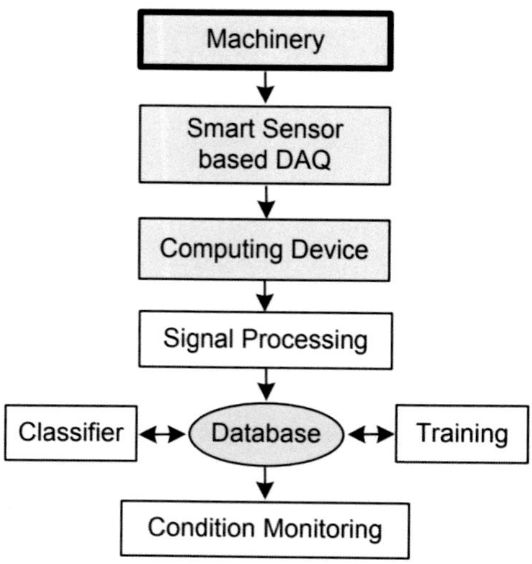

Figure 2. Architecture of the developed intelligent monitor.

Smart Sensor-Based DAQ

A classical data acquisition system consists of hardware such as sensors, amplifiers, antialiasing filters, and a DAQ board. It is not only expensive but also inconvenient for real-world applications [4]. As the microprocessors become cheaper and faster in processing, it is possible to integrate the related DAQ functions onto chips for smart sensors. The authors' research team has developed many smart sensor-based DAQ systems that can collect different types of signals [16] [17]. In this work, wireless smart vibration and distance sensors are used for tower crane condition monitoring. The vibration sensing unit uses ICP-accelerometers (e.g., PCB 320 series) with a bandwidth of 15,000 Hz, acceleration range of ±140 g, sensitivity of 150 mV/g, and measurement range of ±4900 m/s^2 pk. A distance smart sensor is used to measure gear rotation. Signal conditioning functions (e.g., amplification, rectification, and antialiasing filters) are implemented in IC filter chips so as to reduce size and cost. The wireless communication is based on SWAP protocol with frequency of 915 Hz. A user interface is developed to enable modification of DAQ reconfigurable parameters such as sensor calibration, sampling frequency, filter specifications, monitoring time interval, etc.

Adaptive Neurofuzzy Diagnostic Classifier

A novel ANF diagnostic classifier is developed for gear system monitoring. To improve monitoring accuracy, fault diagnosis is performed gear by gear by the use of time synchronous average filtering [15]. The conditions of each gear are classified into three categories: healthy (C_1), possibly damaged (C_2), and damaged (C_3), respectively. Consider L input variables, $\{x_1, x_2, \cdots, x_L\}$; three membership functions (MFs), Small, Medium, and Large, are assigned to each input variable. The diagnostic classification, in terms of the diagnostic indicator y, can be formulated as

$$\begin{aligned} &\Re_j: \text{If } (x_1 \text{ is } A_{1j}) \text{ and } (x_2 \text{ is } A_{2j}) \\ &\text{and } \cdots \text{ and } (x_L \text{ is } A_{Lj}) \Rightarrow (y \subset C_j \text{ with } w_j) \end{aligned} \quad (1)$$

where A_{lj} are linguistic MFs of x_l, $l = 1, 2, \cdots, L$, $j = 1, 2, \cdots, J$; J is the number of rules; w_j is the weight factor (contribution) of the rule $\Re_j$ to the diagnostic operation.

The number of rules is associated with the diagnostic reasoning operations of input variables. For example, if the classifier has three inputs $\{x_1, x_2, x_3\}$ (i.e., L = 3), then diagnostic classification will be performed by the following criteria:

Four rules are associated with healthy condition (C_1) reasoning: If (all three indices are Small), or (two are Small and one is Medium), then (the object is Healthy).

Four rules are associated with damage condition (C_3) formulation: If (all three indices are Large), or (two are Large and one is Medium), then (the object is Damaged).

Except for these two states, the object is possibly damaged (C_2), and the classification of C_2 is performed by 9 rules. In all of these cases, different feature association (rule) corresponds to a different weight grade w_j.

Figure 3 schematically shows the network architecture of this diagnostic classifier. Unless specified, all the network links have unity weights. The input nodes in layer 1 transmit the monitoring indices $\{x_1, x_2, \cdots, x_L\}$ to the next layer. Each node in layer 2 acts as an MF. The nodes in layer 3 perform the fuzzy T-norm operations. If a product operator is used, the firing strength of rule $\Re_j$ is

$$\eta_j = \prod_{l=1}^{L} A_{lj}(x_l) \qquad (2)$$

where $A_{lj}(\bullet)$ denote MF grades.

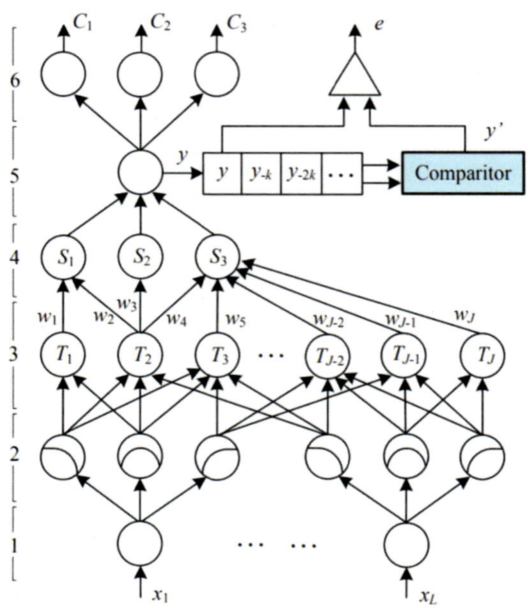

Figure 3. The network architecture of the diagnostic classifier.

Defuzzification is performed in layer 4. If J_1, J_2 and J_3 are the number of rules associated with the classes C_1, C_2 and C_3, respectively, the belonging grade to each class is:

$$S_1 = \frac{\sum_{J_1} \eta_j w_j}{\sum_{J_1} \eta_j}$$

$$S_2 = \frac{\sum_{J_2} \eta_j w_j}{\sum_{J_2} \eta_j}$$

$$S_3 = \frac{\sum_{J_3} \eta_j w_j}{\sum_{J_3} \eta_j} \qquad (3)$$

The diagnostic indicator will be

$$y = S_1 \vee S_2 \vee S_3 . \qquad (4)$$

Diagnostic Decision-Making

In machinery health condition monitoring, if the diagnostic indicator $y \leq \varepsilon_1$, the object is considered healthy; ε_1 is a threshold ($\varepsilon_1 = 0.50$ in this case). If $\varepsilon_1 < y \leq \varepsilon_2$, the monitored object is possibly damaged; ε_2 is another threshold ($\varepsilon_2 = 0.70$ in this case). Otherwise, if $y > \varepsilon_2$, the object is damaged. An appropriate alarm should be given to the user for the occurrence of the defect.

When the indicator y values are formulated, the corresponding historical indicator y' in the database will be used to verify the diagnostic results so as to improve diagnostic reliability. In operation, the objective is to minimize the match error, $e = |y - y'|$, between the current value of the diagnostic indicator y and its comparator value y'. If y and y' belong to different classes, for example C_1 (healthy) and C_2 (possibly damaged), there are three possible reasons: 1) the defect recognized by the ANF classifier is caused by noise instead of real component damage; 2) the historical comparator data sets are not accurate; and 3) the convergence of the ANF classifier deteriorates due to local minima. These problems can be solved by the training process. For the first reason, the error-making rules will be punished to reduce their weight factors w_j to the diagnostic classification. If the database is not accurate, it will be updated using more accurate data sets for future monitoring and training application. If the errors are caused by possible trapping of location minima, the rule boundary properties in the decision space will be modified by appropriate training operations.

Hybrid System Training

Once the ANF classifier is set up, it should be properly trained to improve its performance. In this case, a hybrid training strategy is suggested to update system parameters. The linear parameters will be trained by the use of the classical least squires estimator (LSE).

Many algorithms have been proposed in literature for nonlinear parameter optimization, each having its merits and limitations. The Levenberg-Marquardt (L-M) algorithm possesses quadratic convergence close to a minimum [18]. Its convergence is still reasonably well even if the initial estimates are relatively poor. This characteristic is important in the developed real-time monitoring applications. In addition, the L-M algorithm has been proven globally convergent in many applications by properly choosing the step factors. In this monitoring application, the system uses a less powerful standalone computing device (e.g., a small laptop computer), which has limited RAM space. In using the L-M algorithm, when the number of intermediate variables becomes large in operation, the computation of the inverse of a large size Hessian matrix will take a very long time. The excessive computing load can dramatically slow down or even block the computing operations. To solve this problem, a novel recessive L-M training method is suggested in this work to facilitate its implementation and improve its convergence.

The objective or error function with respect to adjustable parameters θ_k at the current time instant k is defined as

$$E(\boldsymbol{\theta}_k)=\frac{1}{2}\sum_{k=1}^{K}\left[y_k(\boldsymbol{\theta}_k)-y_d\right]^2=\frac{1}{2}\sum_{k=1}^{K}r_k^2(\boldsymbol{\theta}_k)=\boldsymbol{r}_k^{\mathrm{T}}(\boldsymbol{\theta}_k)\boldsymbol{r}_k(\boldsymbol{\theta}_k) \tag{5}$$

where $y_k(\theta_k)$ is the kth output, $k = 1, 2, \cdots, K$; y_d is the desired output. $r_k(\theta_k)$ is the error vector that can be either linear or nonlinear.

To simplify expression, the variable θ_k is dropped in the related terms in these manipulations. By taking the Tayler series expansion and neglecting higher order terms,

$$\begin{aligned}\boldsymbol{\theta}_{k+1} &\approx \boldsymbol{\theta}_k+\lambda_k\left(\boldsymbol{J}_k^{\mathrm{T}}\boldsymbol{J}_k+\eta_k\boldsymbol{I}\right)^{-1}\boldsymbol{J}_k^{\mathrm{T}}\boldsymbol{r}_k=\boldsymbol{\theta}_k+\lambda_k\boldsymbol{H}_k^{-1}\boldsymbol{J}_k^{\mathrm{T}}\boldsymbol{r}_k\\ &=\boldsymbol{\theta}_k+\left(1-\alpha_k\right)\boldsymbol{H}_k^{-1}\boldsymbol{J}_k^{\mathrm{T}}\boldsymbol{r}_k\end{aligned} \tag{6}$$

where $J_k \in \mathbb{R}^{Z\times Z}$ is the Jacobin matrix; Z is dimension of θ_k (or the number of parameters to be trained); $H_k \in \mathbb{R}^{Z\times Z}$ is the Hessian matrix; $I \in \mathbb{R}^{Z\times Z}$ is an identity matrix; $\lambda_k = 1 - \alpha_k$; α_k is the forgetting factor; η_k is the learning that can be adapted by a line search method.

The Hessian matrix can be expressed as

$$\boldsymbol{H}_k = \alpha_k \boldsymbol{H}_{k-1} + (1-\alpha_k)(\boldsymbol{J}_k^{\mathrm{T}}\boldsymbol{J}_k + \eta_k \boldsymbol{I}). \tag{7}$$

In implementation, instead of computing the Z × Z matrix $\eta_t I$ at each time step, η_t is added to one of the diagonal elements of $\boldsymbol{J}_k^{\mathrm{T}}\boldsymbol{J}_k$ at each time instant

$$\boldsymbol{H}_k = \alpha_k \boldsymbol{H}_{k-1} + (1-\alpha_k)(\boldsymbol{J}_k^{\mathrm{T}}\boldsymbol{J}_k + Z\eta_k \boldsymbol{\Lambda}) \tag{8}$$

where $\Lambda \in \mathbb{R}^{Z\times Z}$ has only one nonzero element located at t{mod(Z) + 1} diagonal position, or

$$\Lambda_{ii} \in \begin{cases} 1, & \text{if } i = k\{\text{mod}(Z)+1\}, \text{ and } k > Z \\ 0, & \text{otherwise} \end{cases} \tag{9}$$

Correspondingly, Equation (8) can be rewritten as

$$\boldsymbol{H}_k = \alpha_k \boldsymbol{H}_{k-1} + (1-\alpha_k)(\boldsymbol{U}\boldsymbol{V}^{-1}\boldsymbol{U}^{\mathrm{T}}) \tag{10}$$

where $\boldsymbol{U}^{\mathrm{T}} = \begin{bmatrix} & & & \boldsymbol{J}_k^{\mathrm{T}} & & & \\ 0 & \cdots & 0 & 1 & 0 & \cdots & 0 \end{bmatrix}$, and $\boldsymbol{V}^{-1} = \begin{bmatrix} 1 & 0 \\ 0 & Z\eta_k \end{bmatrix}$.

The computation of $\boldsymbol{H}_k^{-1}$ in Equation (6) is very time consuming, and it not suitable for real-time applications. Here $\boldsymbol{H}_k^{-1}$ is computed by using following approach:

$$(\boldsymbol{A}+\boldsymbol{BCD})^{-1} = \boldsymbol{A}^{-1} - \boldsymbol{A}^{-1}\boldsymbol{B}(\boldsymbol{C}^{-1} + \boldsymbol{DA}^{-1}\boldsymbol{B})^{-1}\boldsymbol{DA}^{-1} \tag{11}$$

where A, B, C and D are matrices such that A and $(C^{-1} + DA^{-1}B)$ are nonsingular matrices.

Based on Equation (11), Equation (6) can be rewritten as

$$\begin{aligned} \boldsymbol{\theta}_{k+1} &= \boldsymbol{\theta}_k + (1-\alpha_k)\boldsymbol{H}_k^{-1}\boldsymbol{J}_k^{\mathrm{T}}\boldsymbol{r}_k \\ &= \boldsymbol{\theta}_k + (1-\alpha_k)\times\Big\{(\alpha_k\boldsymbol{H}_{k-1})^{-1} - (\alpha_k\boldsymbol{H}_{k-1})^{-1}(1-\alpha_k)\boldsymbol{U} \\ &\quad \times\left[\boldsymbol{V} + \boldsymbol{U}^{\mathrm{T}}(\alpha_k\boldsymbol{H}_{k-1})^{-1}(1-\alpha_k)\boldsymbol{U}\right]^{-1}\boldsymbol{U}^{\mathrm{T}}(\alpha_k\boldsymbol{H}_{k-1})^{-1}\Big\}\boldsymbol{J}_k^{\mathrm{T}}\boldsymbol{r}_k \end{aligned} \tag{12}$$

Let $A = \alpha_k H_{k-1}$, $B = (1 - \alpha_k)U$, $C = V^{-1}$, and $D = U^{\mathrm{T}}$. From Equation (12), the recursive L-M algorithm can then be represented by

$$\boldsymbol{\theta}_{k+1} = \boldsymbol{\theta}_k + \boldsymbol{\Phi}_k \boldsymbol{J}_k \boldsymbol{r}_k \tag{13}$$

where $\boldsymbol{\Phi}_k = \frac{1}{\alpha_k}\left[\boldsymbol{\Phi}_{k-1} - \boldsymbol{\Phi}_{k-1}\boldsymbol{U}\boldsymbol{S}^{-1}\boldsymbol{U}^{\mathrm{T}}\boldsymbol{\Phi}_{k-1}\right]$,

$$\boldsymbol{S} = \alpha_k \boldsymbol{V} + \boldsymbol{U}^{\mathrm{T}}\boldsymbol{\Phi}_{k-1}\boldsymbol{U}. \tag{14}$$

S is a matrix with dimension 2 × 2; its inverse computation is simple and can be implemented for real-time monitoring applications. $\theta_0 = 0$; Φ_t is a covariance matrix with initial condition $\Phi_0 = \rho I$, where $\rho \in [10^2, 10^6]$, $\eta_k \in [0.001, 10]$, and $\alpha_k \in [0.95, 1]$.

In hybrid training, as each data sample is inputted to the ANF classifier, the linear consequent parameters w_j are updated by using the LSE. The nonlinear classifier MF parameters are trained by the use of the recursive L-M method. On the other hand, adaptive training is preferred in real-time applications because: 1) it is necessary for time-varying systems; 2) it possesses randomness that may help to escape from a local minimum; and 3) it is useful when the number of training data is large.

PERFORMANCE EVALUATION

The effectiveness of the proposed ANF classifier techniques will be verified in this section by the use of experimental tests.

Experimental Setup

The experimental setup used in this study is schematically shown in Figure 4. The system is driven by a 3-HP induction motor. A pair of gears are tested in this case.

The loading is provided by heavy-duty magnetic breaks connected to the gearbox. The motor rotation is controlled by a speed controller, which allows tested gearbox operation in the range of 20 to 3600 rpm. The vibration is measured using two smart vibration sensors (accelerometers) mounted on the gearbox housing.

A displacement sensor is used to provide gear rotating information, which is used as a reference for time synchronous average filtering. The collected signals are wirelessly fed to a receiver that is connected to a computer through a USB port.

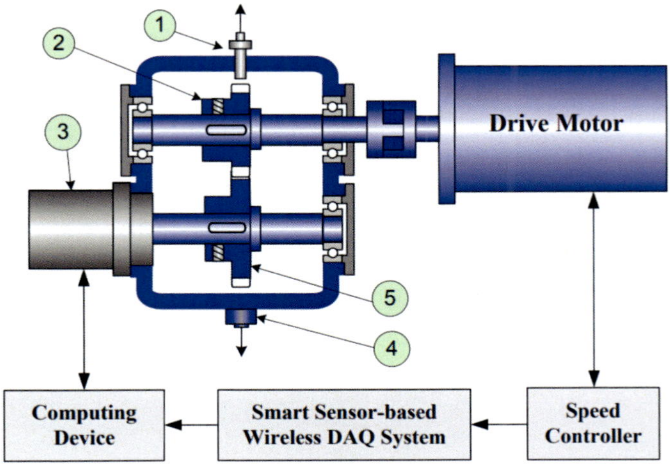

Figure 4. Experimental setup: (1) the smart displacement sensor, (2) the driving gear, (3) the load drive, (4) the smart vibration sensor, (5) the driven gear.

Monitoring Indices

Gear fault consists of localized damage and distributed defects (e.g., wear and pitting). This work focuses on localized gear fault diagnosis because a localized fault can not only generate transmission errors but also may cause sudden failures [15] [19]. For example, a tiny crack may occur at a gear tooth due to dynamic loading, stress concentration, or impacts. As the gear system continues in operation, the crack propagates continuously. Unfortunately, by the time that the crack is visible, the gear may be more than 90% used up. When a tooth is broken out, the following teeth will be damaged quickly due to extra impacts, which could induce severe accidents in tower crane applications.

One of the initiatives of this work is to perform fault diagnosis of the gearbox gear by gear. Because the measured vibration is an overall signal generated from various rotary sources, the first step is to differentiate the signal specific to each gear by using the time synchronous average filter [15]. In this filtering process, all of the signals that are non-synchronous to the rotation of the gear of interest (e.g., those from bearings and shafts) are filtered out. As a result, each gear signal is computed and represented in one full revolution, called the signal average, which will be used for advanced analysis.

Several techniques have been proposed in the literature for gear fault detection; however, each has its own advantages and limitations, and is efficient for specific applications only [13] [14]. As a result, the more features are applied in diagnostic classification, the more possible it is to achieve conflicting diagnostic results. The selected features should be robust, that is, sensitive to component defects but insensitive to noise (i.e., the signal not carrying health information). By investigation, three features are selected in this work from different information domains: energy, amplitude, and phase:

- wavelet energy function, using the overall residual signal that is obtained by bandstop filtering out the gear mesh frequency $f_R N$ and its harmonics, where f_R is the rotation frequency (in Hz) of the gear of interest and N is the number of teeth of the gear;
- beta kurtosis, using the overall residual signal;
- phase demodulation using the signal average. Details of these techniques can be found from [15].

Performance Evaluation by Experiment

To verify the effectiveness of the developed intelligent monitor and the related techniques, systematic tests have been conducted corresponding to different machinery conditions. Three gear cases are tested in this case, as illustrated in Figure 5: (a) healthy gears, (b) cracked gears, and (c) scored gears.

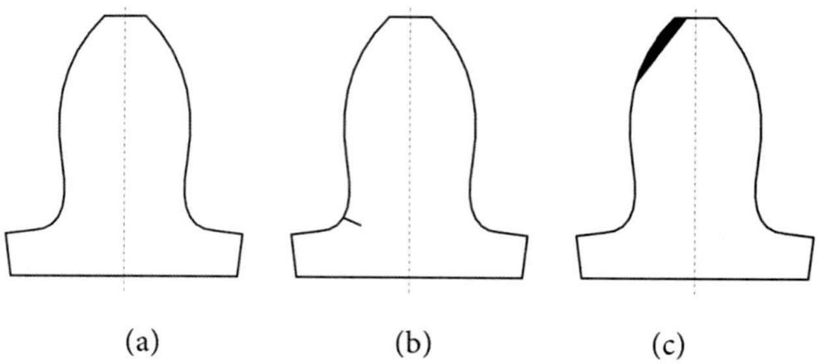

Figure 5. Tested gears: (a) Healthy gears; (b) Cracked gears; (c) Scored gear.

To make a comparison, two other related diagnostic classifiers are considered:

- **Classifier-1**: A general ANFIS classifier, but using the proposed hybrid training technique with the LSE and the recursive L-M for training. It is to compare the proposed ANF classification scheme.

- **Classifier-2**: It has the same network architecture with the ANF classifier, but the training is using the LSE and the classical L-M algorithm. It is to compare training efficiency of the proposed hybrid training based on the recursive L-M algorithm.

A series of tests have been conducted corresponding to different gear conditions. Firstly, both gears are healthy. The tests are undertaken with different load levels and speed. 225 data sets are collected for system training (120 pairs) and testing (105 pairs). Then the driven gear with a simulated crack is installed and tested. 300 data sets are collected; 200 pairs are used for training and 100 pairs for testing. In scored gear testing, similarly, 300 sets of data are collected; 200 pairs are used for training and 100 pairs for testing. Test results are summarized in Table 1, in which FA stands for False Alarms, and MA stands for Missed Alarms.

Table 1. Summary of the testing results using different classifiers (FA: False Alarms, MA: Missed Alarms).

Diagnostic Classifier	Healthy Gears		Cracked Gears		Chipped Gears		Overall Accuracy %
	MA	FA	MA	FA	MA	FA	
Classifier-1	1	3	2	2	2	2	88.4
Classifier-2	0	2	3	1	1	1	94.2
ANF Classifier	0	1	2	1	1	1	98.8

It is seen the developed ANF classifier provides best performance in this test, which has recorded 3 missed alarms and 2 false alarms during the testing periods. The missed alarms are mainly induced when the motor rotates at a very low speed, and the resulting signature modulation due to a small tooth crack is very weak. The false alarms are caused by the dramatic speed and load variations in testing. ANF classifier performs better than Classifier 1 that generates 5 missed alarms and 7 false alarms. It can verify the better classification efficiency of the proposed ANF diagnostic classification scheme. The developed ANF classifier also outperforms Classifier-2 that has recorded 4 missed alarms and 4 false alarms during the testing periods. It can verify the effective convergence of the proposed the recursive L-M algorithm.

Monitoring of the Hoist Gearbox

The purpose of the developed ANF classifier is for hoist gearbox condition monitoring and diagnostics. Figure 6 schematically shows the structure of a hoist drive in a tower crane. The gearbox consists of two pairs of gears. The monitoring is taken gear-by-gear. Vibration signals are collected using the smart vibration sensors. The gear rotation signal is measured by a wireless encoder, which provides a reference for time synchronous average filtering.

Figure 7 shows a demo of an interface for this application example. The processing results are from the developed ANF monitoring system. It shows the current diagnostic indicator values of each gear, as well as the comparator values. To facilitate monitoring, the green color is used to specify the "healthy state", whereas yellow and red colors indicate "possibly damaged" and "damaged" conditions, respectively.

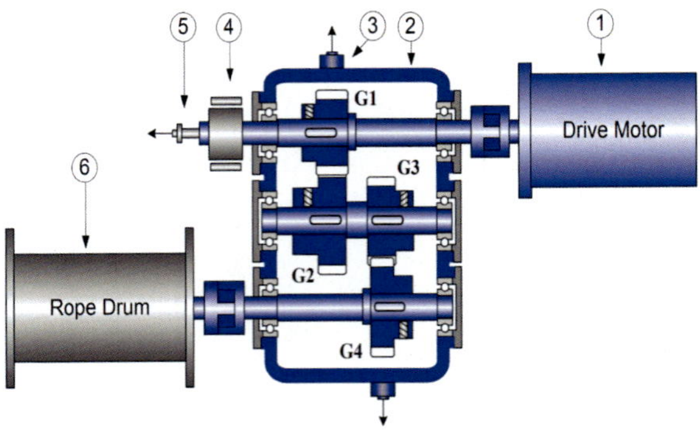

Figure 6. A hoist drive unit: (1) drive motor, (2) gearbox, (3) smart vibration sensors, (4) brake clutch, (5) encoder, (6) rope drum.

	Current Values	Comp Values
Gear 1 :	0.414	0.419
Gear 2 :	0.365	0.371
Gear 3 :	0.512	0.530
Gear 4 :	0.881	0.867

Figure 7. An interface example of the developed intelligent motor.

CONCLUSION

A new ANF monitoring system is developed in this work for health condition monitoring of tower cranes, especially for hoist gearboxes. Vibration signals are measured using wireless smart sensors. Fault detection is performed gear-by-gear. A new ANF diagnostic classifier is developed to integrate strengths of several signal processing techniques for more accurate fault diagnosis. A novel hybrid training method based on a recursive L-M technique is proposed to improve processing efficiency by reducing the matrix size, and enhance the training convergence. The effectiveness of the proposed diagnostic techniques has been verified by experimental tests corresponding to different gear and operating conditions. Test results have demonstrated that the developed ANF monitoring system can provide more reliable fault diagnostics of the gear systems. It has also been implemented for real tower crane hoist drive system monitoring applications.

ACKNOWLEDGEMENTS

This work was supported in part by Natural Sciences and Engineering Research Council of Canada (NSERC), Bare Point Water Treatment Plant, and eMech Systems Inc.

REFERENCES

1. Shaoiro, L.K. and Shapiro, J.P. (2010) Cranes and Derricks. 4th Edition, McGraw-Hill, New York.
2. Zhao, C., Zhang, J., Zhong, X., Chen, S. and Liu, X. (2012) Analysis of Tower Crane Monitoring and Life Prediction. Journal of Adaptive & Dynamic Computing, 2, 12-16. https://doi.org/10.5813/www.ieit-web.org/IJADC/2012.2.3
3. Raviv, G., Shapira, A. and Fishbain, B. (2017) AHP-Based Analysis of the Risk Potential of Safety Incidents: Case Study of Cranes in the Construction Industry. Safety Science, 91, 298-309. https://doi.org/10.1016/j.ssci.2016.08.027
4. Ni, Y., Li, B., Lam, K., Zhu, D., Wang, Y., Lynch, J. and Law, K. (2011) In-Construction Vibration Monitoring of a Super-Tall Structure Using a Long-Range Wireless Sensing System. Smart Structures and Systems, 7, 83-102. https://doi.org/10.12989/sss.2011.7.2.083
5. Ministry of Labor (2011) Safety of Tower Cranes in Construction.
6. Qu, Y., He, D., Yoon, J., Hecke, B., Bechhoefer, E. and Zhu, J. (2017) Gearbox Tooth Cut Fault Diagnostics Using Acoustic Emission and Vibration Sensors—A Comparative Study. Archives of Computational Methods in Engineering, 24, 543-556.
7. Randall, R. (2011) Vibration-Based Condition Monitoring. Wiley, New York. https://doi.org/10.1002/9780470977668
8. Hong, L. and Dhupia, J. (2014) A Time Domain Approach to Diagnose Gearbox Fault Based on Measured Vibration Signals. Journal of Sound and Vibration, 333, 2164-2180. https://doi.org/10.1016/j.jsv.2013.11.033
9. Jardine A. and Banjevic, D. (2008) Condition Monitoring. In: Encyclopedia of Quantitative Risk Assessment, John Wiley & Sons, Hoboken, 341-346. https://doi.org/10.1002/9780470061596.risk0502
10. Vachtsevanos, G., Lewis, F., Roemer, M., Hess, A. and Wu, B. (2006) Intelligent Fault Diagnosis and Prognosis for Engineering Systems. Wile, Hoboken. https://doi.org/10.1002/9780470117842
11. Jing, L., Zhao, M., Li, P. and Xu, X. (2017) A Convolutional Neural Network Based Feature Learning and Fault Diagnosis Method for the Condition Monitoring of Gearbox. Measurement, 111, 1-10. https://doi.org/10.1016/j.measurement.2017.07.017

12. Cerrada, M., Sánchez, R., Pacheco, F., Cabrera, D., Zurita, G. and Li, C. (2016) Hierarchical Feature Selection Based on Relative Dependency for Gear Fault Diagnosis. Applied Intelligence, 44, 687-703. https://doi.org/10.1007/s10489-015-0725-3
13. Wang W. and Kanneg, D. (2009) An Integrated Classifier for Condition Monitoring in Transmission Systems. Mechanical Systems and Signal Processing, 23, 1298-1312. https://doi.org/10.1016/j.ymssp.2008.10.006
14. Modi V. and Wang, W. (2018) A Constrained Evolving Classifier for Gearbox Health Monitoring. Journal of Mechanical and Mechatronics Engineering, 18, 1-7.
15. Wang, W. (2008) An Enhanced Diagnostic System for Gear System Monitoring. IEEE Transactions on Systems, Man and Cybernetics, Part B: Cybernetics, 38, 102-112. https://doi.org/10.1109/TSMCB.2007.908864
16. Wang, W. and Jianu, O. (2010) A Smart Monitor for Machinery Fault Detection. IEEE/ASME Transactions on Mechatronics, 15, 70-78. https://doi.org/10.1109/TMECH.2009.2016956
17. Luong, P. and Wang, W. (2019) Smart Sensor-Based Synergistic Analysis for Rotor Bar Fault Detection of Induction Motors. IEEE/ASME Transactions on Mechatronics. (In Press)
18. Wang, W. and Vrbanek, J. (2008) An Evolving Fuzzy Predictor for Industrial Applications. IEEE Transactions on Fuzzy Systems, 16, 1439-1449. https://doi.org/10.1109/TFUZZ.2008.925918
19. Osman, S. and Wang, W. (2018) A Leakage-Free Resonance Sparse Decomposition Technique for Bearing Fault Detection in Gearboxes. Measurement Science and Technology, 29, Article ID: 035004. https://doi.org/10.1088/1361-6501/aa9d5a

CHAPTER 14

IIoT-Based Intelligent Process Control for Crude Oil Separation: Investigating the Impact of Model-Based Control and Genetic Algorithms

Ali S. Allahloh[1], Mohammad Sarfraz[1], Atef M. Ghaleb[2], and Sobhi Mejjaouli[2]

[1]Department of Electrical Engineering, Aligarh Muslim University, Aligarh, India

[2]Industrial Engineering Department, College of Engineering, Alfaisal University, 11533 Riyadh, Saudi Arabia

ABSTRACT

Integrating the Internet of Things (IIoT) with intelligent process control can improve the performance and efficiency of various industrial processes.

Citation: A. S. Allahloh, M. Sarfraz, A. M. Ghaleb, S. Mejjaouli, "IIoT-Based Intelligent Process Control for Crude Oil Separation: Investigating the Impact of Model-Based Control and Genetic Algorithms", Journal of Sensors, vol. 2023, Article ID 1909835, 20 pages, 2023. https://doi.org/10.1155/2023/1909835.

However, there is a need for continued research on integrating IIoT-based control with advanced techniques such as model-based control and genetic algorithms to fully realize these benefits. This study investigates the impact of such integration on the performance of crude oil separation processes. To this end, we implemented IIoT-based intelligent process control with model-based control and genetic algorithms using specific IIoT devices and platforms, such as SCADAPACK 535E and FactoryTalk. The results of our implementation show that this technique can significantly improve the quality of crude oil separation, with an increase from 95% to 99% over two months of production. Overall, our findings suggest that integrating IIoT-based intelligent process control with advanced techniques such as model-based control and genetic algorithms can significantly improve the performance of crude oil separation processes and may have similar benefits in other industrial processes. Further research is needed to fully understand these techniques' potential in different contexts and identify best practices for their implementation.

INTRODUCTION

Integrating the Internet of Things (IIoT) with intelligent process control has been a topic of increasing interest in recent years, as the use of IIoT technologies has grown in various industrial and other applications. The Industrial Internet of Things (IIoT) is a crucial aspect of Industry 4.0 and Industry 5.0, which are the latest industrial revolutions driven by technology and value, respectively. The IIoT is aimed at improving manufacturing processes by connecting various devices and systems through the Internet, enabling them to communicate and share data. This has led to the creation of smart factories, where machines and systems are able to work together seamlessly and make decisions based on real-time data. However, there are still several gaps in the existing knowledge on the use of IIoT in intelligent process control that need to be addressed through further research, such as the lack of research on the integration of IIoT-based intelligent process control with other advanced control techniques, such as model-based control and genetic algorithms. There are also challenges to be addressed in implementing IIoT in Industry 5.0, such as data privacy and security, standardization of communication protocols, and integration with legacy systems. The specific research question for this study is this: How does integrating IIoT-based intelligent process control with advanced techniques such as model-based control and genetic algorithms impact the performance of crude oil separation processes? This research is aimed at investigating the potential benefits of integrating IIoT-based intelligent process control with model-based control and genetic algorithms using specific IIoT devices and platforms, such as SCADAPACK 535E and FactoryTalk, in the context of crude oil separation processes. The potential benefits of using these advanced control techniques include the ability to tune PID controllers using genetic algorithms, which can reduce the time required to collect the data needed to train a machine learning model. This approach allows for the development of an AI-based prediction model that is based on a larger and more diverse dataset than would be possible to collect manually, which can improve the accuracy and efficiency of process control. The genetic algorithm is used to optimize the parameters of a PID (proportional-integral-derivative) controller in order to improve the control of a process. The prediction model can then use this dataset to make more accurate predictions and improve control of the process over time. IIoT also plays a key role in Industry 5.0, which emphasizes the importance of human creativity and collaboration with machines. The goal of Industry 5.0 is to create highly customized products and increase efficiency in production with the help of advanced technologies such as edge computing, digital twins, collaborative robots, blockchain, and 6G networks. The integration of AI and machine learning technologies in IIoT is also a promising area for future research.

Finally, the IIoT plays a crucial role in the current state of the art for Industry 4.0 and Industry 5.0 and will continue to shape the future of manufacturing and production processes. The integration of IIoT-based intelligent process control with other advanced control techniques, such as model-based control and genetic algorithms, is a promising area for research, with the potential to improve the performance and efficiency of process control systems in the context of crude oil separation processes.

LITERATURE SURVEY

This literature review on Industry 5.0 and the Industrial Internet of Things (IIoT) highlights the key features and technologies of Industry 5.0, such as edge computing, digital twins, collaborative robots, the Internet of Things, blockchain, and 6G and beyond networks. The authors of these studies emphasize the potential applications of Industry 5.0 in the manufacturing industry and how it can transition from traditional to digital systems, while solving the challenges faced by the industry. The literature provides an overview of the future of manufacturing and the required skills, as well as the integration of traditional automation systems with newer technologies in the IIoT. Other notable studies in the field of IIoT include the examination of the future of control and operations, the development of resource service models, the evaluation of IIoT protocols, and the application of IIoT monitoring solutions for advanced predictive maintenance. The literature also highlights the importance of lightweight authentication protocols, monitoring of gas and oil fields using the IIoT, wireless technology trade-offs for the IIoT, machine learning control for nonlinear dynamics and turbulence, progress and challenges in closed-loop turbulence control, and adaptive fuzzy neural network control methods for wastewater treatment processes. The authors of these papers aim to provide insight into the potential applications and supporting technologies of Industry 5.0. Industry 5.0 is seen as the future of manufacturing and is characterized by being value-driven, as opposed to technology-driven like Industry 4.0. The papers outline key features and technologies of Industry 5.0 such as edge computing, digital twins, collaborative robots, the Internet of Things, blockchain, and 6G and beyond networks [1, 2]. The authors of these papers highlight the potential applications of Industry 5.0 in the manufacturing industry, and how it can help to transition from traditional to digital systems. They also discuss the challenges faced by manufacturing industries and how Industry 5.0 technologies can help to solve these problems. The papers provide an overview of the future of manufacturing and the skills that will be required in this new era, emphasizing the importance of humans and robots working together in the manufacturing process [3, 4]. The papers also provide a comprehensive overview of Industry 4.0 and Industry 5.0 and will be useful for those interested in the development and implementation of Industry 4.0 technologies. The authors of these papers encourage further discussion and debate around these topics [5]. One of the papers presents a trust model for assessing the trustworthiness of Industrial Internet of Things (IIoT) devices based on the neutrosophic weighted product method (WPM). The model uses spatial knowledge, temporal experience, and behavioral pattern information from the IIoT devices to calculate their trust scores. The proposed model provides a promising solution for the security and reliability concerns in IIoT [6]. Another paper presents a novel approach called ASTREAM for anomaly detection in data streams in an IIoT environment. The approach combines sliding window, model update, and change detection strategies with LSHiForest to achieve accurate and efficient anomaly detection with better scalability. The approach is evaluated on the KDDCUP99 dataset, and the results show that it outperforms baselines in terms of accuracy and efficiency

[7]. A third paper introduces the concept of advanced machine-metameric dimension (AmD) to analyze the efficacy, efficiency, and effectiveness of updated Heterogeneous Industrial Internet of Things (HetIoT) machines. The authors propose a HetIoT framework for real-time monitoring of the machines' metameric dimension, resulting in higher precision, recall, F1-score, and purity. The proposed framework has the potential to play a crucial role in industrial management [8].

One of the key challenges in the IIoT is the integration of traditional automation systems with newer technologies such as wireless sensor networks and artificial intelligence. Kutasi [9] discusses the use of IIoT capabilities in the hollow cathode plasma nitriding process, highlighting the potential for improved process control and efficiency. Sasajima et al. [10] examine the latest trends in standardization in industrial automation, particularly with regard to the IIoT.

Security is another important consideration in the IIoT, as the integration of new technologies brings with it the risk of cyberattacks. Eden et al. [11] discuss the forensic analysis of SCADA systems within the IIoT, while Huang [12] investigates the use of artificial intelligence algorithms for enhanced security in industrial process control. Gutierrez-Guerrero and Holgado-Terriza [13] propose an automatic configuration method for OPC UA in IIoT environments to improve security and interoperability.

Other notable studies in the field of IIoT include the examination of the future of control and operations in this new era [14], the development of a resource service model for IIoT systems [15], the evaluation of a novel IIoT protocol on a 5G industrial network [16], and the application of IIoT monitoring solutions for advanced predictive maintenance [17]. Overall, the literature on IIoT is diverse and multifaceted, covering a wide range of topics including automation, security, interoperability, and predictive maintenance. As the field continues to grow and evolve, it is likely that we will see even more innovative research in the coming years. Lara et al. [18] present a lightweight authentication protocol for machine-to-machine communications of resource-constrained devices in the industrial internet of things. Aalsalem et al. [19] present a method for monitoring gas and oil fields using reliable wireless sensing and the internet of things. Seferagić et al. [20] presents a survey on wireless technology trade-offs for the industrial internet of things. Duriez et al. [21] discuss machine learning control for taming nonlinear dynamics and turbulence. Brunton and Noack [22] discuss progress and challenges in closed-loop turbulence control. Qiao et al. [23]

present an adaptive fuzzy neural network control method for wastewater treatment processes with multiobjective operation. Zeghlache and Amardjia [24] present a real-time implementation of a nonlinear observer-based fuzzy sliding mode controller for a twin rotor multi-input multioutput system. Bodyanskiy et al. [25] present an adaptive multivariate hybrid neuro-fuzzy system with on-board fast learning. Deepa and Baranilingesan [26] present an optimized deep learning neural network predictive controller for continuous stirred tank reactors. Sabahi et al. [27] present an indirect predictive type-2 fuzzy neural network controller for nonlinear input-delay systems. Patel and Kumar [28] present an artificial neuro-fuzzy logic PID controller based on a BF-PSO algorithm. Tavakoli et al. [29] present a self-constructing fuzzy neural network controller for damping power system oscillations. Kayacan et al. [30] present a self-learning disturbance observer for nonlinear systems in a feedback-error learning scheme. de Jesús Rubio [31] presents the uniform stable neuro-fuzzy inference system. Talkhoncheh et al. [32] present an observer-based adaptive neural network controller for uncertain nonlinear systems with unknown control directions, subject to input time delay and saturation. Karkevandi-Talkhooncheh et al. [33] present an adaptive neuro-fuzzy controller for a class of nonlinear systems with unknown time-varying uncertainties and external disturbances.

DESIGN AND CONFIGURATION

Proposed System

The proposed control system for the crude oil separation facility in this study consisted of several key components, including the following:

- **SCADAPACK 535E devices**: these were used as IIoT sensors and installed at key points in the crude oil separation process to collect data on process variables such as temperature, pressure, and flow rate
- **FactoryTalk**: this was used as the IIoT platform to monitor the process in real-time and to tune the PID controllers using machine learning based on past datasets that included the values of PID parameters set by experts manually
- **Mathematical model of the crude oil separation process**: this was developed as part of the model-based control approach and was used to optimize the PID controller settings using genetic algorithms

- **Genetic algorithms**: these were used to find the optimal values of the PID controller parameters that minimized a cost function defined based on performance criteria such as process stability and efficiency
- **Statistical analysis software**: this was used to analyze the data collected on process variables and the quality of the crude oil separation, as well as on the efficiency of the data collection process
- **Artificial intelligence (AI) model**: this was built using machine learning tools in a program called ML.NET, which is run in a visual studio dot net environment. The model employed a deep learning neural network technique to make predictions about the optimal parameters for the loop controller

The overall architecture of the proposed control system involved collecting data on the crude oil separation process using the SCADAPACK 535E devices, transmitting the data to the FactoryTalk platform, and then using the data to tune the PID controllers and build the AI model. The genetic algorithms were used to optimize the PID controller parameters, and the statistical analysis software was used to assessment the impact of the advanced control techniques on the performance of the crude oil separation. The AI model was built using the deep learning neural network technique to make predictions about the optimal parameters for the loop controller, and the system was designed to overcome the challenge of having insufficient data to train the model by using the IIoT to collect and prepare data for model training.

The proposed control system in this study is an IIoT-based intelligent process control that integrates model-based control and genetic algorithms, which was implemented at a crude oil separation facility. The system has several advantages that provide real-time monitoring and improved control performance, leading to accurate predictions of loop controller parameters. The SCADAPACK 535E devices are used to collect process data which is transmitted to the FactoryTalk platform, allowing for real-time monitoring of the process. By using model-based control and genetic algorithms, the system optimizes the PID controller settings, leading to improved process stability and efficiency. Additionally, the AI model built using the deep learning neural network technique and the IIoT data collection enables accurate predictions of loop controller parameters. However, the implementation of this control system also comes with several disadvantages, including complexity and cost. Developing a mathematical model of the crude oil separation process and training an AI model can be a complex and time-consuming task. The success of the system also relies on the quality of the data collected by the SCADAPACK 535E devices, so proper data quality assessment is necessary. Implementing and maintaining this control system require significant investment in technology and resources.

Implementation

Figure 1 depicts the general architecture for implementing the proposed system. The research design for this study was a case study, in which we implemented IIoT-based intelligent process control with model-based control and genetic algorithms at a crude oil separation facility. We selected this research design because it allowed us to investigate the impact of these advanced control techniques on real-world processes in a controlled environment.

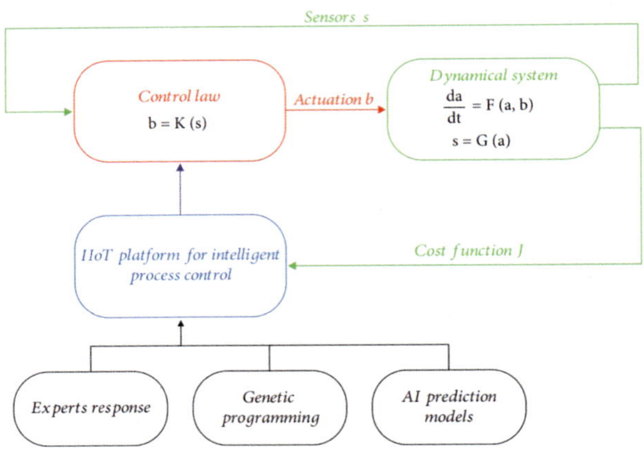

Figure 1. IIoT-based machine learning control.

To implement the IIoT-based intelligent process control, we used SCADAPACK 535E devices as IIoT sensors and FactoryTalk as the IIoT platform. The SCADAPACK 535E devices were installed at key points in the crude oil separation process and were used to collect data on process variables such as temperature, pressure, and flow rate. The data collected by the SCADAPACK 535E devices was transmitted to the FactoryTalk platform, which was used to monitor the process in real-time and to tune the PID controllers using machine learning based on past datasets that included the values of PID parameters set by experts manually.

In addition to the IIoT-based intelligent process control, we also implemented model-based control and genetic algorithms at the crude oil separation facility. The model-based control approach involved developing a mathematical model of the crude oil separation process, which was used to optimize the PID controller settings using genetic algorithms. The genetic algorithms were used to find the optimal values of the PID controller parameters that minimized a cost function, which was defined based on performance criteria such as process stability and efficiency.

To assess the impact of the IIoT-based intelligent process control with model-based control and genetic algorithms on the performance of the crude oil separation process, we collected data on process variables such as temperature, pressure, density, and flow rate over a two-month period of production. We also collected data on the quality of the crude oil separation, as well as on the efficiency of the data collection process. The data were analyzed using statistical analysis software to assess the impact of the advanced control techniques on the performance of the crude oil separation.

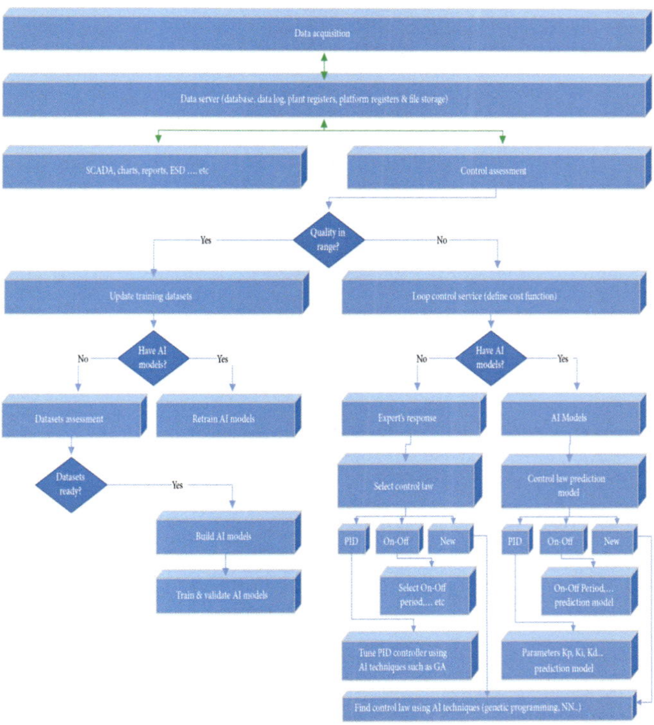

Figure 2. Proposed IIoT platform for intelligent process control flowchart.

Flowcharts in Figures 2 and 3 declare that building an artificial intelligence (AI) model to predict the optimal parameters for a loop controller is a challenging task, as there is no data to train the model on. To address this challenge, the flowchart proposes the use of the industrial internet of things (IIoT) to record and prepare data for model training. This data is then subjected to a quality assessment to ensure that it is suitable for training the model and will result in the most accurate predictions.

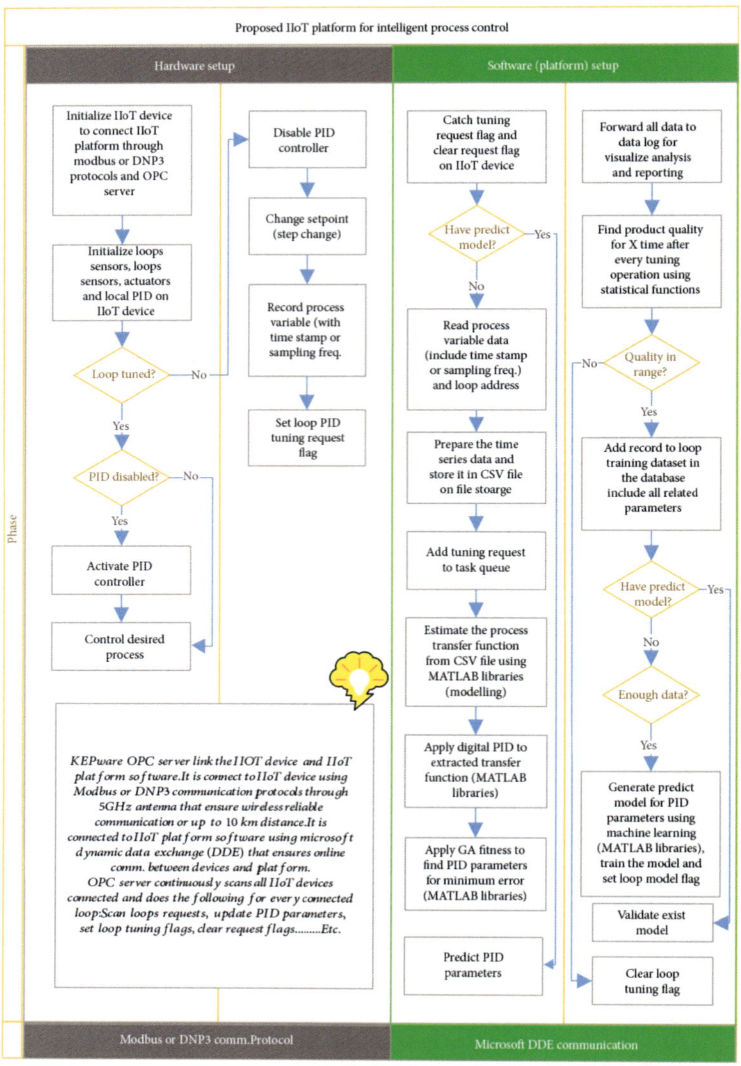

Figure 3. Proposed IIoT platform for intelligent process control.

The AI model is then built using machine learning tools in a program called ML.NET, which is run in a visual studio dot net environment. The model employs a deep learning neural network technique, which is a type of machine learning that involves training a model on large amounts of data in order to recognize patterns and make decisions. This technique is well suited for building highly accurate models and is likely to be effective in predicting the optimal parameters for a loop controller. Overall, the purpose of this process, as declared by the flowcharts, is to build an AI model that can accurately predict the optimal parameters for a loop controller in an industrial setting, using the IIoT to collect and prepare data for model training and the deep learning neural network technique to build the model. By following this process, it is possible to overcome the challenge of having insufficient data to train the model and build a highly accurate AI model for predicting loop controller parameters.

Process Model

The pneumatic control valve regulates and controls various types of processes such as level control in the oil separator, as shown in Figure 4.

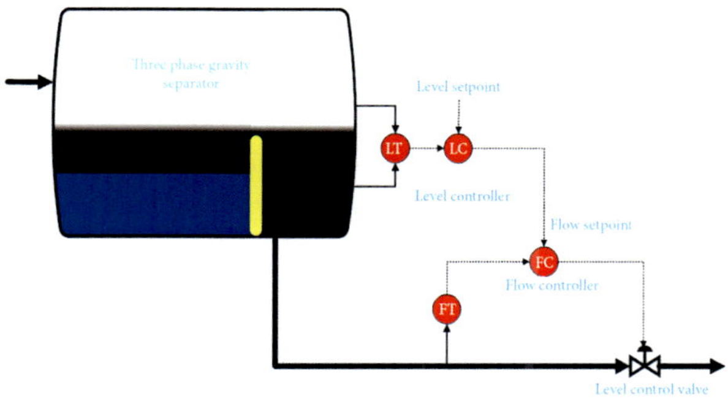

Figure 4. Selected process to control.

The nature of the valve is spring-mass dynamics, and the equation described the time series model of it is:

$$y(t) = G_P\left(1 - e^{\zeta^2\omega_n t}\left[\cos\left(\omega_n\zeta t\sqrt{1-\zeta^2}\right) + \frac{\zeta}{\sqrt{1-\zeta^2}}\sin\left(\omega_n\zeta t\sqrt{1-\zeta^2}\right)\right]\right). \tag{1}$$

Digital PID Controller

The PID in continues form is express by equation as

$$c = K_c\left(e + \frac{1}{T_I}\int edt + T_D\frac{de}{dt}\right). \tag{2}$$

Then convert the continues form to discrete form using sampling interval ΔT as follows:

i. The error e will be e_n

ii. The integral part is nothing but summation of errors

$$\frac{\Delta T}{T_I}\sum e_n \tag{3}$$

iii The derivative is deference between past and current error

$$\frac{T_D}{\Delta T}(e_n - e_{n-1}) \tag{4}$$

Putting all three parts together, the digital algorithm equation is.

$$c_n = K_c\left[e_n + \frac{\Delta T}{T_I}\sum e_n + \frac{T_D}{\Delta T}(e_n - e_{n-1})\right], \tag{5}$$

where K_c, T_I, and T_D are the PID tuning parameters: gain, reset time in sec/repeat, and derivative time in seconds, respectively. For the independent form of PID or parallel form, K_p, K_I, and K_D gains proportional, integral, and derivative, respectively, will be

$$K_P = K_c, K_I = \frac{K_c}{T_I},$$

$$K_D = K_c \times T_D. \tag{6}$$

And the parallel form of PID will express in equation as

$$c_n = K_P e_n + K_I \Delta T \sum e_n + K_D \frac{(e_n - e_{n-1})}{\Delta T}. \tag{7}$$

Since it is required to control the position of the valve, the calculation algorithm programming form upgraded from Equations (5) and (7) to P_n at nth sampling instant as

$$P_n = K_P(S - V) + K_I \Delta T \sum_0^n (S - V) + K_D \frac{\Delta(S - V)}{\Delta T} + P_m, \tag{8}$$

where P_m is a median valve position, S is setpoint, and V is variable. Because of the noisy data sampling, the derivative part in Equation (8) has to be chosen carefully using the technique called four-point difference while the constant setpoint is

$$\frac{\Delta e}{\Delta T} = \frac{1}{6\Delta T}(e_n - e_{n-3} + 3e_{n-1} - 3e_{n-2}), \quad (9)$$

where e = (S – V).

Simulation of IIoT Device and Process

The Internet of Things (IIoT) platform needs to be connected to the crude oil separation process through an IIoT device with a PID controller in order to ensure connectivity between the platform and the process.

It is important to test the IIoT platform before implementing it in the actual process in order to ensure its functionality. One solution for this is to use a virtual or simulated process, which can be achieved using the Proteus software.

The Proteus software includes a range of tools and libraries that can be used to simulate the crude oil separation process and IIoT devices, such as Laplace primitives, Microchip PIC microcontrollers, Arduino microcontrollers, Ethernet controllers, and an oscilloscope. The Laplace primitives require the damping value ζ and natural frequency ω_n in order to simulate the process.

The IIoT device is responsible for sensing and actuating the crude oil separation process by collecting data from sensors and transmitting it to the IIoT platform. The platform then uses this data and the algorithms of intelligent controllers to execute control commands based on the optimal values received in order to achieve the desired control actions.

Figure 5 shows the second-order tools in the Proteus software that are used to simulate the real crude oil separation process based on the damping ratio and natural frequency.

Figure 6 shows the implementation of an IIoT device that serves the crude oil separation process model, which is driven by an industrial IoT device built on an Arduino Mega 2560. The IIoT device includes a digital-to-analog converter (DAC104S085) and a virtual Ethernet network, using an ENC28J60 Ethernet controller and the Modbus TCP protocol.

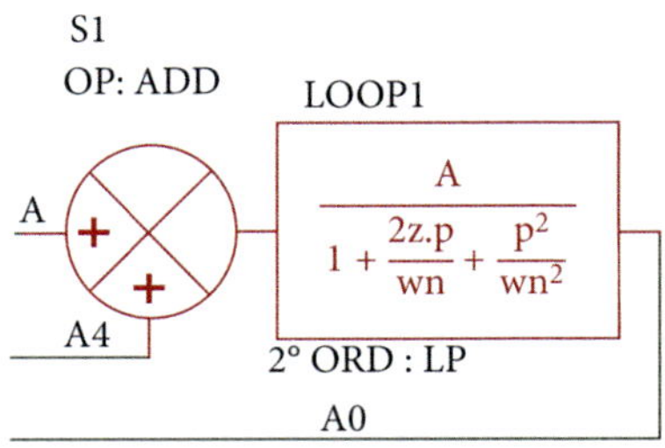

Figure 5. Second-order process simulation tools in the Proteus software.

The entire system is simulated in the Proteus software, with various features such as the ability to start or stop the PID controller, control four loops (only one at a time), select the type of tuning (remote using a genetic algorithm or local using potentiometers), set setpoints using potentiometers for each loop, monitor the network using the Proteus virtual terminal tool (UART), and monitor the process with and without a PID controller using the Proteus oscilloscope tool.

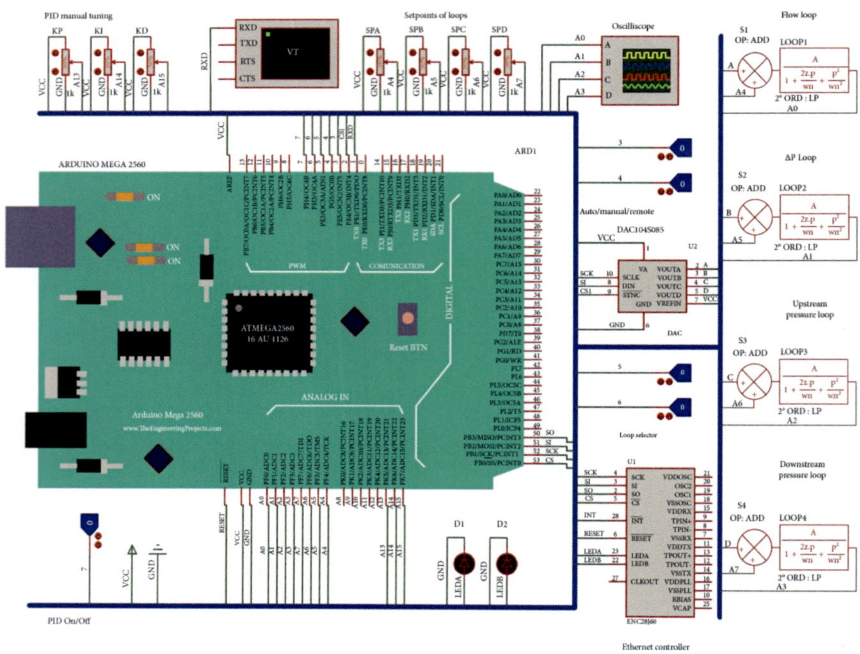

Figure 6. IIoT device and process simulation in Proteus.

The experiment in this paper is divided into five parts:

(1) Plant selection to control the second-order system with a different damping ratio ζ and natural frequency ω_n

(2) Apply the digital PID written on Arduino and tune it manually using potentiometers

(3) Add the Ethernet connectivity to the IIoT device to work as a client and test the connection with the server

(4) Transmit the process's step's response time series data and sampling time to the workstation through Ethernet network

(5) Execute the program on a work station that reads the process response, estimate the model and apply the genetic algorithm to the model, and generate K_P, K_I, and K_D, then send it back to Arduino

RESULTS

Manual Tuning Controller

Figures 7–12 show the Proteus software oscilloscope results of applying the manual tuned local PID controller built on Arduino at different sampling frequencies. Analog potentiometers connected to analog inputs of Arduino board simulator in Proteus software were used to change the values of K_P, K_I, K_D, setpoint, and sampling frequency manually. These process steps are as follows:

(1) Set the sampling frequency and setpoint

(2) Disable the PID controller

(3) Change the setpoint

(4) Monitor the process variable on Proteus's oscilloscope

(5) Enable the PID controller

(6) Change the values of K_P, K_I, and K_D

(7) Change the setpoint up and down

(8) Repeat steps (4), (6), and (7)

(9) Change the sampling frequency and repeat steps (4), (6), and (7)

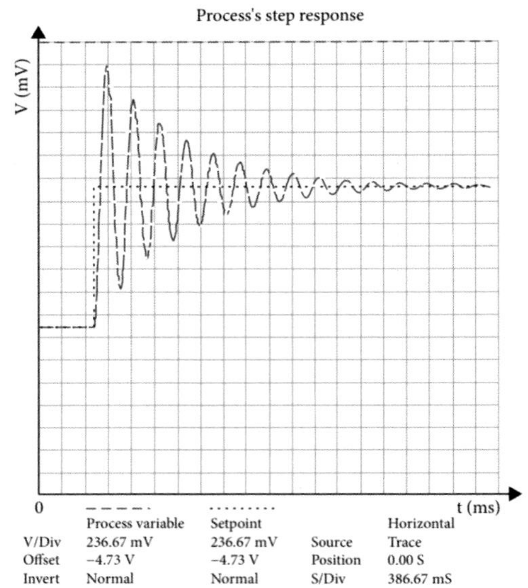

Figure 7. Four-inch short tube valve step's response without controller and setpoint raised up.

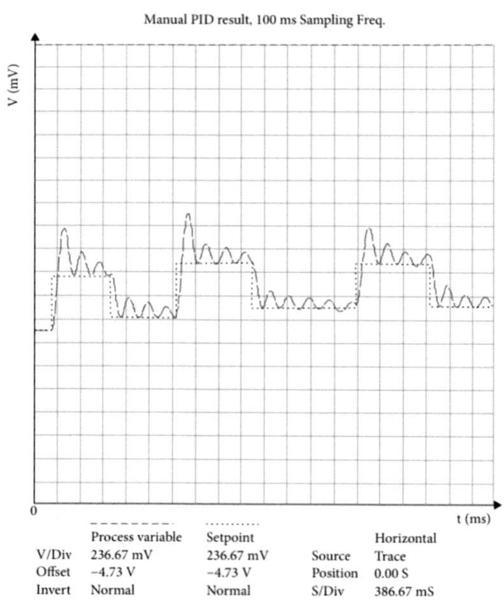

Figure 8. Four-inch short tube valve step's response with manual tuning PID, 100 ms sampling time, and setpoint changed less than 50%.

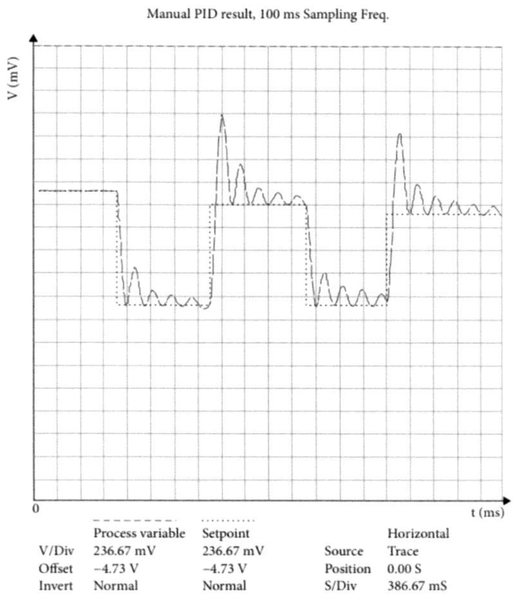

Figure 9. Four-inch short tube valve step's response with manual tuning PID, 100 ms sampling time, and setpoint changed more than 50%.

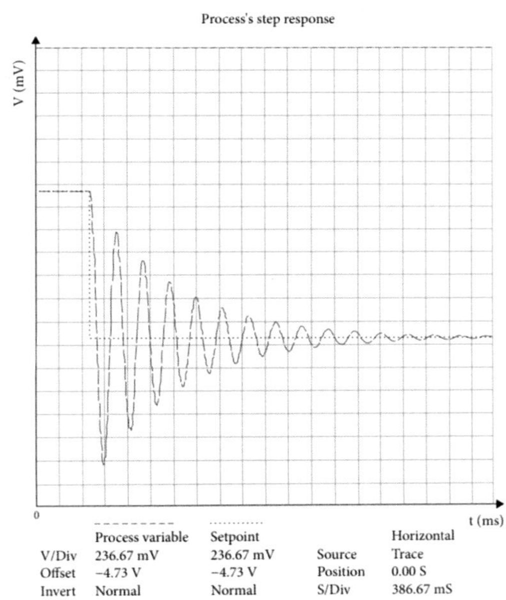

Figure 10. Four-inch short tube valve step's response without controller and setpoint dropped down.

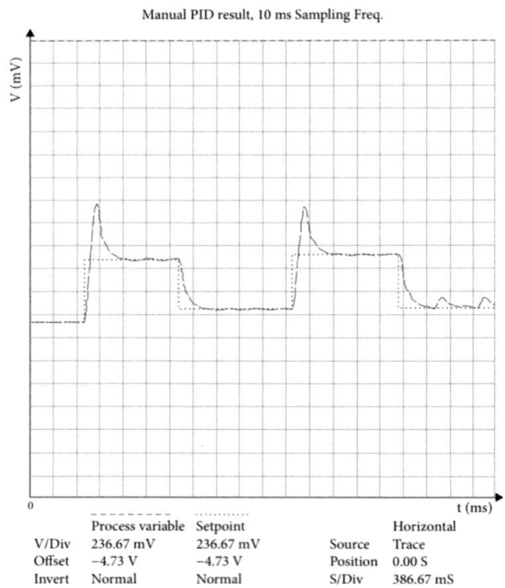

Figure 11. Four-inch short tube valve step's response with manual tuning PID, 10 ms sampling time, and setpoint changed less than 50%.

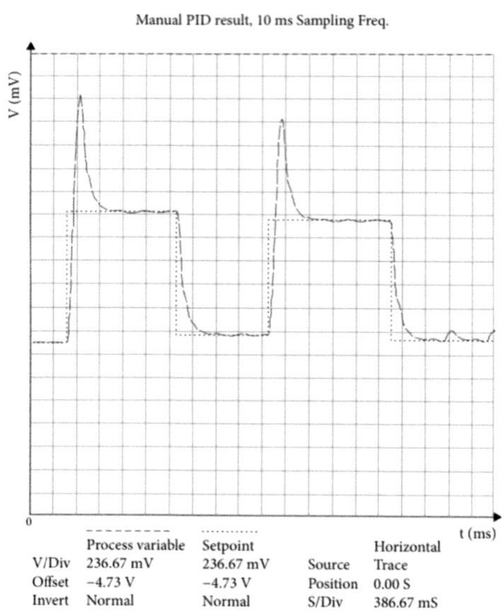

Figure 12. Four-inch short tube valve step's response with manual tuning PID, 10 ms sampling time, and setpoint changed more than 50%.

Process Modeling on Workstation

One of the key functions of the IIoT device in the crude oil separation process is to transfer step response time series data and the corresponding sampling frequency value to the workstation. This data is used to model the process and apply the controller with optimal parameters. In this research, the Arduino simulation in Proteus reads the step response of the crude oil separation process and transfers it, along with the sampling frequency value, to the workstation via the Ethernet network. The workstation receives this data from the IIoT device and stores it in a CSV file using VB.net. This CSV file is then passed to MATLAB, which applies transfer function estimation techniques to the data in order to model the crude oil separation process. The process's response time series data can be seen in Figure 13, which shows the loop CSV file.

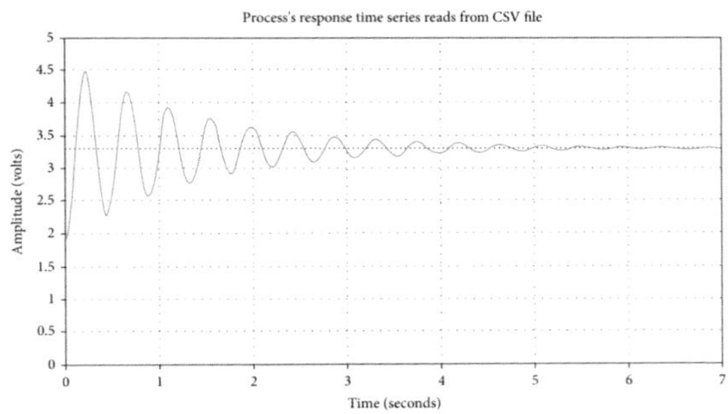

Figure 13. Process's response reads by IIoT platform from CSV file.

The estimated continuous-time identified transfer function of process is

$$\frac{203.9}{s^2 + 1.428s + 203.9}. \tag{10}$$

Optimize the Controller Parameters Using Genetic Algorithm

Optimizing the controller parameters is crucial for achieving the best results in the crude oil separation process. This research uses a genetic algorithm to find effective PID gains as shown in Figure 14 that minimize a cost function. The cost function used in this example is the linear quadratic regulator (LQR) cost function, with Q = 1 and R = 0:001 for a step response of $w_r = 1$. The transfer function of the system being controlled is given by Equation (10).

$$J = \int_0^T Q(w_r - y)^2 + Ru^2 d\tau. \quad (11)$$

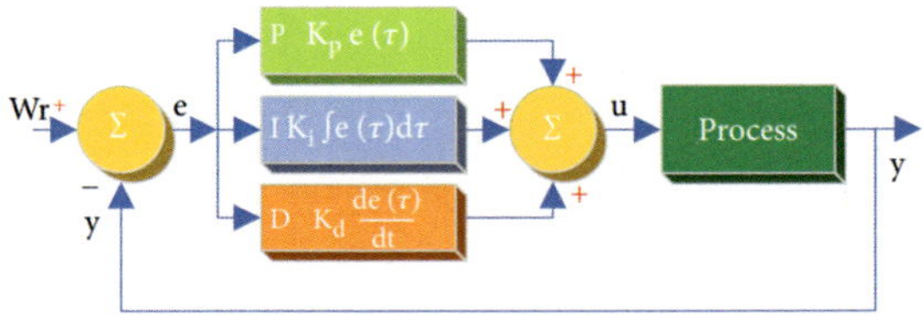

Figure 14. PID controller to define the cost function and optimization.

Figure 15 displays the complete process of utilizing GA to optimize the controller. Once the MATLAB libraries have completed the transfer function estimation, the transfer function is passed back to MATLAB in order to apply the PID controller to the generated model. The optimized values of K_P, K_I, and K_D are then sent back to the Arduino using VB.NET or the Kepware OPC server, using the Modbus TCP communication protocol. The fitness function in MATLAB is used to find the best values for K_P, K_I, and K_D that minimize the error, with a population size of 40 and a maximum of 15 generations. Figure 15 illustrates the role of the IIoT platform in employing the genetic algorithm to optimize the controller parameters in the crude oil separation process. The optimization results can be seen in Figures 16–21.

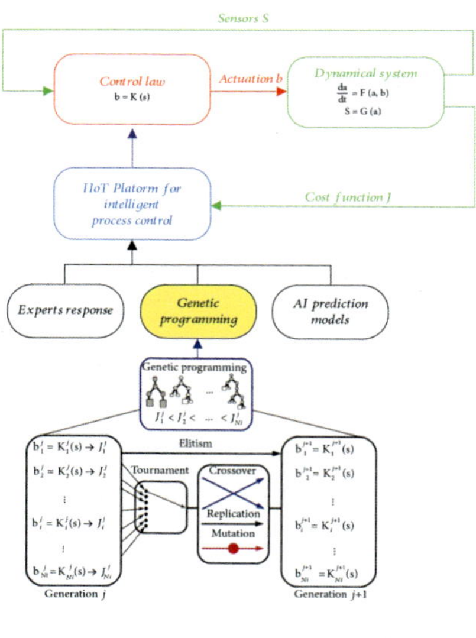

Figure 15. IIoT-based machine learning control using the genetic algorithm.

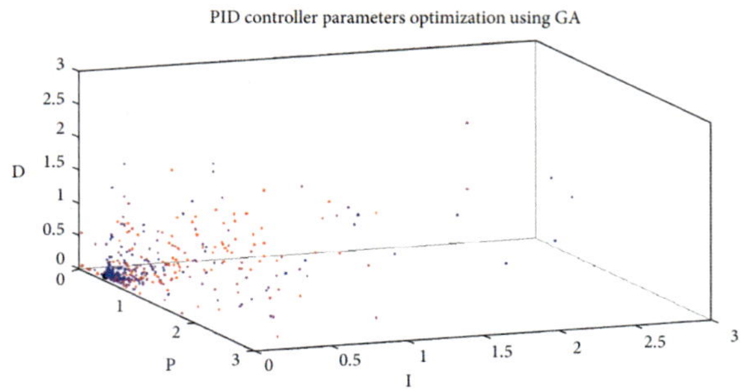

Figure 16. PID controller parameter optimization using genetic algorithm technique.

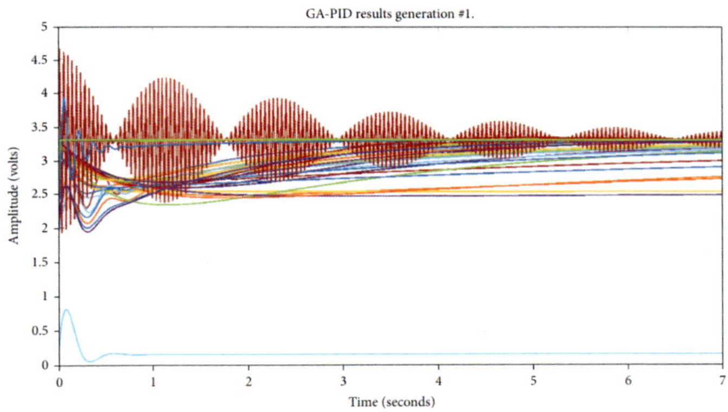

Figure 17. Generation No. 1 result.

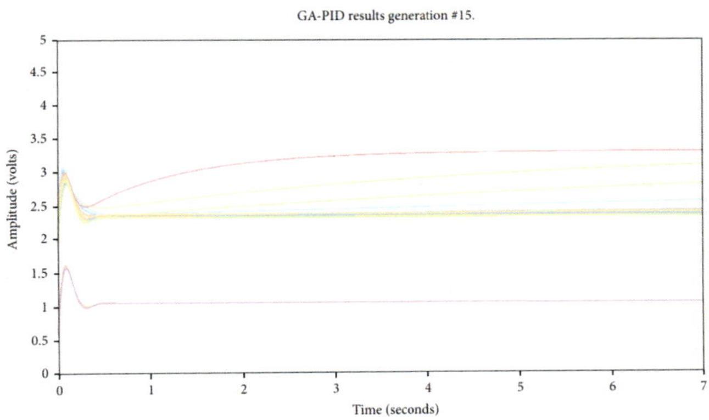

Figure 18. Generation No. 15 result.

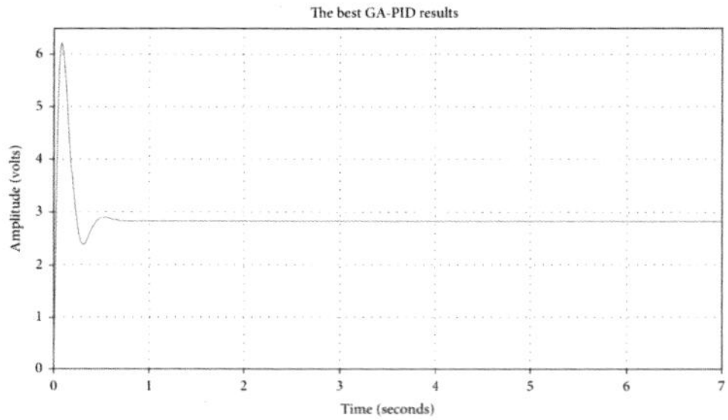

Figure 19. The best of result $K_P = 0{:}4596$, $K_I = 0$, and $K_D = 0{:}0840$.

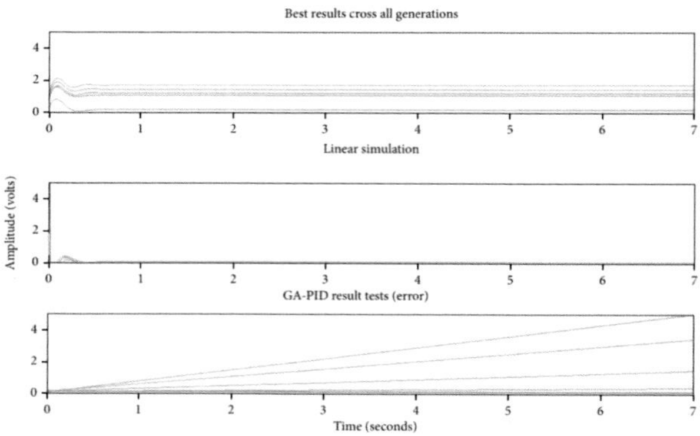

Figure 20. The best of generations result.

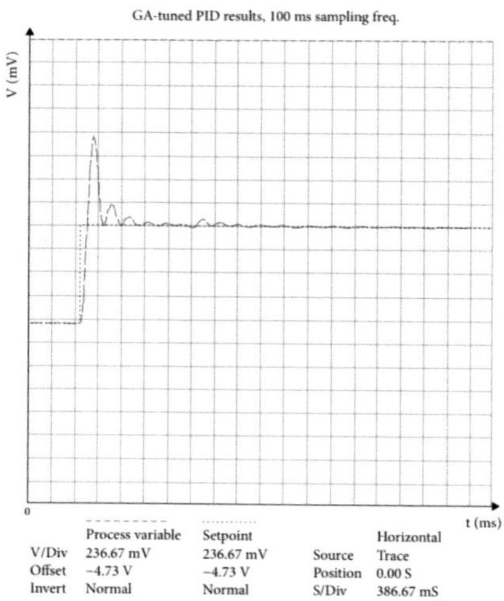

Figure 21. Proteus's oscilloscope results when best result applied as obtained by GA $K_P = 0{:}4596$, $K_I = 0$, and $K_D = 0{:}0840$.

Experimental Results and Discussion

In this research, the SCADAPACK 535 remote controller serves as the IIoT device for the crude oil separation process shown in Figure 22. With its extremely high performance features, including a 600 MHz 32-bit Dual Core Cortex A9 CPU, 128 MB NAND Flash, 128 MB DDR3 Ram, 6 analog inputs, 2 analog outputs, 18 digital inputs, 3 Ethernet ports, Modbus TCP, Modbus 485, DNP3 communication protocol, and 9 digital outputs, the SCADAPACK 535 is exceptionally well suited for control in the crude oil separation facility.

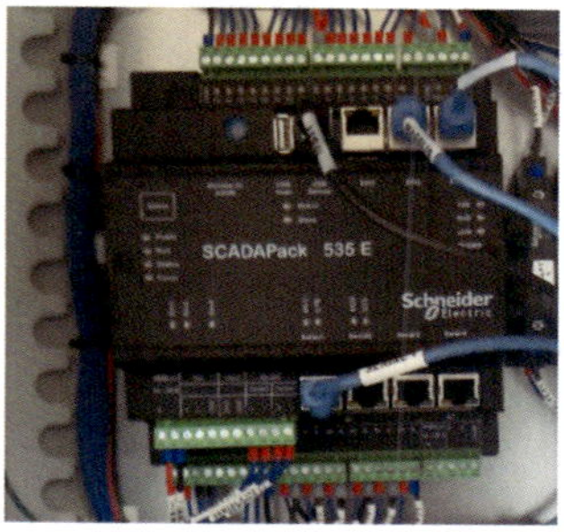

Figure 22. Remote terminal programmable logic controller SCADAPACK 535 IIoT device.

The process variable (level) is measured by a displacer level transmitter, and the target of the PID controller is to maintain the level at the setpoint by manipulating the flow of oil at the outlet. Meanwhile, the density of the oil at the outlet indicates the quality of the separation process. The oil flow and density are measured by an Emerson Micromotion Coriolis flow meter, which transmits this data to the SCADAPACK 535 through Modbus RTU over RS485.

Once installed, the new IIoT device effectively controls the pneumatic valve actuator and regulates the oil flow, as shown in Figure 23. This figure illustrates the behavior of oil flow rate and density for two months before using the new IIoT platform. The density of the oil at the outlet is a key indicator of the separation quality, so a normal distribution statistical approach is used to monitor it. The results show that the mean density is μ = 0:84 g/cm^3 and the standard deviation is σ = 0:02, indicating that 95% of the oil outlet density falls within the range of 0.80 to 0.88.

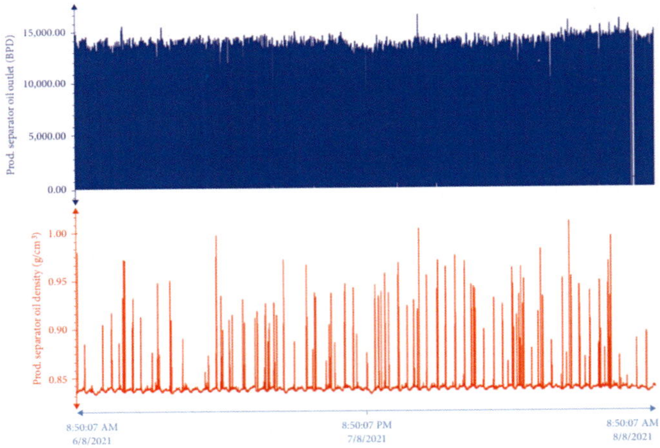

Figure 23. Oil flow and density chart with manual tuning.

The oil phase in the crude oil separator is located between the water and gas phases, which means it can be affected by the water phase. Figure 24 shows the water flow rate and density chart for the same two-month period as Figure 23. The statistical results of the water density are $\mu = 1:06$ g/cm^3 and $\sigma = 0:01$, indicating that 95% of the water outlet density falls within the range of 1.04 to 1.08. This range indicates that the water level controller is performing well.

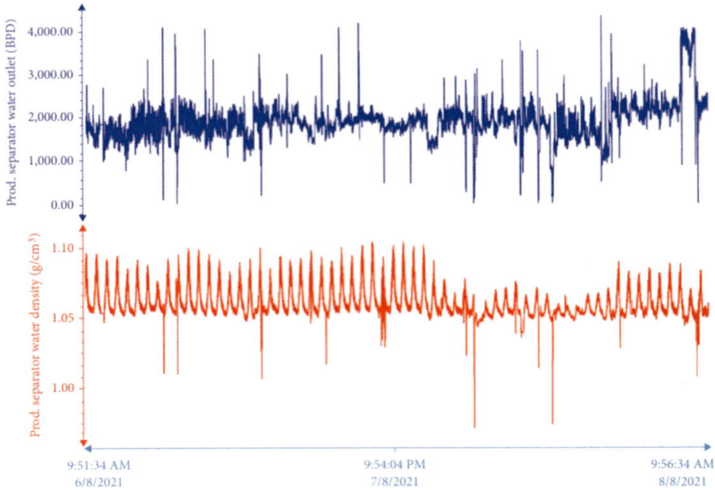

Figure 24. Water flow rate and density chart with manual tuning.

Once the IIoT device is installed and connected to the IIoT platform, the process model is generated from the time series step response, and the PID controller is tuned with the help of MATLAB. Figure 25 shows the chart of oil flow rate and density for two months after using the new IIoT platform. The statistical results show that the mean density is $\mu = 0{:}84$ g/cm^3 and the standard deviation is $\sigma = 0{:}01$, indicating that 95% of the oil outlet density falls within the range of 0.82 to 0.86. This improvement in the separation process demonstrates the effectiveness of the IIoT platform in tuning the PID controller.

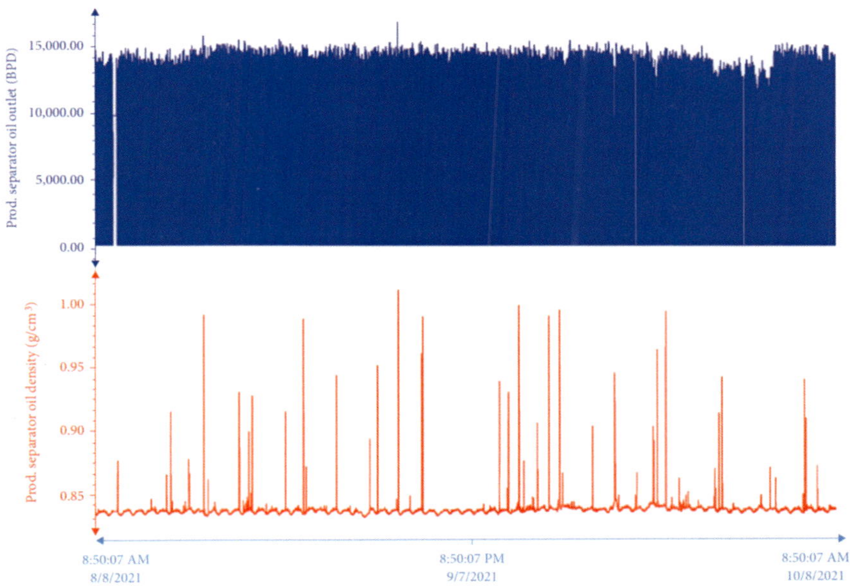

Figure 25. Oil flow rate and density chart with the new IIoT platform.

The improvement in the oil outlet shown in Figure 25 should not affect the water phase. Figure 26 shows the water outlet flow rate and density for the same two-month period as Figure 25. The statistical results show that the mean density is $\mu = 1{:}06$ g/cm^3 and the standard deviation is $\sigma = 0{:}01$, indicating that 95% of the water outlet density falls within the range of 1.04 to 1.08. This demonstrates that there is no impact on the water outlet due to the new IIoT platform.

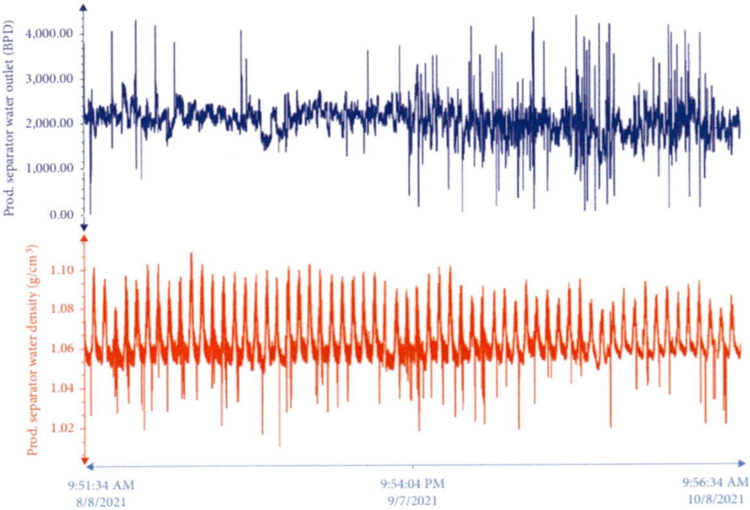

Figure 26. Water flow rate and density chart with new IIoT platform.

The new IIoT platform successfully reduced the standard deviation of the oil outlet density from 0.02 to 0.01. This means that the 95% range of oil density improved from 0.80 to 0.88 g/cm^3 to 0.82 to 0.86 g/cm^3, resulting in a higher quality oil with better API and lower water content. The 99% range of oil density also improved from 0.78 to 0.9 g/cm^3 to 0.81 to 0.87 g/cm^3, indicating a significant improvement in the separation process. However, in both cases, the minimum value of the oil outlet density is 0.83 g/cm^3, indicating that there is no gas present in the oil. The 99% range of oil outlet density is therefore 0.83 to 0.87 g/cm^3. These results demonstrate the effectiveness of the IIoT platform in improving the crude oil separation process.

CONCLUSION

In conclusion, this research has demonstrated the effectiveness of implementing an IIoT platform and advanced control techniques in optimizing the performance of a crude oil separation process. Using the genetic algorithm to tune the PID controller and the machine learning model to predict optimal control parameters improved the separation quality from 95% to 99%. The integration of IIoT technology and advanced control techniques resulted in a reduction of the oil outlet density standard deviation and an overall improvement in the efficiency of the separation process. These findings highlight the potential of IIoT and advanced control techniques to maximize process efficiency and improve the quality of oil production. As we move towards Industry 5.0, it is crucial that we adopt innovative technologies and approaches to improve the performance of industrial processes. Integrating IIoT and advanced control techniques in the crude oil separation process is a step towards realizing Industry 5.0. The development of 6G networks will provide even greater opportunities for real-time monitoring, control, and optimization of industrial processes. This will further improve efficiency, quality, and sustainability in the oil production industry. Finally, this research provides valuable insights into the benefits of combining IIoT and advanced control techniques in optimizing industrial processes. The findings have the potential to drive innovation and drive the adoption of Industry 5.0 and 6G technologies in the oil production industry.

DATA AVAILABILITY

Data supporting figures are not publicly available according to company policy. However, these datasets can be accessed on request from Mr. Ali S. Allahloh, upon the completion of a Data Usage Agreement, according to policies from the Yemen Petroleum Exploration and Production Authority (PEPA).

ACKNOWLEDGMENTS

This work is supported by Alfaisal University, Riyadh, Saudi Arabia.

REFERENCES

1. P. K. Maddikunta, Q. V. Pham, B. Prabadevi et al., “Industry 5.0: a survey on enabling technologies and potential applications,” *Journal of Industrial Information Integration*, vol. 26, article 100257, 2022.
2. N. Fazal, A. Haleem, S. Bahl, M. Javaid, and D. Nandan, “Digital management systems in manufacturing using Industry 5.0 technologies,” in *Advancement in Materials, Manufacturing and Energy Engineering, Vol. II. Lecture Notes in Mechanical Engineering*, P. Verma, O. D. Samuel, T. N. Verma, and G. Dwivedi, Eds., Springer, Singapore, 2022.
3. X. Xu, Y. Lu, B. Vogel-Heuser, and L. Wang, “Industry 4.0 and Industry 5.0Inception, conception and perception,” *Journal of Manufacturing Systems*, vol. 61, pp. 530–535, 2021.
4. R. Sanchez-Iborra and M. D. Cano, “State of the art in LP-WAN solutions for industrial IoT services,” *Sensors*, vol. 16, no. 5, p. 708, 2016.
5. M. A. Khan and K. A. Abuhasel, “An evolutionary multi-hidden Markov model for intelligent threat sensing in industrial internet of things,” *Journal of Supercomputing*, vol. 77, no. 6, pp. 6236–6250, 2021.
6. M. A. Khan and N. S. Alghamdi, “A neutrosophic WPM-based machine learning model for device trust in industrial internet of things,” *Journal of Ambient Intelligence and Humanized Computing*, pp. 1–15, 2021.
7. Y. Yang, X. Yang, M. Heidari et al., “ASTREAM: data-stream-driven scalable anomaly detection with accuracy guarantee in IIoT environment,” *IEEE Transactions on Network Science and Engineering*, 2022.
8. M. A. Khan and K. A. Abuhasel, “Advanced metameric dimension framework for heterogeneous industrial Internet of things,” *Computational Intelligence*, vol. 37, no. 3, pp. 1367–1387, 2021.
9. D. N. Kutasi, “Process control with iiot capabilities of the hollow cathode plasma nitriding,” in *2019 20th International Carpathian Control Conference (ICCC)*, pp. 1–5, Krakow-Wieliczka, Poland, 2019.
10. H. Sasajima, T. Ishikuma, and H. Hayashi, “Notice of removal: future iiot in process automationlatest trends of standardization in industrial automation, iec/tc65,” in *2015 54th Annual Conference of the Society of Instrument and Control Engineers of Japan (SICE)*, pp. 963–967,

Hangzhou, China, 2015.

11. P. Eden, A. Blyth, K. Jones et al., "SCADA system forensic analysis within iiot," in *Cybersecurity for Industry 4.0. Springer Series in Advanced Manufacturing*, L. Thames and D. Schaefer, Eds., Springer, Cham.
12. Z. Huang, "Application of the artificial fish school algorithm and particle filter algorithm in the industrial process control particle filtering algorithm for industrial process control," *Mathematical Problems in Engineering*, vol. 2020, Article ID 3070539, 2020.
13. J. M. Gutierrez-Guerrero and J. A. Holgado-Terriza, "Automatic configuration of opc ua for industrial internet of things environments," *Electronics*, vol. 8, no. 6, p. 600, 2019.
14. I. Harjunkoski, "Future of control and operations in the era of industrial internet of things," *Computer Aided Chemical Engineering*, vol. 40, pp. 2275–2280, 2017.
15. W. Li, B. Wang, J. Sheng, K. Dong, Z. Li, and Y. Hu, "A resource service model in the industrial IoT system based on transparent computing," *Sensors*, vol. 18, no. 4, p. 981, 2018.
16. G. Hatzivasilis, K. Fysarakis, O. Soultatos, I. Askoxylakis, I. Papaefstathiou, and G. Demetriou, "The industrial internet of things as an enabler for a circular economy HY-LP: a novel IIoT protocol, evaluated on a wind parks SDN/NFV-enabled 5g industrial network," *Computer Communications*, vol. 119, pp. 127–137, 2018.
17. F. Civerchia, S. Bocchino, C. Salvadori, E. Rossi, L. Maggiani, and M. Petracca, "Industrial Internet of Things monitoring solution for advanced predictive maintenance applications," *Journal of Industrial Information Integration*, vol. 7, pp. 4–12, 2017.
18. E. Lara, L. Aguilar, M. A. Sanchez, and J. A. García, "Lightweight authentication protocol for m2m communications of resource-constrained devices in industrial internet of things," *Sensors*, vol. 20, no. 2, p. 501, 2020.
19. M. Y. Aalsalem, W. Z. Khan, W. Gharibi, and N. Armi, "An intelligent oil and gas well monitoring system based on internet of things," in *2017 International Conference on Radar, Antenna, Microwave, Electronics, and Telecommunications (ICRAMET)*, pp. 124–127, Jakarta, Indonesia, 2017.

20. A. Seferagić, J. Famaey, E. De Poorter, and J. Hoebeke, "Survey on wireless technology tradeoffs for the industrial internet of things," *Sensors*, vol. 20, no. 2, p. 488, 2020.
21. T. Duriez, S. L. Brunton, and B. R. Noack, *Machine Learning Control-Taming Nonlinear Dynamics and Turbulence*, vol. 116, Springer, 2017.
22. S. L. Brunton and B. R. Noack, "Closed-Loop turbulence control: progress and challenges," *Applied Mechanics Reviews*, vol. 67, no. 5, 2015.
23. J. F. Qiao, Y. Hou, L. Zhang, and H. G. Han, "Adaptive fuzzy neural network control of wastewater treatment process with multiobjective operation," *Neurocomputing*, vol. 275, pp. 383–393, 2018.
24. S. Zeghlache and N. Amardjia, "Real time implementation of non linear observer-based fuzzy sliding mode controller for a twin rotor multi-input multi-output system (TRMS)," *Optik-International Journal for Light and Electron Optics*, vol. 156, pp. 391–407, 2018.
25. Y. Bodyanskiy, O. Vynokurova, G. Setlak, D. Peleshko, and P. Mulesa, "Adaptive multivariate hybrid neuro-fuzzy system and its on-board fast learning," *Neurocomputing*, vol. 230, pp. 409–416, 2017.
26. S. N. Deepa and I. Baranilingesan, "Optimized deep learning neural network predictive controller for continuous stirred tank reactor," *Computers and Electrical Engineering*, vol. 71, pp. 782–797, 2018.
27. K. Sabahi, S. Ghaemi, J. Liu, and M. A. Badamchizadeh, "Indirect predictive type-2 fuzzy neural network controller for a class of nonlinear input - delay systems," *ISA Transactions*, vol. 71, Part 2, pp. 185–195, 2017.
28. R. Patel and V. Kumar, "Artificial neuro fuzzy logic PID controller based on BF-PSO algorithm," *Procedia Computer Science*, vol. 54, pp. 463–471, 2015.
29. A. R. Tavakoli, A. R. Seifi, and M. M. Arefi, "Designing a self-constructing fuzzy neural network controller for damping power system oscillations," *Fuzzy Sets and Systems*, vol. 356, pp. 63–76, 2019.
30. E. Kayacan, J. M. Peschel, and G. Chowdhary, "A self-learning disturbance observer for nonlinear systems in feedback-error learning scheme," *Engineering Applications of Artificial Intelligence*, vol. 62, pp. 276–285, 2017.
31. J. de Jesús Rubio, "USNFIS: uniform stable neuro fuzzy inference system," *Neurocomputing*, vol. 262, pp. 57–66, 2017.

32. M. K. Talkhoncheh, M. Shahrokhi, and M. R. Askari, "Observer-based adaptive neural network controller for uncertain nonlinear systems with unknown control directions subject to input time delay and saturation," *Information Sciences*, vol. 418-419, pp. 717–737, 2017.
33. A. Karkevandi-Talkhooncheh, S. Hajirezaie, A. Hemmati-Sarapardeh, M. M. Husein, K. Karan, and M. Sharifi, "Application of adaptive neuro fuzzy interface system optimized with evolutionary algorithms for modeling CO_2-crude oil minimum miscibility pressure," *Fuel*, vol. 205, pp. 34–45, 2017.

SECTION 4
AUTOMATION APPLIED IN DIFFERENT SECTORS

CHAPTER 15

Simulation of the Industrial Oil Adsorption Purification Process for Automation Tasks

Liudmyla Yaroshchuk[1] and Yevheniia Tiurina[1]

[1]Automation Hardware and Software Department, National Technical University of Ukraine “Igor Sikorsky Kyiv Polytechnic Institute”, Kiev, Ukraine

ABSTRACT

Objective reasons associated with an increase of transport quantity and the volume of industrial goods and energy service production lead to the accumulation of used oils and greases. These substances can be reused in the case of their purification. Such resource saving is facilitated by the development of technologies, in particular, adsorption purification and

Citation: Zishan Huang, “Automatic Intelligent Control System Based on Intelligent Control Algorithm”, Journal of Electrical and Computer Engineering, vol. 2022, Article ID 3594256, 10 pages, 2022. https://doi.org/10.1155/2022/3594256.

efficient systems for their automation. The article carried out a detailed analysis of the continuous adsorption purification technological system and indicated its significant differences from the point of view of control systems with special preparation of raw materials. The conclusion about the presence of nonstationarity and stochasticity sources of continuous adsorption as a control object is substantiated, which made it possible to impose requirements on the properties of the control model. The existing methods of adsorption mathematical description are analyzed within their use for continuous control. Structures of dynamic models based on a combination of analytical and experimental methods are proposed. Analytical models are based on mass balances of substances, known forms of approximations for describing equilibrium conditions and adsorption kinetics are studied, and experimental and statistical modeling is carried out to determine possible structures of the connection between technological variables during control. A structural-parametric scheme of the model and a scheme for its adaptation in control systems have been formed. An example of an adsorption simulation model formation using *MATLAB+Simulink* is shown. The results can be used to develop software for control systems of continuous adsorption and to test control algorithms.

INTRODUCTION

Waste industrial oils and greases (OG) are substances that cannot decompose under normal environmental conditions and are hazardous environmental pollutants. At the same time, they contain a significant number of substances that can be used in various spheres of human life. The creation and operation of purification facilities for the waste oil and grease (WOG) recovery is now a perspective direction for the industrial development, which is aimed at improving the Earth's ecosystem state through resource conservation and biosphere protection. Continuous adsorption technology of WOG is the most effective way of their purification. The creation of software for control systems of such industries is an urgent task due to the complexity of processes, raw material instability, and poor information support.

The efficiency of automatic control systems (ACS) largely depends on the type and parameters of the control algorithm used. Modern technical means of automation (TMA) make it possible to implement various control algorithms, so there is a need for preliminary testing of competing algorithms on mathematical models of the technological control object (TCO). Equally important is the task of maintaining parameters of the ACS controllers in accordance with the adsorption process properties as the control object. This task is related to the necessity to adapt the software of control systems.

Justification for the choice and adaptation of ACS algorithms to a greater extent relate to complex technological objects of large capacity, especially those that are at the implementation stage [1]. Such objects can be considered continuous adsorbers designed for WOG recovery in large volumes.

The above problems are primarily related to the properties of the TCO mathematical models. For use in control systems, they must satisfy conditions for the reproduction accuracy of the objects properties and be convenient for adaptation. To study various control algorithms, models should be able to simulate the dynamic properties of input material flows and external and internal disturbances.

These conditions require a reasonable choice of the type for the adsorption process mathematical model and the mathematical processor for its creation. Important tasks are the formation of criteria for selecting models and the formation of plans for their testing.

Given the processes' nonstationarity occurring in industrial adsorbers during the industrial oil's continuous purification, as well as the significant difficulties in organizing and conducting experiments in production conditions, the importance and relevance of modeling these TCOs should be noted.

The aim of the study is to create a model of the adsorption process, designed both to compare alternative control algorithms for industrial adsorbers and for use in control systems for these adsorbers. The objectives of the study are as follows:

i. Determine a modeling method that is convenient for correcting the model based on information from a real adsorber
ii. To carry out structural identification of the adsorption model based on experimental studies
iii. Create a structural-parametric scheme of the model
iv. Implement the model using *MATLAB+Simulink*

LITERATURE REVIEW

A lot of attention is devoted to the modeling of technological processes. Models are performed in different ways, depending on the TCO complexity, the information support of the automation system, and the purpose of the model. In [2, 3], mathematical expressions for modeling adsorption processes as nanosystems and analysis of these expressions from the point of view of practical application are presented. Methods based on the heat and mass transfer theory of the adsorption process are given in [4–11]. These works shows that the use of analytical models is possible only under assumptions about the properties of substances and simplifications regarding the behaviour of these substances in apparatuses. An example of modeling for a one-component adsorbate (water vapor) is given in [12]. The authors of this work investigated mass transfer in a fixed layer of a solid phase using silica gel. In [5], mathematical dependences are given for describing single-component adsorption of copper on different adsorbents. In [6], various aspects of the analytical modeling of adsorption processes are considered in search of the model that adequately describes the experimental data.

Consideration of multicomponent mixtures is found in a few works [13–15], which use complex analytical dependences to describe heat and mass transfer in heterogeneous systems. The work [16] describes adsorption processes of several components from the standpoint of determining the highest selectivity degree of the adsorbent, i.e., what quantity of the studied mixture components will be most adsorbed on the proposed adsorbents. The studies carried out in [17] generally describe multicomponent adsorption, but from the position of a mixture of components that must be removed, without taking into account adsorption properties of each separately. In [18], the authors developed the mathematical model of the dynamic mode of adsorption of inclusions on the interface. Such a model takes into account the physicochemical properties of media in contact with each other but contains many assumptions that can introduce significant errors into the model, depending on the properties of the adsorbent and the raw material.

There are works devoted to computer simulation of adsorption processes in specialized mathematical packages. In particular, in [19], the authors use the *MATLAB+Simulink* environment to simulate the dynamic mode of the catalytic cracking process, and in [20], the creation of a 3D model of the adsorption process in the *ANSYS CFX* software product is described. In [21], the mathematical model was developed to solve the problem of pollutant transport in groundwater based on the stationary incompressible two-dimensional Navier-Stokes equation for the fluid flow and the two-dimensional convection-diffusion equation for the pollutant (one type of pollutant was used). This model was implemented using *FreeFEM* and *MATLAB* environments. Since the adsorption process of multicomponent mixtures is difficult for its behaviour description and adequate reproduction in models that will be used in control systems, it is necessary to provide such a mathematical apparatus that will have the functionality to respond to changes in the properties of raw materials, adsorbent, and technological parameters (such as temperature and pressure directly inside the installation) of the process as a whole.

An analysis of the studies described above shows that the use of exclusively analytical methods is limited for the separation of multicomponent mixtures with an unstable composition. The methods require knowledge of many variables that characterize the properties of material flows and the state of technological equipment, which is impossible in production conditions. Sophisticated computer models are effective in scientific research to determine the form of relationships between adsorption process variables. Since control tasks require models that can be satisfied only by the capabilities of the information subsystem of the production automation system, it is useful to work with the above laboratory studies and computer modeling ideas.

MATERIALS AND METHODS

We consider a general technological system (TS) for the WOG property recovery and outline the boundaries of the part that should be taken into account when modeling the adsorption process.

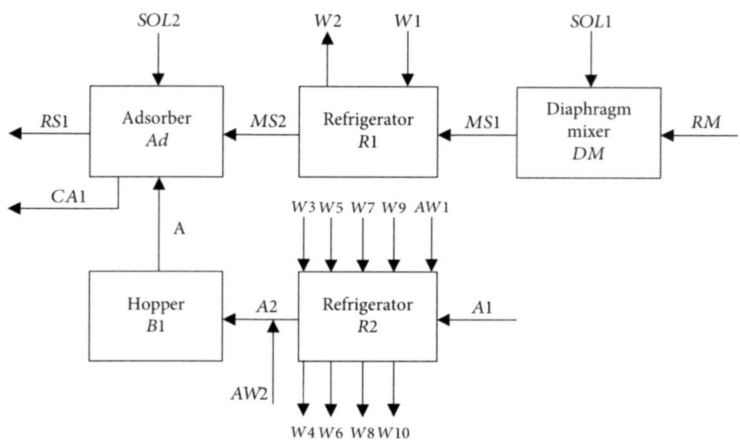

Figure 1. Structural scheme of the TS for the preparation processes of raw material and adsorption and cooling of the adsorbent. RM: contaminated raw material; *SOL*1: solvent at the diaphragm mixer inlet; *MS*1: mixture of raw ma-

terial and solvent; *W*1, *W*2: water at the refrigerator *R*1 inlet and outlet, respectively; *SOL*2: solvent for the water seal; *MS*2: cooled mixture of raw material and solvent; *A*1, *A*2: regenerated adsorbent hot and cooled, respectively; *W*3, *W*5, *W*7, *W*9: water at the coil inlet of the refrigerator *R*2; *W*4, *W*6, *W*8, W10: water at the coil outlet of the refrigerator *R*2; *AW*1: air to create a fluidized bed in the refrigerator *R*2; *AW*2: air for the pneumatic transport system; *A*: the adsorbent at the adsorber inlet (hereinafter, referred to as the adsorbent); *RS*1: purified product; *CA*1: contaminated adsorbent.

Contaminated raw materials (WOG) and a circulating adsorbent enter the input of the common TS (in one part of the TS, it becomes contaminated, adsorbing substances from the raw materials, and in the other part, it is regenerated). The general technology of WOG recovery involves several stages: preparation of WOG (raw materials), adsorption, and adsorbent recovery (desorption, drying, regeneration, and cooling). Figure 1 shows a structural scheme of part of the TS that contains the adsorber and others most closely associated with its apparatuses.

Preparation of raw materials: the diaphragm mixer *DM* receives contaminated raw material *RM* and solvent *SOL*1 used to reduce the viscosity of *RM*. The resulting mixture *MS*1 is cooled in the refrigerator *R*1 to the adsorption temperature with water *W*1 fed through the coils.

Adsorption: the adsorbent *A* (from the intermediate hopper *B*1) and the cooled mixture of raw material and solvent *MS*2 are simultaneously fed countercurrently into the adsorber *Ad*. The adsorbent saturated with aromatic hydrocarbons and resins leaves the adsorber in the suspension form. Solvent *SOL*2 is fed into the adsorber water seal to prevent the removal of raw materials with the suspension. The purified product—the raffinate solution *RS*1—exits the top of the adsorber.

Contaminated adsorbent *CA*1 from the adsorber enters the technological recovery units. It can be considered regenerated (adsorbent *A*1) after the stages of contaminant desorption, drying and burning resins out of it.

Cooling of the regenerated adsorbent: the regenerated hot adsorbent stream *A*1 is cooled with water in a fluidized bed refrigerator *R*2 to a predetermined adsorption temperature. The cooled adsorbent *A*2 enters the hopper *B*1 and from it to the adsorber *Ad* (adsorbent *A*). Desorption, drying, and burning are not considered, since the regenerated adsorbent properties are finally determined after it has been cooled in the refrigerator *R*2.

Each of the material flows shown in Figure 1 has several properties that form the properties of the reduced raw material *RS*1 at the adsorber outlet.

Determining the model type requires the analysis of these properties as information flows. Figure 2 shows a technological-parametric scheme for the contaminated oil and grease recovery.

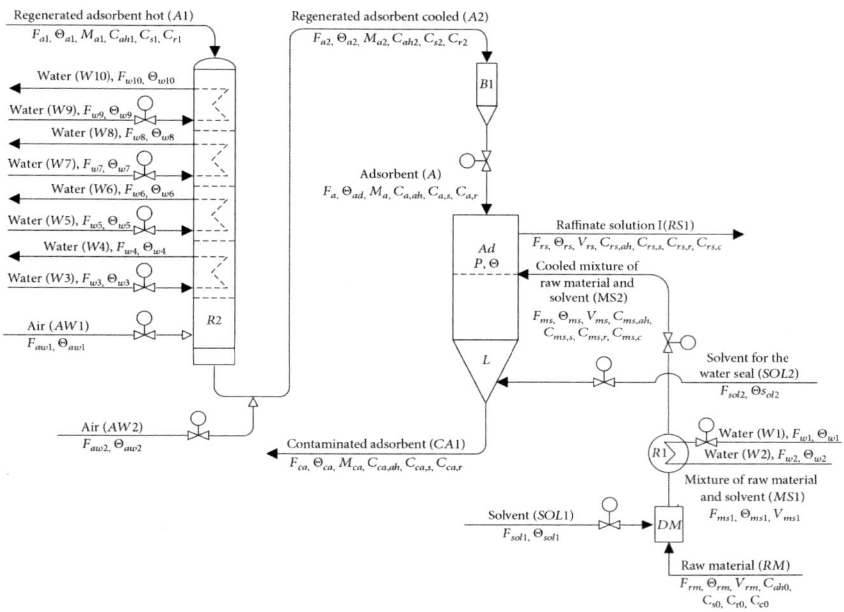

Figure 2. Technological and parametric scheme of the processes of raw material preparation, adsorption, and adsorbent recovery with information flow designation.

The designation of material flow properties that are indicated in Figure 2 is revealed.

(i) Raw material (*RM*): F_{rm}: consumption; Θ_{rm}: temperature; V_{rm}: viscosity, C_{ah0}: aromatic hydrocarbon (AH) content; C_{s0}: sulphur content; C_{r0}: resin content; C_{c0}: content of other impurities

(ii) Solvent (*SOL*1): F_{sol1}: consumption; Θ_{sol1}: temperature.

(iii) Mixture of raw material and solvent (*MS*1): F_{ms1}: consumption; Θ_{ms1}: temperature; V_{ms1}: viscosity

(iv) Water (*W*1) at the refrigerator *R*1 inlet: F_{w1}: consumption; Θ_{w1}: temperature

(v) Water (*W*2) at the refrigerator *R*1 outlet: F_{w2}: consumption; Θ_{w2}: temperature

(vi) Solvent for the water seal (*SOL*2): F_{sol2}: consumption; Θ_{sol2}: temperature

(vii) Cooled mixture of raw material and solvent (*MS*2): F_{ms}: consumption; Θ_{ms}: temperature; V_{ms}: viscosity; $C_{ms,ah}$: aromatic hydrocarbon content; $C_{ms,s}$: sulphur content; $C_{ms,r}$: resin content; $C_{ms,c}$: content of other impurities

(viii) Regenerated adsorbent hot (*A*1): F_{a1}: consumption; Θ_{a1}: temperature; M_{a1}: moisture content; C_{ah1}: aromatic hydrocarbon content; C_{s1}: sulphur content; C_{r1}: resin content

(ix) Inlet water flows (*W*3, *W*5, *W*7, *W*9) in the refrigerator R2: F_{w3}, F_{w5}, F_{w7}, F_{w9}: consumption; Θ_{w3}, Θ_{w5}, Θ_{w7}, Θ_{w9}: temperature

(x) Outlet water flows (*W*2, *W*4, *W*6, *W*8, *W*10) from the refrigerator *R*2: F_{w4}, F_{w6}, F_{w8}, F_{w10}: consumption; Θ_{w4}, Θ_{w6}, Θ_{w8}, Θ_{w10}: temperature

For the simulation model, we will take into account only the flows *W*3, *W*4.

(i) Regenerated adsorbent cooled (*A*2): F_{a2}: consumption; Θ_{a2}: temperature; M_{a2}: moisture content, C_{ah2}: aromatic hydrocarbon content; C_{s2}: sulphur content; C_{r2}: resin content

(ii) Adsorbent (*A*): F_{a}: consumption; Θ_{a}: temperature; M_{a}: moisture content, $C_{a,ah}$: aromatic hydrocarbon content; $C_{a,s}$: sulphur content; $C_{a,r}$: resin content

(iii) Adsorber (*Ad*): *P*: pressure in the apparatus (or the pressure difference between the upper and lower parts of the adsorption column); Θ: adsorption temperature; *L*: solvent level (*SOL*2) in the water seal

(iv) Raffinate solution I (*RS*1): F_{rs}: consumption; Θ_{rs}: temperature; V_{rs}: viscosity; $C_{rs,ah}$: aromatic hydrocarbon content; $C_{rs,s}$: sulphur content; $C_{rs,r}$: resin content; $C_{rs,c}$: content of other impurities

(v) Contaminated adsorbent (*CA*1): F_{ca}: consumption; Θ_{ca}: temperature; M_{ca}: moisture content, $C_{ca,ah}$: aromatic hydrocarbon content; $C_{ca,s}$: sulphur content; $C_{ca,r}$: resin content

The *control action* [1] ACS adsorber can be the consumption of the adsorbent (F_a), the mixture of raw material and solvent (F_{ms2}), and water (F_{w1}, F_{w3}) at the stages of cooling the input streams and the solvent (F_{sol1}). The most important possible *controlled variables* are the concentrations of contaminants in the product ($C_{rs,ah}$, $C_{rs,s}$, $C_{rs,r}$), pressure (*P*) and adsorption temperature (Θ), and level (L); *controlled values* are the concentrations of aromatic hydrocarbons, sulphur, and resins in the contaminated adsorbent ($C_{ca,ah}$, $C_{ca,s}$, $C_{ca,r}$). Disturbances are the concentrations of pollutants, i. e. AH, sulphur, resins, in the raw material ($C_{ms,ah}$, $C_{ms,s}$, $C_{ms,r}$) and adsorbent ($C_{a,ah}$, $C_{a,s}$, $C_{a,r}$), temperatures of the raw material (Θ_{ms}) and adsorbent (Θ_a). The final structure of the automation system depends on many factors.

Figure 2 shows that the preparation processes and adsorption itself are complex multiparameter control objects with a very small number of control actions. Therefore, the model concept is as follows: when applying the same control actions and basic disturbances to the model and the adsorber, the model must be adequate to the object and satisfy the given accuracy indicator. There are no requirements for the application of specific mathematical formalisms to parts of the model or the model as a whole.

First, explore the possibilities of analytical modeling for the processes occurring in the adsorber with a continuous operating principle. Consider the general patterns of transfer of the energy amount, mass, and momentum between different parts of the technological system [22]:

$$\frac{\partial \Theta}{\partial t} = -\left(w_x \frac{\partial \Theta}{\partial x} + w_y \frac{\partial \Theta}{\partial y} + w_z \frac{\partial \Theta}{\partial x} \right) + k_\Theta \nabla^2 \Theta + \frac{q_r}{\rho c_p},$$

$$\frac{\partial c_i}{\partial t} = r_{iv} - \left(w_x \frac{\partial c_i}{\partial x} + w_y \frac{\partial c_i}{\partial y} + w_z \frac{\partial c_i}{\partial z} \right) + D_i \nabla^2 c,$$

$$\frac{\partial w_x}{\partial t} = -\left(w_x \frac{\partial w_x}{\partial x} + w_y \frac{\partial w_x}{\partial y} + w_z \frac{\partial w_x}{\partial z} \right) + v \nabla^2 w_x + \left(X - \frac{1}{\rho} \frac{\partial p}{\partial x} \right), \tag{1}$$

where Θ is the temperature; t is the time; x, y, and z are the designation of Cartesian coordinates; w_j is the fluid velocity along the j-th coordinate; $k_\Theta \nabla^2 \Theta + q_r/\rho c_p$ is the Fourier equation describing the pattern of heat propagation in the system due to heat conduction; k_Θ is the thermal diffusivity; $\nabla^2 = \partial^2/\partial x^2 + \partial^2/\partial y^2 + \partial^2/\partial z^2$, e.g., $\nabla^2\Theta - \partial^2\Theta_i/\partial x^2 + \partial^2\Theta_i/\partial y^2 + \partial^2\Theta_i/\partial z^2$; q_r, ρ, and c_p are power of the heat source per unit volume, density, and heat capacity, respectively; i, c_i, r_{iv}, and D_i are the component number, concentration, rate of formation per unit volume, and diffusion coefficient of the i-th component, respectively; v is the volume; p is the pressure; and X is the projection of the mass force on the x coordinate.

Multicomponent adsorption in the case of WOG (n-component) will require consideration of a system of n-1 equations of System 1.

For the mathematical formalization of adsorption, equations relating to the concept of phase equilibrium and adsorption kinetics are also used. Equilibrium conditions are equations whose factors are the temperature and the concentration c of the substance adsorbed by the adsorbent in the balanced moving phase, and the initial variable is the adsorption capacity a (the amount of substance adsorbed by a unit mass or volume of the adsorbent):

$$a = f(c, \Theta_i). \tag{2}$$

The adsorption theory considers the external (to the adsorbent surface) and internal (inside adsorbent particles) transfer of an adsorbed substance. The rate of this process characterizes the kinetics of external adsorption. It is given using the following expression:

$$\frac{da}{dt} = \beta(c - c^*), \tag{3}$$

where β is the coefficient of external mass transfer per unit volume of the adsorbent and c and c^* are the adsorbed substance content in the volume of the mixture and on the surface, respectively.

Adsorption inside particles is the most complex process in terms of modeling. The equations that can be applied to describe it should contain the parameters of the geometry of the adsorbent internal chambers and the physicochemical properties of the adsorbent, adsorbate, and adsorptive as nanoparticles. For multicomponent mixtures, theoretical equations with such exponents are highly general [23–25].

The analysis of the above mathematical expressions shows that even models of the type as in Equations (1)–(3) require significant amount of information and can be adequate to TCO only if all the assumptions regarding the physicochemical properties and methods of mutual movement of the interacting substances are performed. Existing models of molecular adsorption can only be considered as possible ways of mathematical description of the substance's behaviour under certain conditions. Therefore, the use of complex analytical models also requires adaptation to production conditions when changing the properties of the WOG. However, due to the bulkiness, multiparametric nature, and lack of the necessary TMA, the process of model adaptation will be very difficult. For these reasons, the model is based on a system of simplified dynamic equations for pollutants [26, 27]:

$$F_m y_{ah0} - F_m y_{ah1} - SK_{m,ah} \frac{(y_{ah0} - y_m(x_{ah1})) + (y_{ah1} - y_m(x_{ah0}))}{2} = V_m \rho_m \frac{d}{dt}\left(\frac{y_{ah0} + y_{ah1}}{2}\right) + V_a \rho_a \frac{d}{dt}\left(\frac{x_{ah0} + x_{ah1}}{2}\right),$$

$$F_a x_{ah0} - F_a x_{ah1} + SK_{a,ah} \frac{(x_a(y_{ah1}) - x_{ah0}) + (x_a(y_{ah0}) - x_{ah1})}{2} = V_a \rho_a \frac{d}{dt}\left(\frac{x_{ah0} + x_{ah1}}{2}\right) + V_m \rho_m \frac{d}{dt}\left(\frac{y_{ah0} + y_{ah1}}{2}\right),$$

$$F_m y_{s0} - F_m y_{s1} - SK_{m,s} \frac{(y_{s0} - y_m(x_{s1})) + (y_{s1} - y_m(x_{s0}))}{2} = V_m \rho_m \frac{d}{dt}\left(\frac{y_{s0} + y_{s1}}{2}\right) + V_a \rho_a \frac{d}{dt}\left(\frac{x_{s0} + x_{s1}}{2}\right),$$

$$F_a x_{s0} - F_a x_{s1} + SK_{a,s} \frac{(x_a(y_{s1}) - x_{s0}) + (x_a(y_{s0}) - x_{s1})}{2} = V_a \rho_a \frac{d}{dt}\left(\frac{x_{s0} + x_{s1}}{2}\right) + V_m \rho_m \frac{d}{dt}\left(\frac{y_{s0} + y_{s1}}{2}\right),$$

$$F_m y_{r0} - F_m y_{r1} - SK_{m,r} \frac{(y_{r0} - y_m(x_{r1})) + (y_{r1} - y_m(x_{r0}))}{2} = V_m \rho_m \frac{d}{dt}\left(\frac{y_{r0} + y_{r1}}{2}\right) + V_a \rho_a \frac{d}{dt}\left(\frac{x_{r0} + x_{r1}}{2}\right),$$

$$F_a x_{r0} - F_a x_{r1} + SK_{a,r} \frac{(x_a(y_{r1}) - x_{r0}) + (x_a(y_{r0}) - x_{r1})}{2} = V_a \rho_a \frac{d}{dt}\left(\frac{x_{r0} + x_{r1}}{2}\right) + V_m \rho_m \frac{d}{dt}\left(\frac{y_{r0} + y_{r1}}{2}\right), \quad (4)$$

where F_m and F_a are the flow rates of the mixture of raw material and solvent and adsorbent, respectively; y_{i0} and x_{i0} are the input concentrations in the mixture of raw material and solvent and adsorbent, respectively; y_{i1} and x_{i1} are the output concentrations in the mixture of raw material and solvent and adsorbent, respectively; $i = ah, s, r$; ah is the aromatic hydrocarbons; s is the sulphur; r is the resins; $y_m(x_{i0})$, $y_m(x_{i1})$, $x_a(y_{i0})$, and $x_a(y_{i1})$ are the equilibrium concentrations of aromatic hydrocarbons, sulphur, and resins in the mixture of raw material and solvent and adsorbent, respectively; $K_{m,i}$ and $K_{a,i}$ are the mass transfer coefficients expressed in terms of the component molar particles in the mixture of raw materials and solvent and adsorbent, respectively; V_m and V_a, are the column volume filled with the mixture of raw material and solvent and adsorbent, respectively; and $_m$ and ρ_a are the densities of the mixture of raw material and solvent and adsorbent, respectively.

The simplification of this system consists, in particular, in the following assumptions: the adsorber is a control object with lumped parameters; mass transfer coefficients are constant values; the motive force for the transfer of pollutants from adsorbate to adsorbent is directly proportional to the average value of the differences in the equilibrium concentrations of these substances in adsorbate and adsorbent; mass accumulation is proportional to the average concentrations: $(x_{ah0} + x_{ah1})/2$, $(y_{ah0} + y_{ah1})/2$, $(x_{s0} + x_{s1})/2$, $(y_s + y_{s1})/2$, $(x_{r0} + x_{r1})/2$, and $(y_{r0} + y_{r1})/2$.
These simplifications will reduce the number of unknown parameters and, thus, the duration of the model adaptation in production conditions.

The mass of each of the pollutants that have passed from the adsorbate to the adsorbent per unit time, in Equation (4), is represented by expressions that are identical in structure. For example, an expression for aromatic hydrocarbons

$$SK_{m,ah} \frac{(y_{ah0} - y_m(x_{ah1})) + (y_{ah1} - y_m(x_{ah0}))}{2}. \tag{5}$$

In the case of WOG, the input properties of the raw material are usually unstable, so even Equation (5) cannot provide the necessary information in real time with the required accuracy.

A significant list of factors and disturbances even in such a simplified model complicates its practical application in ACS. Based on this conclusion and recognizing the convenience of the current model adaptation as a very important requirement, based on System 4 and taking into account the transport delay in the TCO, it is proposed to present the models of the main control channels in the form

$$W_n(s,t) = \frac{k_n(t)}{T_{2n}(t)s^2 + T_{1n}(t)s + 1} e^{-\tau_n s}, \tag{6}$$

where n is the control channel number, $n = 1, 2, 3$; (ah, S, r); T_{1n} and T_{2n} are the adsorber time constants; s is the Laplace transform symbol; t is the time; k_n is the transmission coefficient; and τ_n is the transport delay time of the control object.

Since the above simplifications distort actual relationships between the variables of heat and mass transfer processes in the adsorber, it is proposed to represent each of the parameters of Equation (6) by the function $_{wk}$, which will be adjusted when comparing the values of the selected variable at the outputs of the adsorber and its model.

To select the form of the functions $_{wk}$ nonlinearity consider experimentally confirmed mathematical expressions that characterize various aspects of adsorption processes [1, 22], in particular, formulas for calculating the index *a*. Since there are no recommendations for calculating *a*, for the adsorption WGO purification, several formulas are used that can be taken as a basis. Among them are the formulas of Langmuir, Freundlich, Redlich-Peterson, BET (the theory of polymolecular (multilayer) adsorption by Brunauer, Emmett, and Teller), Dubinin-Radushkevich, Boyd, and Lagergren [22, 27, 28]. Each of the formulas is based on a certain set of various assumptions related both to the properties of material flows and to the process conditions.

So, in the Dubinin-Radushkevich formula [22], the adsorption index *a* depends nonlinearly on the properties of material flows and regime parameters as follows:

$$a = \frac{a_m BP}{1 + BP}, \tag{7}$$

where a_m is the capacity of a monolayer of adsorbed substance in adsorbent, *P* is the partial pressure, and *B* is the parameter, which is proposed to be calculated by the formula

$$B = \frac{\alpha_m \exp\left(q_m / kT\right)}{k_0 \sqrt{(2\pi mkT)}}, \tag{8}$$

where *m* is the mass of the molecule, *k* is Boltzmann's constant, *T* is the absolute temperature, k_0 is the entropy coefficient, and q_m is the adsorption heat of one molecule.

The Freundlich formula (for energetically inhomogeneous surfaces) has the following form:

$$\ln a = \ln K_F + \frac{1}{n} \ln C, \tag{9}$$

where K_F and n are the Freundlich adsorption constants characterizing the adsorption capacity and the interaction intensity of substances "adsorbent-adsorbate," respectively.

The structure of these expressions has common features—they are nonlinear with exponential, logarithmic, and power components of the adsorbate concentration, temperature, and pressure in the adsorber. Based on this, it is necessary to substantiate which structures of the functions $_{wk}$ should be introduced into the model designed to control and study algorithms.

The definition of the form φ_{wk} was made using the results of studies given in [6, 14, 15].

Thus, the adsorption of tetrafluoroethane isomers on NaY zeolite was studied in [6]. Figure 3 shows experimental data displaying the dependence of the adsorbed substance quantity on pressure and its logarithmic and polynomial approximations.

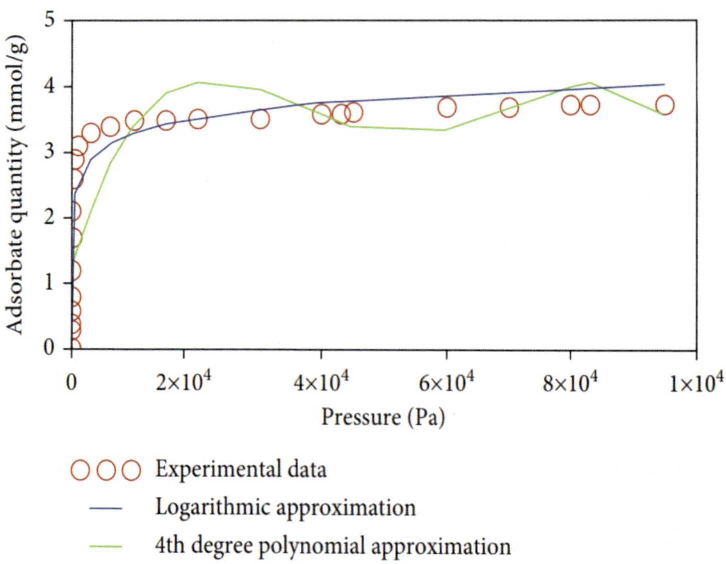

Figure 3. Approximation of the dependence of the adsorbed substance quantity on pressure during the adsorption of tetrafluoroethane on NaY zeolite.

The work [14] considers the separation of Krypton (Kr) from Xenon (Xe) on different adsorbents. Figure 4 shows the dependence of Kr quantity on relative pressure using FAU LSX K—faujasite zeolite with Si/Al = 1—and, accordingly, its approximation by exponential and quadratic parabolas.

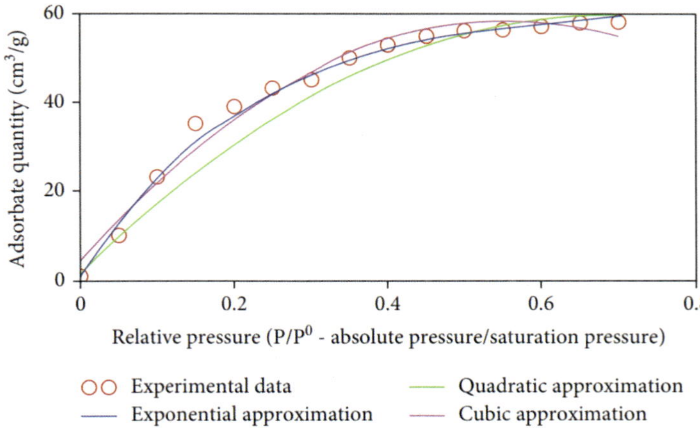

Figure 4. Approximation of the dependence of Kr quantity on relative pressure during the adsorption on FAU LSX K zeolite.

The authors in [15] considered the problem of used fluorine-containing compound utilization by extracting from them and separating fluorocarbons by adsorption methods. Figure 5 shows the dependence of adsorbed difluoromethane R32 (adsorbate) quantity on pressure during adsorption on 5A zeolite with modified pore sizes and the approximation of the obtained dependence of adsorbate amount on pressure by an exponent, quadratic, and cubic polynomials.

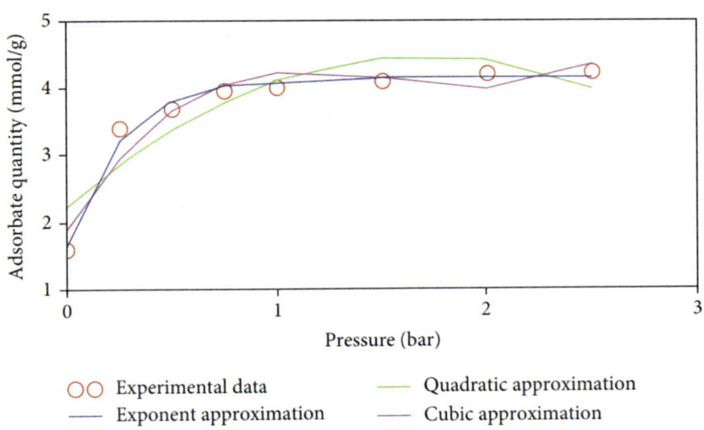

Figure 5. Approximation of the dependence of difluoromethane R32 quantity on pressure during adsorption on 5A zeolite.

Figures 6 and 7 show dependencies of fluorocarbon adsorbate quantity on temperature at various pressures [15]: dependence of R32, R22 (chlorodifluoromethane), and R125 (pentafluoroethane) quantity on temperature during adsorption on zeolite 5A and a pressure of 1.5 bar (Figure 6); dependence of R134a (1,1,1,2-tetrafluoroethane), R125, and R143a (1,1,1-trifluoroethane) quantity on temperature during adsorption on zeolite 13X and a pressure of 1 bar (Figure 7).

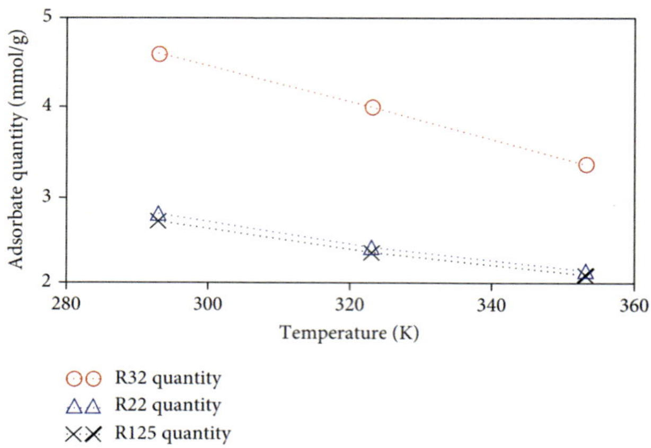

Figure 6. Dependence of the adsorbate quantity on temperature using zeolite 5A (pressure 1.5 bar).

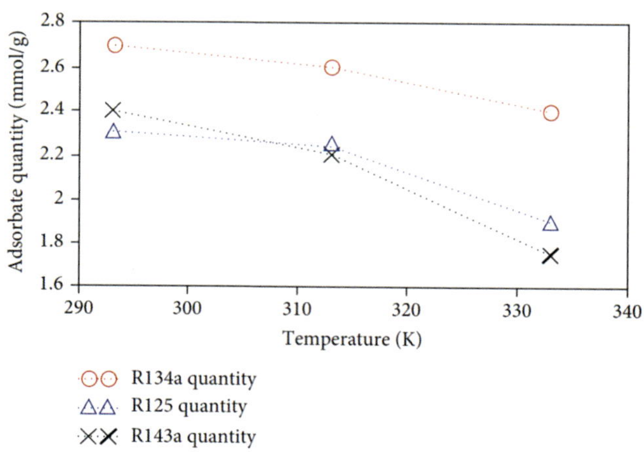

Figure 7. Dependence of the adsorbate quantity on temperature using zeolite 13X (pressure 1 bar).

Figure 8 shows results of fluorocarbon components R134a, R125, and R143a adsorption on zeolite 13X, obtained from the two-level Langmuir-Freundlich equation at temperatures of 293 K, 313 K, and 333 K.

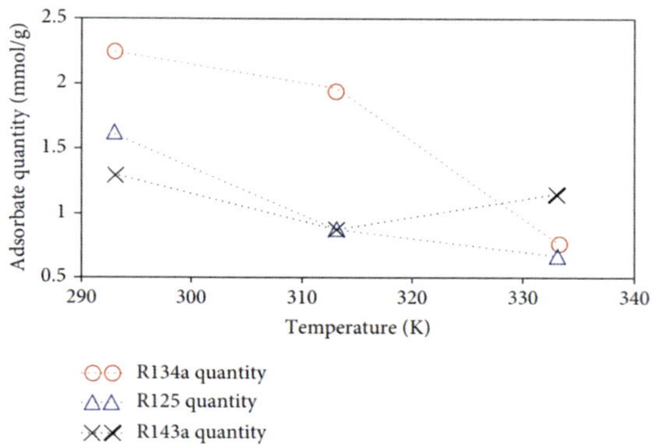

Figure 8. Dependence of the adsorbate quantity on temperature during adsorption on zeolite 13X.

Figure 9 shows the dependencies of the fluorocarbon adsorbate R32, R22, and R125 quantity on temperature at a pressure of 1 bar on HKUST-1 zeolite, which has a three-dimensional porous structure with three types of cages formed by the connection of dimeric copper clusters coordinated with four carboxylate groups of trimesic acid [15].

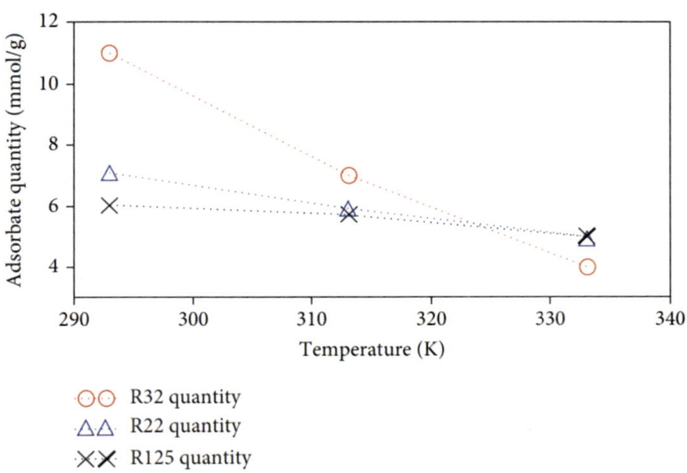

Figure 9. Dependence of the adsorbate quantity on temperature using zeolite HKUST-1 at a pressure of 1 bar.

Based on the studies on structural identification, which are illustrated in Figures 3–9, for the control channel transmission coefficient, it is proposed to choose two competing structures $\varphi_{n,kn}$—exponential and polynomial—of the following types:

$$\varphi_{n,kn}(t) = a_0 + a_1 e^{a_2 \Theta(t)} + a_3 e^{a_4 P(t)},$$

$$\varphi_{n,kn}(t) = \varphi_{n,kn,\Theta}(\Theta(t)) + \varphi_{n,kn,P}(P(t)) + \varphi_{n,kn,c}(c(t)), \tag{10}$$

where a_i is the exponential model parameters; $\varphi_{n,kn,\theta}(\Theta(t))$, $\varphi_{n,kn,P}(P(t))$, and $\varphi_{n,kn,c}(c(t))$ are polynomials of the 2nd or 3rd degree for temperature, pressure, and concentration.

Correction of functions 10 and 11 when adapting the model should not be difficult.

Consider the main control channel $\ll F_a \longrightarrow C_{rs,ah} \gg$. Its transfer function, for example, is as follows:

$$W_1(t) = W_{F_a \longrightarrow C_{ah}}(t) = \frac{k_{11}(t)}{T_{21}(t)s^2 + T_{11}(t)s + 1} e^{-\tau_1 s}, \tag{11}$$

$$k_{11}(t) = a_{10} + a_{11} e^{a_{12}\Theta(t)} + a_{13} P(t)^{a_{14}}, \tag{12}$$

$$T_{11}(t) = b_{11}, \quad T_{21}(t) = b_{21}. \tag{13}$$

It is appropriate to choose the values T_{11}, T_{21}, and k_{11}, obtained from simplified material balance 4 or from the available experimental data as the starting values of the parameters for corrective functions.

Figure 10 shows a generalized structural scheme of the system for adsorption model adapting through several control channels and channels of disturbances.

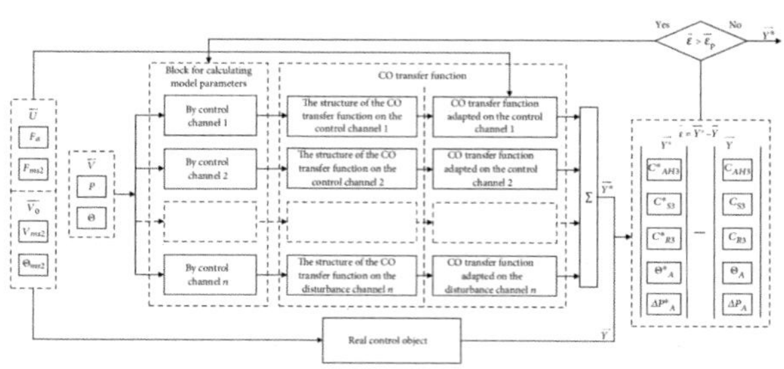

Figure 10. Structural scheme of the adsorber model adaptation system.

It allows to create a model adaptation algorithm. According to the algorithm, vectors of input signals $\vec{U}$ (control actions) and $\overrightarrow{V_0}$ (mode parameters) are fed to the inputs of each control channel and the real TCO. The adaptive model implies the presence of an adaptation block to determine the transfer function coefficients of the model, the structure of which is known. The vector of mode parameters $\vec{V}$ is fed to the input of the same block. The vector of the adaptive model $\overrightarrow{Y*}$ is compared with the vector of the output signals of the adsorber $\vec{Y}$. If the error value $\vec{\varepsilon}$ is not acceptable, then in the adaptation block, the transfer function coefficients are tuned until the condition $\vec{\varepsilon} \leq \overrightarrow{\varepsilon_p}$.

Figure 11 shows an example of a structural-parametric scheme of the model with the proposed corrective functions for the control channel "$F_a \rightarrow C_{rs,ah}$."

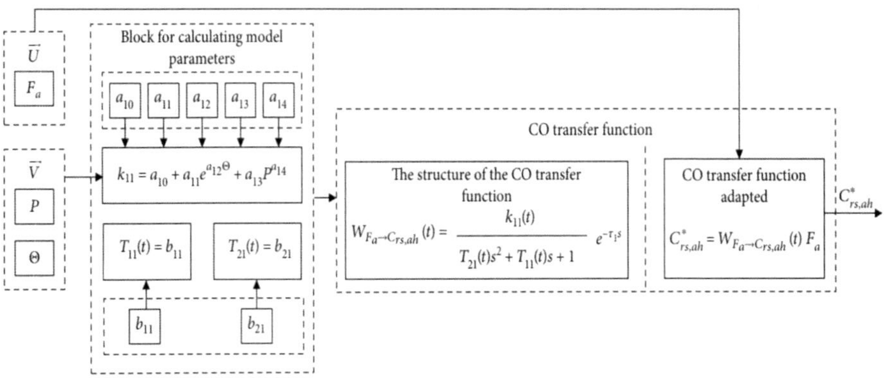

Figure 11. Structural-parametric scheme of the model.

Figure 12 shows the structural scheme of the adsorption model by control channel "$F_a \rightarrow C_{rs,ah}$," implemented by means of *MATLAB+Simulink*. The structure of the transfer function in Equation (12) is represented by two blocks: the first has form of the aperiodic part $WF_a \rightarrow C_{rs,ah}$; the second is the transport delay $e^{-\tau 1s}$, where τ_1 is the transport delay time, s. Control action F_a and the parameters calculated in the corresponding block for calculating model parameters (BCMP) are fed to the input of this function. Figure 13 shows the structure of the BCMP subsystem, which provides calculation of model parameters according to Equation (12).

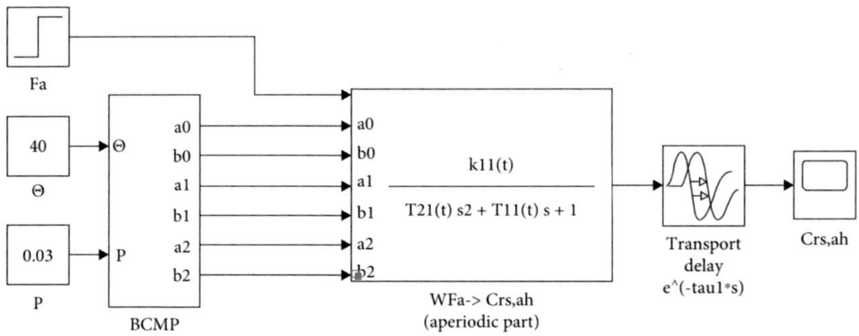

Figure 12. Structural *Simulink* model of adsorption by control channel "$F_a \rightarrow C_{rs,ah}$."

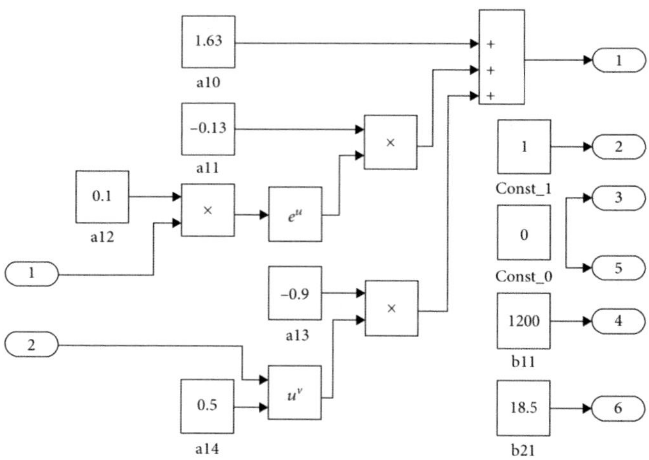

Figure 13. Subsystem structure of the BCMP of *Simulink* adsorption model.

Such a *Simulink* model can be used to compare adsorber control algorithms, supplementing it with disturbance blocks typical for production conditions.

RESULTS AND DISCUSSION

An analysis of the technological system for purification (regeneration) of industrial (industrial and transport) oils and lubricating oils showed that it is characterized by two types of changes in the properties of substances (i.e., types of nonstationarity of processes in the FEC). The first of them is systemic in nature, since it occurs every time when raw materials are received from another supplier. This type of nonstationarity can be more or less predictable (periodic) with one supplier of raw materials or unpredictable (random) due to its receipt from several suppliers.

The second type of change (nonstationarity) is purely random. It is related to the properties of the adsorbent (the impact of adsorbent regeneration processes) and raw materials (even from one supplier). The properties of both substances are random variables; we consider them as external perturbations. The indicators of the state of the internal elements of the design of the adsorber and other devices are considered internal perturbations [23].

The study of the chemical-technological system and the analysis of existing methods for modeling the adsorption process made it possible to determine the type of models and the algorithm for their use in computer systems for controlling the purification processes, namely, the PTO, taking into account the named nonstationarity.

Processing the results of open access studies made it possible to determine the structures of approximation equations that can be applied in control systems for continuous adsorbers.

The obtained results allow software developers to use polynomial 1-3 orders and exponential models of control channels. Substantiation of the type of models designed to control the process of adsorption for PTO purification, based on theoretical provisions and experimental studies, can be considered a scientific novelty.

Taking into account the complexity of experimental work in production conditions and the inefficiency of research in laboratory conditions, such results are of practical importance.

CONCLUSIONS

As a result of the studies presented in the article, were obtained following:

- It is substantiated that the adsorption process can be considered nonstationary due to the receipt of contaminated raw materials from different sources and the lack of systems for stabilizing the properties of contaminated raw materials in each source
- It is determined that nonstationarity of processes has two causes and two types of nonstationarity of processes; this should be taken into account in process control algorithms

- The analysis of the structure and material flows of the technological purification system made it possible to determine the information flows
- It has been determined that the following should be taken into account when modeling: the adsorption purification process is a multiparameter TCO, the values of most of the process variables cannot be determined continuously, and under production conditions, they will not be measured even in a laboratory
- Mathematical model of each control channel is proposed in the form of a second-order transfer function with a transfer coefficient that is a nonlinear function of temperature and pressure in the adsorber
- The study of data from open sources has identified the types of functions that can reflect the dependence of the transfer coefficients on the regime parameters; these include exponentials, polynomials of 2-3 orders, and logarithms
- A structural-parametric diagram of the model and a diagram of the system for adapting the adsorber model were created
- A simulation model of adsorption processes was created using *MATLAB+Simulink*, which corresponds to the concept of modeling

These results were obtained through the use of system analysis methods, in particular, the analysis of the technological system and its components and the analysis of methods for describing technological processes and adsorption, in particular, taking into account the requirements for the mathematical model of processes as control objects.

Such control-focused modeling research work related to continuous adsorption purification of used oils and greases based on the generalization of other adsorbents and adsorbates has not previously been reported in the literature. The achieved results can be considered as they have both scientific novelty and practical value. They can be used to create software for control systems for adsorption purification processes not only of WOG but also of other substances.

Proposed structures of the model allow solving the following problems related to the control of adsorption purification process, namely,

- Determining the list of controlled variables necessary to control the adsorption process using the recommended model

- Determining the dynamic properties of controlled variables to synchronize their entry into the process model
- Choosing and investigating a computer method of the adsorption model adaptation to the conditions of this process in production conditions
- Creating and investigating the computer operation of an adaptive control system for the quality indicators of the contaminated industrial oil and grease purification both in the mode of unchanged raw materials and when the supplier of contaminated raw materials is changed
- Coordinating functioning of separate control systems of the production system in order to save resources for implementation
- Developing plans for experimental studies of individual technological variables in production conditions to check the properties of controlled variables as random variables and to reconcile them when using models
- Developing plans for experimental studies of disturbance and control channels to check the model adequacy

The relevance of this project is in the fact that a process model was proposed to control the process of used oil and grease purification, the behaviour of which would correspond to the behaviour of this process in production conditions; it would be simple enough after the formula and in terms of the number and accuracy of the input variable measurement. The lack of mathematical models reduces effectiveness of control systems for complex technological apparatuses.

Also, the obtained model gives an opportunity to conduct simulation experiments to study various algorithms for controlling production processes, reducing time for designing and implementing control systems.

REFERENCES

1. L. D. Yaroshchuk and Y. O. Tiurina, "Hierarchy of control tasks for continuous process of adsorption regeneration of lubricants," *Bulletin of Cherkasy State Technological University*, vol. 2, no. 2, pp. 49–62, 2021.
2. A. I. Klyndjuk, *Surface Phenomena and Dispersed Systems: Manual for Students of Chemical Engineering Specialties*, Belarusian State Technological University, 2011.
3. M. R. Petryk, O. M. Khimich, I. V. Boiko, D. M. Mykhalyk, M. M. Petryk, and V. I. Kovbashyn, "Mathematical modeling of heat transfer and adsorption of hydrocarbons in nanoporous zeolite catalysts of waste gas neutralization systems," *The National Academy of Sciences of Ukraine, V. M. Glushkov Institute of Cybernetics*, 2017.
4. V. V. Sabadash, "Theoretical bases of sorption processes on natural and synthetic sorbents, [Ph.D. thesis]," *Processes and the Equipment of Chemical Technology*, Ministry of Education and Science of Ukraine, Lviv Polytechnic National University, 2019.
5. X.-W. Wu, H.-W. Ma, L.-T. Zhang, and F.-J. Wang, "Adsorption properties and mechanism of mesoporous adsorbents prepared with fly ash for removal of Cu(II) in aqueous solution," *Applied Surface Science*, vol. 261, pp. 902–907, 2012.
6. M. Z. Kvasko, M. Z. Kvasko, O. S. Zhurakovsky, and Y. Y. Zhurakovsky, "Analysis of parameters of the mathematical model of tetrafluoroethane adsorption on NaY Zeolite," in *Computer modeling in chemistry and technologies and systems of sustainable development - KMHT- 2012: collection of scientific articles of the Third International Scientific and Practical Conference*, pp. 87–89, Rubizhne, Kyiv, 2012, https://ela.kpi.ua/handle/123456789/18821.
7. O. K. Ye, "Sorption and biosorption of pharmaceutical substances on activated carbon for their effective removal from water, [Ph.D. thesis]," Tech. Rep., The National Academy of Sciences of Ukraine, A. V. Dumanskyi Institute of Colloid Chemistry and Water Chemistry, 2017.
8. N. Ayawei, A. N. Ebelegi, and D. Wankasi, "Modelling and interpretation of adsorption isotherms," *Journal of Chemistry*, vol. 2017, Article ID 3039817, 11 pages, 2017.

9. A. J. Poloz, N. V. Nikolenko, and J. R. Jebich, "Peculiarities of adsorption of oligomeric components of epoxy compositions on aluminium powder," *Questions of chemistry and chemical technology*, vol. 1, pp. 49–60, 2019.
10. T. G. Myers, F. Font, and M. G. Hennessy, "Mathematical modelling of carbon capture in a packed column by adsorption," *Applied Energy*, vol. 278, article 115565, 2020.
11. R. Ortmann, "Untersuchungen zur adsorption und desorption von aldehyden und aminen im spurenbereich," 2016, https://nbn-resolving.org/urn:nbn:de:hbz:464-20160606-150124-5.
12. D. V. Bashkirov, A. V. Klinov, and A. I. Razinov, *Mathematical modeling of the adsorption process on the example of dehydration of humid air with a fixed layer of silica gel*, Bulletin of Kazan Technological University, vol. 17, no. 19, 2014.
13. I. M. Ivanenko, T. A. Dontsova, and Y. M. Fedenko, "Adsorption, adsorbents and catalysts based on them," *Igor Sikorsky KPI*, 2018.
14. B. Alagappan, "Assessing different zeolitic adsorbents for their potential use in Kr and Xe separation, [Ph.D. thesis]," *Professional Papers, and Capstones*, 2013.
15. J. A. D. K. Wanigarathna, "Adsorption-based fluorocarbon separation in zeolites and metal organic frameworks, [Ph.D. thesis]," Tech. Rep., Nanyang Technological University, Singapore, 2018.
16. M. D. Myshljavceva, "Mathematical modeling of difficult adsorption systems on a surface of solid bodies: a transfer matrix method. [Ph.D. thesis]," *Mathematical modeling, numerical methods and complexes of programs, Federal State Budgetary Educational Institution of Higher Professional Education*, Omsk State Technical University, 2013.
17. M. L. M. Navarro Cerutti, F. V. Hackbarth, D. Maass et al., "Copper-exchanged Y zeolites for gasoline deep-desulfurization," *Adsorption*, vol. 25, no. 8, pp. 1595–1609, 2019.
18. O. Malko and A. Malko, "Modeling of the process of adsorption of inclusions on the interface," in *Information technologies and computer modeling: materials of articles of the International Scientific and Practical Conference*, pp. 358–361, Ivano-Frankivsk, 2018.
19. D. Ahmed and S. Ateya, "Modelling and simulation of fluid catalytic cracking unit," *Journal of Chemical Engineering & Process Technology*, vol. 7, no. 4, 2016.

20. L. T. Fajzullina and E. J. Tumanova, "Mathematical modeling of the adsorption process for various methods of adsorbent packing," in *5th All-Russian Student Scientific and Technical Conference Intensification of Heat and Mass Transfer Processes, Industrial Safety and Ecology*, pp. 280–283, Kazan National Research Technological University, 2018.
21. J. Nyende, I. Enyogoi, J. Mango, and H. Kasumba, "Numerical simulation of a two-dimensional groundwater pollute transport problem using incompressible steady-state Navier-Stokes equations and diffusion-convection equations," *Modelling and Simulation in Engineering*, vol. 2022, Article ID 7419502, 20 pages, 2022.
22. V. B. Kogan, "Theoretical basics of typical processes of chemical technology," *Himija*, 1977.
23. L. D. Yaroshchuk and Y. O. Tiurina, "Verbal modeling of microprocesses of continuous oil and grease adsorption," *Bulletin of NTUU Igor Sikorsky Kyiv Polytechnic Institute*, no. 2, pp. 78–88, 2020.
24. F. G. Quintero-Álvarez, C. K. Rojas-Mayorga, D. I. Mendoza-Castillo, I. A. Aguayo-Villarreal, and A. Bonilla-Petriciolet, "Physicochemical modeling of the adsorption of pharmaceuticals on MIL-100-Fe and MIL-101-Fe MOFs," *Adsorption Science & Technology*, vol. 2022, article 4482263, pp. 1–14, 2022.
25. S. Rodríguez-Narciso, J. A. Lozano-Álvarez, R. Salinas-Gutiérrez, and N. Castañeda-Leyva, "A stochastic model for adsorption kinetics," *Adsorption Science & Technology*, vol. 2021, article 5522581, pp. 1–21, 2021.
26. As. M. Gumerov, N. N. Valeev, Az. M. Gumerov, and V. M. Emelianov, "Mathematical modeling of chemical-technological processes," *5th All-Russian Student Scientific and Technical Conference Intensification of heat and mass transfer processes, industrial safety and ecology*, Kazan' State Technological University, pp. 23–25, 2018.
27. J. I. Dytnerskij, "Processes and Apparatuses of Chemical Technology: a Textbook for Universities," *Part 2. Mass transfer processes and apparatuses*, Himija, 1995.
28. A. V. Oboturov, *Surface phenomena and dispersed systems, lecture notes*, Educational institution "Mogilev State University of Food", 2011.

CHAPTER 16

Opportunities and Challenges of Robotics and Automation in Offshore Oil & Gas Industry

Heping Chen[1], Samuel Stavinoha[1], Michael Walker[1], Biao Zhang[2], and Thomas Fuhlbrigge[2]

[1]Ingram School of Engineering, Texas State University, San Marcos, USA

[2]ABB Corporate Research Center, Windsor, USA

ABSTRACT

The oil and gas industry will continue to boom in the coming few decades. Obtaining oil and gas from conventional and non-conventional resources will become more and more challenging. This intensifying need will impose very considerable demands on work force, financial and technology capabilities. Since the future supplies of oil and gas are to expand, advanced technology will become increasingly necessary to obtain access to more challenging conventional and non-conventional resources. Therefore oil and gas technologies will be very costly to operate in the coming future due to

Citation: Chen, H. , Stavinoha, S. , Walker, M. , Zhang, B. and Fuhlbrigge, T. (2014), "Opportunities and Challenges of Robotics and Automation in Offshore Oil & Gas Industry". Intelligent Control and Automation, 5, 136-145. doi: 10.4236/ica.2014.53016.

hostile, hard-to-reach environments. The offshore oil industry will become a complicated myriad of advanced equipment, structures, and work force. Our objectives are to identify potential applications and research directions of robotics and automation in the oil & gas field and explore the obstacles and challenges of robotic and automation applications to this area. This study performs the necessary survey and investigation about the work conditions of robotics and automation equipment in the oil and gas industry, especially offshore oil rigs. The oil & gas industry processes are first investigated. The personals and tasks are then explored. Furthermore, this paper reviews the current robotic automation technology. The challenges and requirements are identified for robotics and automation equipment in the oil and gas industry. The requirements of robotics and automation in the oil & gas industry are presented. Future research opportunities are discussed from a technical perspective.

Keywords: Oil & Gas, Robot, Automation, Hazardous Environment, Teleoperation, Offshore Oil & Gas Industry

INTRODUCTION

The oil and gas industry will continue to boom in the coming few decades. The oil and gas demand will grow rapidly in the next two decades [1] .

The intensifying need to obtain oil and gas from more hostile, hard-to-reach environments will increase the operation cost rapidly in the coming future. Hence, the oil and gas industry keeps looking for lower-cost solutions.

To be competitive and to improve their profit margins, oil & gas companies are committed to cost reduction. They also look for ways to minimize employee costs and improve manufacturing efficiencies and quality besides seeking lower-cost suppliers and less-expensive raw materials. Because of the rising cost of employee salary and benefits like health care, the cost reduction effort in oil & gas companies is offset. Also high employee turnover adds the costs of retraining. Therefore, the oil and gas companies are looking for new technologies to reduce the labor cost. Also safety is a big concern in the oil and gas production. Using robotics in inspection, maintenance and repair could greatly improve the safety and efficiency.

The oil & gas industry's presence is evident in its global networks of market supply and demand relationships. When there are fluctuations, regardless of their origins, consumers are affected in all over the world.

Prices respond to changing markets with upward volatility because of an inelastic demand for oil and petroleum products. The period of high oil prices from 2004 through 2008 led to a steady demand for petroleum. As this market trend persists, oil & gas companies have a window of opportunity to maximize efficiency and productivity to moderate the petroleum market.

One solution to both the need for efficiency and maximum production and the capabilities required to further exploration is to implement robotics and automation in offshore oil & gas environments. Because the offshore oil & gas processes require advanced technologies, offshore environments will deploy the safest, most secure and consistent operations by utilizing industrial robotics and automation, and the latest software and mechanical devices.

In order to investigate the challenges of robotics and automation in oil and gas industry, the necessary survey and investigation about the oil & gas industry processes, the personals and tasks should be explored first. The work conditions must be discussed to explore the requirements of robotics and automation equipment in the oil and gas industry, especially on offshore rigs. To meet the requirements and develop robotics and automation equipment in such work conditions, this paper reviews the current technology that has been developed and discusses the future research opportunities in the oil and gas industry.

OIL AND GAS INDUSTRIAL PROCESSES

The offshore oil & gas industry is a complicated myriad of advanced equipment, structures, and work force. With a proper knowledge of offshore oil and gas rig environments, the applications of industrial robotics and automation are less abstract. Before any real vision of the potential roles robotics and automation in offshore oil processes can emerge, those processes must be enumerated appropriately.

There are many products and services related to oil and gas with an equally substantial potential for markets within the industry. There are three stages through which petroleum products pass: upstream, downstream and midstream models.

The upstream oil sector commonly refers to the searching for, recovery and production of crude oil and natural gas. It is also known as the exploration and production (E & P) sector, including searching for potential oil and gas fields, drilling of exploratory wells, and subsequently operating the wells to recover and bring the crude oil and/or natural gas to the surface [2] .

The midstream oil and gas sector is the relay point for the upstream sector's products. Midstream processes commonly refer to processing, transport, and storage of these products [3] . Because it is possible to produce pipeline quality gas for direct sale to an interstate or intrastate natural gas pipeline in the midstream sector, some treatment or processing of natural gas may occur in the midstream sector and bypass the downstream oil and gas sector completely. The midstream typically links the supply of the oil industry to the demand for energy commodities [3] .

The downstream oil sector refers to the refining of crude oil and the selling and distribution of natural gas, as well as other products derived from crude oil such as liquefied petroleum gas (LPG), gasoline or petrol, jet fuel, diesel oil, other fuel oils, asphalt and petroleum coke [4] [5] . The downstream industry touches consumers through thousands of products. These products include petrol, diesel, jet fuel, heating oil, natural gas and propane to asphalt, lubricants, synthetic rubber, plastics, fertilizers, antifreeze, pesticides, and pharmaceuticals [4] [5] .

This paper provides a fundamental knowledge base to utilize in understanding the environment in which offshore industrial robotics and automation would operate, as well as cover the challenges that exist within that environment. The processes described here relate primarily to the upstream oil and gas sector, as that particular sector hosts the majority of opportunities for robotics and automation. The oil and gas processes and the three sector model they fall into tend to parallel across the onshore and offshore industries, however the processes will be distinguished when necessary if a distinct observation is being made.

The major oil and gas extraction processes include the materials and equipment used and the processes employed. There are four major processes in the oil and gas extraction industry: (1) exploration, (2) well development, (3) production, and (4) site abandonment [6] . After these processes are completed, the production process enters. It is likely the process in which robotics and automation have the largest potential to increase efficiency and create a safer environment for offshore oil and gas rigs, all while cutting construction costs for human necessitated rig designs [6] .

After the Deep Water Horizon oil spill in 2010, the Bureau of Ocean Energy Management, Regulation and Enforcement (BOEMRE) has implemented new mandatory regulations to replace old protocols for the offshore oil & gas industry. Consequences of these new regulations are evident in the day-to-day operations in the oil & gas industry, especially regarding inspection and maintenance of equipment. This is one area of operation that robotics and automation can dramatically improve efficiency, precision, safety, and decrease costs to companies. It is no stretch of imagination to suggest that robots and automation will soon be the primary means to effectively satisfy many of these new regulations.

CHALLENGES AND REQUIREMENTS

In this section, the environments that the robotics and automation will be deployed are investigated. The hardware and software requirements are discussed.

Challenges in Oil and Gas

The deep waters of the Gulf of Mexico, the frigid regions of Russia, and the hot, dusty, undeveloped deserts of the Middle East are merely the geographic challenges facing today's oil and gas exploration and production industry [7] . The work conditions on offshore installations are the first thing to look at when analyzing the environments. The most important ones are as follows:

- **Atmosphere**: The atmospheric conditions on offshore platforms are quite unfriendly. Due to the substances used and generated during the processing of hydro carbon resources, the following three types of gases can occur separately and combined: explosive, toxic and corrosive.
- **Unsheltered maritime environment**: Except for the living quarters and a few technical rooms offshore platforms are partially sheltered and unsheltered. This means there is no sufficient protection against saltwater spray and direct sun light which is also reflected from the sea surface.
- **Heavy weather**: Wind with high speed and squalls, rain, hail and snow. All these weather conditions occur more often and more intense offshore than onshore.
- **Extreme ambient temperature**: Depending on the region the platform is located there can be extreme high and low temperatures. Humidity is also ranging from lower values up to condensing.
- **Constraint space and/or walkways**: The width of typical walkways is about 0.7 - 0.75 m.

Offshore rigs have further logistical issues: (a) it is highly expensive to have people working on the rig as they must be housed and protected; (b) in the case of emergency, it must be possible to evacuate personnel quickly. As oil and gas exploration pushes into more hostile and remote regions, these difficulties become serious obstacles to the financial viability of an offshore installation.

With a basic comprehension of the processes and the personnel tasks involved in the offshore oil and gas industry, the lingering factor is the role that robotics and automated systems could potentially fill. While this is a relatively young market, there are a few examples of robotics and automation used for oil processes, but only a few. Reason for such hesitation derives from the logistical challenges that come with the implementation of a robot or an automated system in an offshore environment, as well as from a general lack of prototypes on the market.

The disaster at the Macondo Prospect, which is called the Deepwater Horizon oil spill [8] , has spawned a great deal of movement and some legislation. Much of the legislation requires an increase in inspection and maintenance at offshore rigs, where robotics and industrial automation will find a role in enabling those in the industry to meet these new standards.

To generate application scenarios for mobile robots, the operations carried out on these types of platforms must be understood. There are scheduled and occasional operations. The scheduled operations are tasks planed in the daily operation schedule. The occasional operations are those triggered by external influence on a more or less random basis. The most important scheduled operations are:

- **Inspection**: gauge readings and valve and lever position readings
- **Monitoring**: gas level, check for leakage, acoustic anomalies, surface condition and check for intruders.
- **Maintenance**: gas and fire detector test, sampling, pigging, cleaning, refilling and pipelines.

The most frequent occasional operations are:

Valve and lever operation: change pressure, change flow rate and start or stop equipment operation.

Gas leakage: identify and locate, stop dangerous operations (welding, cutting, …) and secure area and stop leakage and monitor concentration drop.

Fire: identify and locate fire.

The robot will have to operate at various levels of automation: fully automatic, semi-automatic and manual. Fully automatic operations require no human intervention. There will be various tasks using semi-automatic operations, which will require varying degrees of human interaction. This is quite different from the more traditional industrial robotic applications because human decision makers must be within the control loop to collaborate with the robots and the control system. The successful application of robotics and automation in the oil and gas industry will rely on the seamless integration of man, technology, and organization. Compared to the fully automated offshore robots, the inspection robots are the simplest because they may need constant human involvement. The manipulation robot is more complex than the inspection robot because it has to make decisions while performing different tasks. In the short term, the inspection robot could have many applications in the oil & gas industry.

Requirements for Hardware Development

In order to be suitable for dependable and useful offshore operations, the following basic requirements must be met by the hardware of the mobile inspection and manipulation robots:

- The robot must be certified—or as a prototype be certifiable. The robotic system must be explosive-proof, weather-proof and salt-water-proof [9] .
- The electronics must be suitable for harsh environments.
- The drive systems of the robot must be suitable for the hostile environments and the very special floor conditions.
- The robot has to maneuver in confined spaces. Therefore its size must be adapted to the previously defined reference passage.
- The robot must be equipped with highly reliable sensors to perceive its surroundings, especially to detect obstacles.
- If the robot is supposed to navigate in different levels, the robot must be able to move vertically on simple ladder-type steel profiles.
- The robot must be equipped with sensors to track its position for autonomous motion.
- The robot must be equipped with a manipulator to handle objects and position sensors.

- The robot must have appropriate application sensors and tools to execute inspection and manipulation tasks autonomously, semiautonomously or manually.
- The robot must be able to communicate with a central control station, for example by Wireless LAN.

Requirements for Communication

A major challenge for teleoperation within the oil and gas industry is the remote nature of offshore installations. The offshore rigs can be located hundreds of miles away from land, conducting complex and dynamic operations in harsh environments.

Operation failures in such installations may result in major consequences for human operators, the environment and process equipment. Safe and efficient teleoperation is critical for such unmanned facilities, securing benefit and optimal productivity at remote locations.

Requirements for Software Development

A mobile robot for inspection and manipulation in offshore environments may only be acceptable if it can be used without expert knowledge but rather easily and intuitively as a daily-used tool. This implies that:

- the robot can be controlled manually, semi-automatically and autonomously;
- new inspection and manipulation tasks can be programmed quickly and without the assistance of specialists;
- anyone working next to the robot can interact with it safely.

Therefore, the control software for a mobile offshore robot should provide functions to

- navigate on the platform without collisions both in remote-controlled mode and in automatic mode, for example move to a given target autonomously;
- easily program typical inspection, monitoring and/or manipulation tasks;
- execute pre-programmed inspection, monitoring and/or manipulation tasks automatically;
- enable supervision and control of the robot from a remote location;

- display sensor data such as camera images, the current gas concentration or other sensor information on the remote screen;
- alert the remote operators when abnormal sensor values are detected;
- review sensor recordings of past autonomous inspection tasks at the remote screen.

Requirements for Robotic Systems

Because it is installed at remote and isolated places, offshore oil and gas platforms pose a challenging environment for their human operators due to the unsheltered maritime environment, heavy weather and unfriendly, often explosive, toxic and corrosive atmosphere. In order to apply mobile robotics technologies in offshore environments, a number of challenges that do not exist in any other application area of mobile robots must be overcome [10] :

- **Complex environments**: Offshore installations contain complex structures such as pipes, flanges, tanks, steel frames, stairways and many more. These structures can be very hard to detect by sensor systems applied typically in mobile robotics. The installed sensors on the mobile offshore robot must be able to distinguish relevant structures in the environment such as obstacles and free passages. The process equipment on offshore platforms is distributed among different levels, often including intermediate or mezzanine levels. only stairs or ladders are available to move from one level to the other. An appropriate means to move the robot from one level to another is required for a single robot to cover all levels.
- **Floors**: Offshore platforms on the contrary mainly consist of plain steel floor and gratings. There are many small holes, sharp edges, slopes and steps up to several centimeters in height.
- **Explosive atmospheres**.
- **Corrosive environments**: splash salty water, salty air and corrosive chemicals.
- **Temperature**: the ambient temperature on offshore platforms shows significant variations, depending on the platform's location. The mobile offshore robot must thus be operable in temperature ranges between −30°C and +50°C.
- **Humidity**: The relative humidity up to 100% and condensing.

- **Other conditions**: possibly highly radiant heat from process equipment, heavy precipitation, storms, and direct sunlight.

CURRENT ROBOTICS TECHNOLOGIES

There are different kinds of robots in the oil & gas industry. The remotely operated underwater vehicles are discusses first. Since there are many underwater pipeline problems, underwater pipeline repair robotic systems are developed to repair the pipelines. Mobile robot platforms for topside oil & gas platform inspection and operations are investigated by academia and industry. Several systems are presented to understand their features and specifications.

Remotely Operated Underwater Vehicles (ROVs)

ROVs are unoccupied, highly maneuverable underwater robots operated by an operator aboard a surface vessel. They are linked to the ship by different cables that carry electrical signals back and forth between the operator and the ROV. Most are equipped with video cameras and lighting systems. Sometimes additional equipment is added to expand the vehicle's capabilities, including still cameras, manipulators or cutting arms, water samplers, and instruments that measure water clarity, light penetration, and temperature. ROVs were developed for industrial purposes, such as internal and external inspections of pipelines and the structural testing of offshore platforms; however, they are now used for many applications including scientific ones. They have proven extremely valuable in ocean exploration. They are also implemented for educational programs at aquaria.

ROVs vary in size from that of a bread box to a small truck. Deployment and recovery operations range from simply dropping the ROV from a small boat to complex deck operations using large winches for lifting and Aframes to take the ROV back onto the deck. In most cases, however, ROV operations are simpler and safer to conduct than many types of occupied-submersible or diving operations.

The disadvantages of using a ROV include: (a) the human presence is lost, making visual surveys and evaluations more difficult; (b) the lack of freedom from the surface due to the ROV's cabled connection to the ship. An ROV operator controls the vehicle from a system on board the ship using a joystick, a camera control, and a video monitor. The operator moves the vehicle and the camera to desired locations. The vehicle's depth, direction, and geographic position (latitude and longitude) are also recorded. ROVs are often kept aboard vessels mounting submersible operations for several reasons. In the event that a submersible becomes entangled or otherwise incapacitated, an ROV can investigate the scene to help the operators make decisions. If appropriate, cutter blades can be deployed to the manipulator arm to free the sub. If a sub loses power and is not able to surface, the ROV's manipulator arm can grab onto the sub. The deck crew can then bring it to the surface. ROVs also support exploration and science objectives. When the submersible cannot be used because of weather or maintenance problems, the ROV often can take its place. It can also be used to investigate questionable dive sites before a sub is deployed, limiting risk to the expensive subs and their pilots.

There are different kinds of ROV manipulators developed [11] -[13] . One example is the TITAN manipulator [11] . Constructed from titanium, the TITAN is uniquely capable of withstanding the industry's harsh and repetitive needs. Some features are:

- Seven degrees of freedom;
- Durable and reliable in harsh subsea environments;
- Large operating envelope;
- High lift-to-weight ratio;
- Depth rating from 4000 m to 7000 m;
- Titanium construction;
- Hydraulic.

The master controller of the ROVs contains function keys for selecting configuration options and a display for viewing diagnostic and status information. The controller's advanced operational features include individual joint operation, position scaling (altering the ratio of master arm movement to manipulator arm movement), programmable routines, incremental gripper movement, individual joint diagnostics, and automatic error checking.

Deep Water Pipeline Repair Robotic Systems

Since there are many requests to repair deep water pipelines from the international oil and gas companies, StatOil [14] and Cheron [15] have developed pipeline repair robotic systems. The pipeline repair robots can work for water depths down to 1000 meters.

SINTEF Topside Robotic System

Because it is difficult to access subsea installations, normally-unmanned automated topside platforms may be an alternative through increased accessibility for large maintenance operations. Moreover, topside platforms may statistically recover up to 22 percent more of the oil or gas than a subsea alternative in a reservoir [7] . Due to these reasons, SINTEF worked with Norwegian energy producer StatoilHydro on a remotely operated offshore topside platform [8] [10] [16] in the robotic lab facility in Trondheim, Intelligent and reliable robotic and instrumentation systems has been developed to enable onshore operators to monitor and control all of the platform's processes.

Compared to their current manned counterparts, the normally-unmanned oil platforms offer potentially significantly lower commissioning and operation costs. Remote inspection and maintenance (I&M) operations can be performed on offshore oil and gas platforms as an alternative to traditional offshore platforms. The platforms separate the work area accessible by human operators from a closed permanently unmanned area (PUA) serviced by robots. Some important scheduled I & M operations inside the PUA can be performed by robots to replace human operators such as gauge readings, valve and lever operations and monitoring gas level, leakage, acoustic anomalies and surface conditions. Standard 6-axes robot manipulators (Kuka KR-16) are mounted on a mobile platform to automatically connect to custom-built tools and sensors such as vibration-measurement sensors, a valve-operating tool, and a lid operation tool. The operators will remain on land and are able to perform different tasks, reducing both risks and costs.

Fraunhofer Inspection Robot

The Fraunhofer inspection robot [17] was developed and tested by the Fraunhofer Institute of Manufacturing Engineering and Automation (IPA) for inspection. The robot uses specifically shaped objects such as pipes and poles as well as stripes of reflective tape applied to the environment for localization. It was tested continuously for 12 hours per day in hazardous locations and in tropical environment with 35+°C ambient temperature, up to 90% rel. humidity, and direct sunlight. A laser scanner is used to perceive its environment. A six-axis lightweight arm installed on the robot carries a camera to perform visual inspections. The robot platform has various application-specific sensors, such as a stereo microphone as well as gas and fire sensors. Wireless LAN and Bluetooth are deployed to enable the robot to communicate with the central control PC and with a mobile operator control device.

The robot is capable of safe navigation in offshore environments, which enables the robot to autonomously record sensor data at key locations or continuously monitor sensor data along a predetermined path. A manual operator control device, such as an explosion-proof PDA, can be used to teach the robot to perform inspection tasks. The robot autonomously executes the inspection tasks after the inspection tasks have been taught. The sensor data are recorded and displayed at the central control PC. The operators in the central control room are able to supervise all relevant sensor data in real time. In addition, the robot can be teleoprated to assist the user by analyzing data of its environment sensors. The user drives the robot around the oil & gas platform in order to get close to objects that need to be inspected or manipulated. A graphical user interface can display the objects detected by the robot through its sensors. The robot can then perform the planned tasks automatically such as executing movements or grasping operations to solve tasks, e.g. positioning its camera in front of a gauge or turning a hand wheel.

CMU Inspection Robot

The sensabot [18] was developed by Carnegie Mellon University, supported by Shell. Sensabot was designed for severe weather, pack ice, and temperatures that range from over 100˚F (40˚C) to below −31˚F (−35˚C). On top of that, the robot should work in an environment with a corrosive, toxic, foul-smelling gas and explosive methane gas. Sensabot meets the International Electrotechnical Commission (IECEx) standards for electronic equipment and ANSI safety standards for guided industrial vehicles. The safety systems on Sensabot reduce the risks of environmental hazards and operating around personnel.

The mobile base of the Sensabot can deal with concrete, gravel, and steel gratings and tackles ramps, light snow, and slippery surfaces. It can go through human-sized walkways. It can access to multiple levels of a facility via ramps, elevators, and a cog rail systems.

There are different inspection sensors on Sensabot including (a) detectors to measure concentrations of gasses such as methane; (b) a powerful pan/tilt/zoom camera for remote operators to visually inspect pipes, valves and fittings for corrosion and other types of damage; (c) temperature sensors to help operators spot overheating equipment; (d) vibration sensors to enable operators to monitor the condition of pumps, motors, and bearings; (e) a microphone to help operators detect audible problems with machinery; (f) ladar and video to detect obstacles in its vicinity (such as equipment that lies in its path).

FURTHER RESEARCH OPPORTUNITIES

Because of the challenges and special requirements in the oil & gas industry, there are many exciting research opportunities for robotics and automation. This section discusses the research areas in robot manipulators, mobile platforms, teleoperation and automated equipment.

Manipulator

Currently, pipe handling robots and ROV arms are two types of manipulators used in oil and gas. As people are looking for more automated oil rigs, specialized robot arms should be developed. The robot arms should be able to tolerate the harsh environments in the oil and gas areas. Besides that• The inspection robots should have a large workspace since the oil and gas equipment is typically much bigger than that in automotive manufacturing;

- The maintenance and/or repair robots should be able to handle big payload;
- The operation robots should have different tools;
- Light weight manipulator with big workspace for inspection;
- Heavy duty manipulator for operations.

Mobile Platform

ROVs are widely used in subsea operations to install equipment, perform maintenance and repair tasks etc. However there are many disadvantages for the current ROVs:

- Cables: the cables with the ROVs are easily tangled because they extend from the vessel to where the ROV is located. Since there are many objects in the sea, the cables could be tangled. Once ROV is tangled, the cables have to be cut.
- ROVs are not easy to operate. Even though there are some training systems to help the operators, it is a difficult task for the operators.

For the future research in the ROV, the opportunities could be:

- ROVs without using long cables from a vessel.
- Semi-autonomous ROVs. The ROVs can guide itself to avoid obstacles and locate the tasks.
- Autonomous ROVs. The ROVs are preprogrammed to perform given tasks before operations.

The ROVs with intelligence can make decisions and perform given tasks. Besides ROVs, mobile platforms will have many applications in oil and gas:

- Topside inspection and monitoring;
- Topside maintenance and repair;
- Topside operations.

The future research for the mobile platforms could be:

- Mobile platforms should be able to tolerate harsh environments;
- Mobile platforms should be able to anchor themselves automatically;
- Teleoperated mobile platforms could be more feasible in a short term.

Due to the development of unmanned platforms on the surface and teleoperated seafloor drilling rigs, mobile manipulators will parallel deep sea ROVs in maintaining platforms, where it is reasonable to assume the industry will see many more unmanned platform projects in the very near future.

Teleoperation

Teleoperation will still be the enabling technology in a short term because the oil & gas industry hesitates about using fully automated robotic system on the oil and gas operations. Different teleoperation technologies should be developed for different applications, for example, mobile platforms on the rig and subsea mobile platforms.

Subsea Robots

Half of the world's remaining oil reserves lie beneath the oceans' floors. Getting to them presents distinct engineering challenges, ranging from weatherand ice-related issues to the environmental and human safety risks. Automated, robot-operated, unmanned drilling rig that sits on the bottom of the sea would resolve the problems related to weather-tossed seas, darkness, floating icebergs, and pack ice. Instead of building an offshore surface platform to drill for oil and gas at the seabed, up to 3000 meters below, exploration companies can use the Seabed Rig system to drill on the sea floor using robotic systems, with only a support vessel at the surface. The Seabed Rig also eliminates the potential for human error, is safer, and costs less than conventional drilling platforms.

There are many opportunities for robotic system in seabed operations. Besides drilling, other possibilities are:

- Automated seabed maintenance and repair system;
- Automated seabed inspection system;
- ROVs.

Automated Equipment

Because the oil and gas industry does not require highly flexible systems, hard automation could be better solutions for the industry. For example, seabed drilling system is more like a hard automation system, instead of robotic system. Furthermore, Robotic drilling systems, such as the one developed by Seabed Rig, would appear to be extremely useful systems for simplifying the drilling process and ensuring that its accuracy is to the utmost degree. Most drilling processes are extremely labor intensive and require continual human supervision, which could be easily remedied by robotic systems.

Related Research Topics

Since an offshore platform is quite complex, there are many wires used to transmit signals or transfer power. Data communication is a critical issue. Some potential research areas are:

- Sensor network;
- Wireless data communication;
- RFID applications;
- Data management.

Since the paper focuses on exploring robotic applications in the oil & gas industry, the above research areas will be discussed in our future research.

CONCLUSIONS

There is little doubt that the oil & gas companies would benefit greatly from the use of more intelligent technologies, not only increasing their efficiency, but also cutting down on human risk factors that are ever present in rig environments.

There is a clear incentive for oil and gas companies to automate their oil and gas facilities, starting with isolated operations, such as pipe handling and assembly for drilling and tasks related to rig operations. These examples represent high-risk operations for humans and therefore provide opportunities to improve health, safety and environmental (HSE) performance.

In addition to productivity and efficiency gains, robots used for high-risk tasks will also lead to improvements in HSE performance. Such tasks are not necessarily always predictable and represent unusual robot activities. The robot will therefore require features that extend the "eyes, ears, and hands" of the human decision makers as they carry out inspections and maintenance operations on the process infrastructure.

Reduced commissioning and operation costs, together with improved Environmental, Health and Safety (EHS) are some of the potential benefits of having normally unmanned topside oil platforms. However, such oil & gas platforms require advanced methods and tools for remote control and monitoring of inspection and maintenance operations.

In this paper, a brief introduction to the oil & gas industry processes is performed. The challenges and requirements for the robotics and automation equipment are explored. Future research opportunities including robot manipulator, mobile platform, teleoperation, and subsea robotics are discussed. Overall, we postulate that there are many opportunities in the oil and gas industry and some research is currently in progress to develop robotic and automation applications. Therefore, it is an optimal time to develop robotic and automation systems that can satisfy the oil & gas requirements.

ACKNOWLEDGEMENTS

Heping Chen, Samuel Stavinoha, g##Bracket#/Michael Walker, Biao Zhang, Thomas Fuhlbrigge The authors would like to thank ABB Corporate Research Center, ABB Inc. Windsor, Connecticut, for supporting the study.

REFERENCES

1. World Nuclear Association (2012) Uranium, Electricity and Climate Change. http://www.world-nuclear.org/info/Energy-and-Environment/Uranium, Electricity-and-Climate-Change/
2. Upstream (Petroleum Industry). (2014). http://en.wikipedia.org/wiki/Upstream_(petroleum_industry
3. Upstream Oil and Gas. (2014). http://www.upstreamoilandgas.com/
4. Gary, J.H. and Handwerk, G.E. (1984) Petroleum Refining Technology and Economics. 2nd Edition, Marcel Dekker, Inc., New York.
5. GE Energy. (2011). http://www.ge-energy.com/solutions/index.jsp#tabs-industries
6. US Environmental Protection Agency (2000) US EPA Office of Compliance Sector Notebook Project: Profile of the Oil and Gas Extraction Industry. US Environmental Protection Agency, Washington DC.
7. Meeting the Challenges of Today's Oil and Gas Exploration and Production Industry. http://www-935.ibm.com/services/us/gbs/bus/pdf/g510-3882-meeting-challenges-oil-gas-exploration.pdf
8. Transeth, A.A., Skotheim, O., Schumann-Olsen, H., Johansen, G., Thielemann, J. and Kyrkjebo, E. (2011) A Robotic Concept for Remote Maintenance Operations: A Robust 3D Object Detection and Pose Estimation Method and a Novel Robot Tool. The International Conference on Intelligent Robots and Systems, Taipei, 18-22 October 2010, 5099-5106.
9. The Robotized Field Operator. http://www.isa.org/InTechTemplate.cfm?template=/ContentManagement/ContentDisplay.cfm&ContentID=79568
10. Kyrkjebo, E., Liljebaäck, P. and Transeth, A.A. (2009) A Robotic Concept for Remote Inspection and Maintenance on Oil Platforms. Proceedings of ASME 28th International Conference on Ocean, Offshore and Arctic Engineering (OMAE 2009), Hawaii, 31 May-5 June 2009, 667-674.
11. Schilling Robotics. (2014). http://www.schilling.com/products/manipulators/Pages/default.aspx
12. Atlas Hybrid Manipulator. (2014). http://www.oceaneering.com/rovs/rov-technologies/atlas-hybrid-manipulator/

13. ECA Robotics. (2014). http://www.eca-robotics.com/en/control-command-security/robotics/2.htm
14. Statoil Developing Deepwater Pipeline Repair Robot. (2014). http://www.offshore-mag.com/articles/2007/03/statoil-developing-deepwater-pipeline-repair-robot.html
15. Chevron's Deepwater Pipeline Repair System Revealed. (2013). http://oil-geneva.blogspot.com/2013/02/chevrons-deepwater-pipeline-repair.html
16. Robots Taking over the Job on Offshore Drilling Platforms. (2014). http://www.sciencedaily.com/releases/2007/12/071221230852.htm
17. MIMROex. Fraunhofer Institute of Manufacturing Engineering and Automation. http://www.ipa.fraunhofer.de/MIMROex.2297.0.html?&L=2
18. SensaBot Inspection Robot. Robotics Institute, Carnegie Mellon University. http://www.ri.cmu.edu/research_project_detail.html?project_id=752&menu_id=261

CHAPTER 17

The Impact of Automation in Painting Furniture Parts

Alexandrine H. Karusine, Chenhuan Cui, Nathanael O. Larsey, and Kanyarugano Tanguy-Nolain

School of Engineering, Information Engineering, Huzhou University, Huzhou, China.

ABSTRACT

Technology and Woodwork industries have been going through a phase of perfection over a period of hundred years, up until today, the wood work traditions are still exist and improving with the help of modern technology. Looking at history and current studies with regards to production of furniture, many factories are producing millions of chairs for entire continents and to maintain the market lead position requires a lot of human dedication and intelligent investments in technology. Using automation in painting

Citation: Karusine, A. , Cui, C. , Larsey, N. and Tanguy-Nolain, K. (2022), "The Impact of Automation in Painting Furniture Parts". Open Journal of Applied Sciences, 12, 1995-2003. doi: 10.4236/ojapps.2022.1212139.

means to automatically control the process including industrial machines like robots, conveyors and others which save time, money and also reduce human efforts. Most automation companies are specialized in industrial automation, industrial robots, communications, electronics, and software for developing complex projects to automate desirable task by integrating robots. This study is purposefully conducted to shed more light on the benefits or impacts of automation in painting furniture parts. One of the benefits that most of companies gain from automating the painting process with robots is the productivity growth. The recruitment process for specialized personnel in the painting processes has also been greatly simplified due to the fact that the robots have taken over the difficult' tasks and protection from areas with high risk inhalation of harmful substances is assured. In this research, Automation is shown to be efficient and effective in the painting system by reviewing different studies. Automated dyeing processes in factories have made progress due to the successful results achieved by automated painting robots. In that regard, the impact of automation in painting furniture parts is evident in the production of fitments that are in almost every home through the solutions offered by automation to the consumer.

Keywords: Painting Robot Vision, Automation Impact, Robot Imaging System

INTRODUCTION

As shown by historical accounts and the purpose of this paper is to show the impact and the concept of an automatic gadget to assist humans or do tasks for them that has existed for a very long time. Examples include concepts like automatically opening doors. Fast forward several hundred years leading up to the Industrial Revolution, this resulted in increased production quest and, consequently, impetus for automation. The introduction of (NC) Numerically Controlled machinery, the eminence of the computer (the 1950s), and the circuitry (1970s) all played a great role in the development of the first, but primitive, industrial robot. The earliest were capable of replacing people in heavy, risky, and tedious activities. They were not able to sense the external surroundings, and were just employed for tasks that are easy like pick and place [1]. Later, engineers were able to include more complex motions, exterior sensors, and operations like deburring, welding, grinding and assembling. Since then, industrial robots have helped replace risky or harmful operations while also lowering costs, increasing productivity, and improving quality.

The Unimate, created in 1954 by George Charles Devol, is widely referred to as the first industrial robot. A few years later, entrepreneur Joseph F. Engelberger and Devol were talking about mutual interests, and they founded Unimation Company [2]. How would you describe an industrial robot then? A mechanical system that can be programmed to perform a variety of jobs, is automatically regulated, capable of using an arm with a variety of tools, works quickly, and has different levels of flexibility [3]. Unimate, the first version, which was used at GM's factory for handling die castings and spot welding, was constructed in 1961. In 1967, the initial instance of a company successfully designing a particular robot to meet their demands was reported. This company developed a robot to complete a spray painting task and afterwards changed its name to ABB group. It is only in this particular case that big businesses started creating industrial robots on their own. By the middle of the 1970s, it was expected that the use of industrial robots would increase at a rate of around 30% annually. After officially taking off, the industrial robotics sector never looked back. The continuing inventive development and upgrading of industrial robots resulted in a significant increase in demand in 2010. Global sales of robots increased by 29% by 2014. As the world grows more and more aware of the great advantages of industrial robots, it is an exciting time to be a part of the robotics industry, most industrial robots are like this Figure 1.

Figure 1. Arm robot with control panel.

A powerful workhorse of a high acceleration and speed robot with the ability to handle huge cargo allows shorter cycle duration and increased output which may position the process equipment near the applicator because of the large load capacity [4].

IMPACT OF VISUAL IDENTIFICATION IN THE PAINTING PROCESS

Racks are loaded onto the OnRobotics overhead conveyor via a load/unload robot, scanned and transferred to the PaintMate Robot for coating. The scanner will identify each part and validate its position in the rack. Once positioned in front of the robot the OnRobotics patented turning system synchronizes with the robot part path, rotating the parts individually as the parts are rotated. The racks are then conveyed back to the load/unload, where a robot unloads them. Allowing the workpiece to be mixed and hung without generating work chaos. Of course, planning ahead of time ensures that all workpieces to be hung have a corresponding spraying program [5].

The visual recognition system is installed on the spraying robot's front end as on Figure 1, and the robot will automatically invoke the trajectory program for the workpiece of this shape. If robotic vision is not introduced, robots will continue to operate as blind machines that follow their scripting. They rigidly adhere to the rules that govern their behavior, which makes them suitable for routine work that would be mechanically demanding and complicated for humans. Robots are changing as we approach the advent of Industry 4.0. It will enable companies to stay up with the fourth industrial revolution's needs and tendencies [6]. There is another method that where the painter holds the manipulator to teach according to the normal spraying operation. During the teaching process, the whole operation process is recorded and stored by the computer, and the computer automatically completes all the operation processes of spraying by imitating the manual teaching process by the manipulator according to the stored information [7]. The advancement of robotics depends on the development of an interactive robot imaging system. A key element of this advancement is machine or robot vision, which provides new accuracy and preciseness levels in intelligent computerized operations. Vision aids are useful. Collaborative robots can inspect, identify, count, measure, or read barcodes, among other things. Multi-operations in one process are made possible by ultra-high-speed imaging and high-quality lenses [8].

The Robotic Vision of the Painting Robot

One or more built-in cameras allow the collaborative robot to "see". At least one robot vision camera, acting as the machine's eye, will be installed on the robotic arm itself. Additional cameras are sometimes put at considered areas within the collaborative robot working space. This configuration enables the camera to collect as much visual data as is necessary for it to work alongside humans to fulfill its function [9] [10]. Prior to deployment, the machine is taught how to recognize the things with which it should interact through programming. Within Figure 2 workspace and a close look on Figure 3, hanged objects are to be scanned in 2D or 3D by the collaborative robot camera. After that the image will be saved in the database for collaborative robots and configured to move and carry out the designated task. Once the programming is finished, the collaborative collaborative robot may be put on the assembly line [11] [12].

As new tasks such as making real-time decisions and triggering actions become more complicated, Accuracy must be increased, and cooperation with humans is essential. Visual feedback is an important aspect for such goals because it is the same thing that vision systems provide to robots [13].

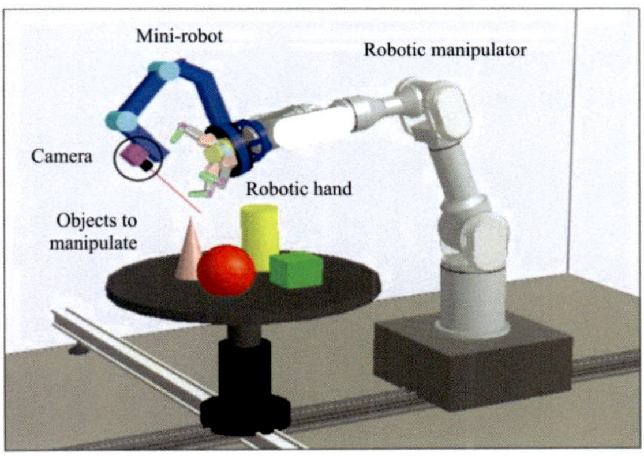

Figure 2. Robotic system.

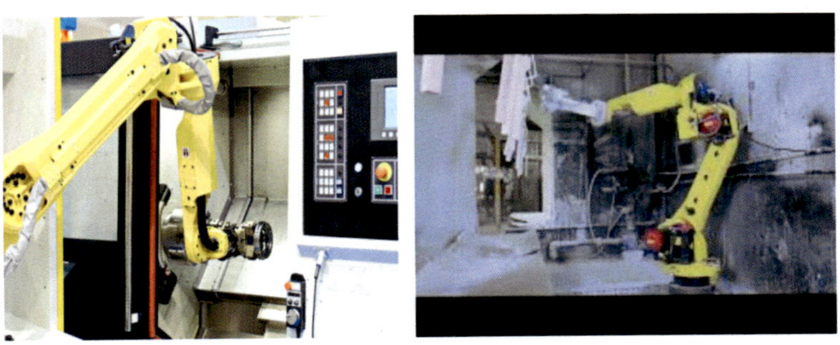

Figure 3. Robot workshop.

The Process of Robot Vision

Hanging conveyor belt has hooks that supports the rotation function mounted on it and this carry different shapes of furniture as it is shown in Figure 4, it moves in front of a painting robot to be sprayed, first the robot with a built-in camera identifies an object moving in front of it in 3D then portray it in 2D send the signals for the program to be identified and be executed then the robot takes the command and starts spraying the part.

The vision system (C) on Figure 5 is an advanced one it detects the parts on the line in 3D and then in 2D, identify them and communicate their exact shape and position to ensure uniform paint coating for the parts produced and then the robot (R_b) start the process (P) painting. This is the immediate issue. Because it defines a set of values that categorize the modeling among 3D points in the world co-ordinate system and 2D image precise location, this method is also referred to as camera calibration. The pinhole camera model is commonly used to model world coordinates projected from a perspective onto an image [14].

Figure 4. Robotic system with hanging conveyor belt.

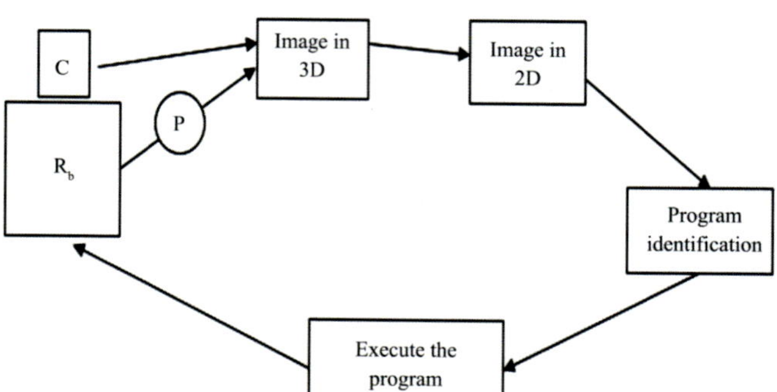

Figure 5. Painting process by an arm robot.

Significant Sectors of Robot's Imaging System

Footage Capturing

Cameras record images of things entering the collaborative robot's workspace. Conveyor systems are likely to deliver the items before the robot if the setup is midway through the assembly line. At a predetermined distance, camera will begin collecting data visually. The machine will then assess the video or film and make necessary improvements to create a detailed image.

Image Treatment

Images will be processed further and examined pixel by pixel. The framework will compare the colors and observable shape of the object to the image stored in its directory.

Connectedness and Reaction

When the device specifies that the item in the image corresponds that which was programmed before, it will take the appropriate action on the object in front of it.

This whole process occurs in a matter of trice.

THE INFLUENCE OF AUTOMATION IN THE PAINTING PROCESS OF FURNITURE PARTS

The above is a rudimentary illustration of how collaborative robot with vision systems may be integrated into a manufacturing line. Automation in the painting process is not only efficient but also represents a technological innovation by applying a liquid paint on wooden surfaces using a robot. Automated dyeing processes in factories have made progress due to the successful results achieved by automated painting robot. Among those successful results there are also some examples that can be illustrated as the automation advantages in the painting process:

- **Increases efficiency**: Images captured by a robot vision camera may be used to train your personal AI model, and the picture data may be gathered automatically. Robots in factories may then utilize the training module to better recognize various furniture flaws.

- **Assures consistency of the product**: Although employees are able to frost a cake, however it would be unreasonable to anticipate that anybody could maintain accuracy for an extended period of time say, six to eight hours. This is an illustration of a repetitious work that robots are a great alternative for. Additionally, industries may benefit from the visual recognition function and be more adaptable in the production lines using robot vision. If they let Collaborative Robot respond to elements that come into their places of vision, they can generate more.
- **Reliable**: Robots that are led by vision are more trustworthy than those that cannot sight. After all, they don't operate in a blind manner. They can be set to bypass an item that is significantly dissimilar from the pre-programmed 2D or 3D pictures in their system and continues to those that pass the initial quality assessment. Like this, manufacturers may increase productivity by using their components or raw materials on the conveyor.
- **Promotes workplace safety**: Outmoded, automatic machines operate in accordance with the directions and speeds set forth in their programming.

  If you put a human employee in its pathway, it won't stop whether they move out of the way or not. Collaborative robot with AI-enabled vision, moreover, have sensors that can identify obstacles in their pre-planned pathways and movements. Instead of starting up at full speed, they stop when their sensors detect an object in their path. The danger of on-site accidents can be considerably decreased in factories with vision-guided collaborative robots.
- **Lowers operational expenses**: Collaborative robot with systems of robotic vision requires a significant financial funding. Despite the fact that returns on investments of any kind are never guaranteed, businesses can be certain of benefits like time and resource savings, increased rates of production, effective and more regular quality of the products, and more reinvigorated workers. Lower operating costs and higher sales are two potential long-term results.
- **Increase in productivity**: The vision system identifies the parts on the conveyor and continuously and automatically adapts the conveyor speed, the robots use the line tracking option meaning they follow the parts as they move and can paint a lot of parts in

real-time. After there will be an evident reduction in the paint consumption.

- **Speed up the operation**: By automating this process it is easy to speed the process up. In fact, the study has proven that automation can handle about 45% of repetitive work [15]. That way the time is perfectly spent, the amount of time spent on the task can't be compared to the production result, means in a matter of hours you will accomplish more than it was going to be if you were using man power to do the task.

DISCUSSION

Painting different furniture parts using an automated process, has proven to be more functional and beneficial in the companies and factories that has chosen this method, rather than the old fashioned way which was the use paint, brush and a lot of workers. Automation can bring many benefits to an industrial finishing operation that cannot be compared to the costs associated with it. Though most of the systems on today's robot can be programmed to identify several shapes or patterns designs which allows some variation in the painting procedure, when it comes to the programmable side of a robot then the programming process can be overwhelming and time consuming. On the other hand In order to avoid any mix-up, wastage of paint and time while using this automatic painting method, like any other device either automated or not automated it has to be kept under a consistent maintenance, to prevent any issues from arising, should this automated system have a breakdown it might cause unnecessary chaos, in that case an engineer is needed to constantly check if everything is running perfectly well. The biggest challenge for industrial companies today lies in improving productivity and operational efficiency, mitigating risk and reducing costs. Automating industrial finishing processes makes it possible to remain profitable in the market demand.

CONCLUSION

Automation is an important method to use in industry nowadays, besides the fact that it can save you hours of time and a lot of money. Using automated system as a robot to spray or paint is one of the effective and efficient ways to accomplish the task given. Robots have emerged as a key component of Industry 4.0 and in order to attain more functionalities plus requirements of the applications that are new, sensor technologies and vision systems can be improved to reach certain flexibility. In companies that deal with the preparation of wood into the finished products like chairs, tables, beds and many more are evolving from a tiresome method to an automated technique, and the outcome has proven to be admired by users. Furthermore, Conditions in the industrial environment and robotics need to be taken into account because without proper environmental management systems and objectives, automation alone can have significant adverse impacts, especially on ecosystems and energy use or resource use. Robots and their applications require a particular vision solution. Robot executions are regularly subject to precise troubles, such as regulator and introduction of objects with different geometrical shapes to the ones that are already in the system. The concept of taking into account, automation systems as an indispensable piece in the manufacturing process is not novel and as time goes new techniques and inventions are still being unveiled to facilitate and speed up the painting process.

REFERENCES

1. Chen, Q. and Wu, Z. (2017) A Brief Comment on the History of Global Industrial Robot. Machinery, 55, 1-4, 19. (In Chinese)
2. Soysal, O. (2005) A Brief History of Robotics. https://stemrobotics.cs.pdx.edu/sites/default/files/History_of_Robotics.pdf
3. Yazawa, T., 矢澤隆之, et al. (2012) Industrial Robot. WO 2013073379A1.
4. Zhang, X. (2019) Teaching Design of ABB Robot palletizing Workstation Based on Robot Art. Electronic Test, No. 22, 112-114. (In Chinese)
5. Hongbo, L.I., Liu, K. and Zheng, Y. (2020) Goods Picking System and Method. US Patent No. 20200109007A1.
6. Peng, K.D., Zheng, L., Yuan, J. and Luo, X. (2017) Analysis and Solution of Industrial Robot Talent Based on Industry 4.0 Background. Value Engineering, 36, 171-174. (In Chinese)
7. Song, K., Wang, F. and Sun, C. (2002) Analysis of Production Line of Spray Coating of Chairs with Intelligence. Furniture, No. 1, 14-17. (In Chinese)
8. Liu, H., Zhao, B. and Zhu, H. (2012) Research of Sorting Technology Based on Industrial Robot of Machine Vision. Proceedings of the 2012 Fifth International Symposium on Computational Intelligence and Design, Washington DC, October 28-29 2012, 57-61.
9. Carlisle, B. and Shimano, B. (2018) Collaborative Robot. US Patent No. 10029369B1.
10. Fechter, M., Seeber, C. and Chen, S. (2018) Integrated Process Planning and Resource Allocation for Collaborative Robot Workplace Design. Procedia CIRP, 72, 39-44. https://doi.org/10.1016/j.procir.2018.03.179
11. Zhang, F. and Pavan, T.S. (2020) Vision Calibration System for Robotic Carton Unloading. EP3706079A1.
12. Švaco, M., Šekoranja, B. and Jerbic, B. (2011) A Multiagent Framework for Industrial Robotic Applications. Procedia Computer Science, 6, 291-196. https://www.sciencedirect.com/science/article/pii/S1877050911005199
13. Pérez, L., Rodríguez, í., Rodríguez, N., Usamentiaga, R. and García, D.F. (2016) Robot Guidance Using Machine Vision Techniques in Industrial Environments: A Comparative Review. Sensors, 16, Article No. 335. https://doi.org/10.3390/s16030335

14. Nguyen, K.V., Banks, D.R. and Tyagi, N.K. (2020) Three-Dimensional Computer Vision Based on Projected Pattern of Laser Dots and Geometric Pattern Matching. US Patent No. 10753736B2.
15. Where Machines Could Replace Humans—And Where They Can't (Yet). https://www.mckinsey.com/capabilities/mckinsey-digital/our-insights/where-machines-could-replace-humans-and-where-they-cant-yet

CHAPTER 18

Application of Electrical Automation Technology in Power System

Kai Wang and Wanqing Li

Department of Electrical Engineering and Automation, College of Electrical Engineering & New Energy, China Three Gorges University, Yichang, China.

ABSTRACT

With the continuous development and progress of science and technology in China, automation technology has occupied an important position in many fields while its application in power system is increasingly widespread. Therefore, the application of electrical automation technology in power system is of great significance for power supply stability and work efficiency.

Citation: Wang, K. and Li, W.Q. (2019), "Application of Electrical Automation Technology in Power System". Journal of Power and Energy Engineering, 7, 8-13. doi: 10.4236/jpee.2019.75002.

In this paper, the author analyzes the application of electric automation technology in power system and makes contributions to the sustainable and stable development of power enterprises.

Keywords: Power System, Electrical Automation Technology, Application

INTRODUCTION

With the rapid development of science, technology and social economy in the new era, people's living standards are constantly improving, whose demand for electricity is also increasing. Since the traditional power supply mode cannot meet people's power demand anymore, relevant enterprises need to innovate the power system. Electrical automation technology has the characteristics of intellectualization and multiple functions, which can effectively solve various problems existing in power system. The introduction of electrical automation technology into the power system can greatly improve work efficiency, reduce the consumption of manpower and material resources in power enterprises, and effectively reduce the probability of human error.

APPLICATION DIRECTION

Simulation System

Because of the development and progress of science and technology, electrical automation technology has gradually been widely used due to the increasingly perfect computer technology. On the basis of sufficient experimental data provided by the simulation system, electrical automation technology can conduct synchronous experiments for steady state and transient state of varieties and different power systems, assist researchers in testing new devices.

A variety of different control devices can form a closed loop system with them, flexibly providing high quality conditions for transmission systems and experiments on intelligent protection control strategies [1] . The introduction of real-time simulation system in the power system provides convenient conditions for the in-depth study of real-time simulation modeling of power system and dynamic load characteristic monitoring, so as to build a laboratory with real-time simulation conditions.

Intelligent Protection

Nowadays, Chinese strength of experimental research has reached the advanced level in the integrated automation field. At the same time, in the field of intelligent protection of automation technology, its leading position is obvious in the worldwide. The device of hierarchical integrated automation, developed by China, is suitable for power stations with different voltage levels. In the electric automatic protection device, the latest network communication, artificial intelligence, integrated automatic control theory, adaptive theory, computer new technology and so on are applied to it. A new study on the automatic protection principle of electrical system was carried out to continuously improve its safety level, so that the new protection device has the advantage of intelligent control [2] .

Artificial Intelligence

Design planning, operation analysis, fault diagnosis of power system and its components are carried for practical application research on the aspects of evolutionary theory, expert systems, and fuzzy logic. Combined with the development needs of relevant enterprises, the intelligent control and application of power system are studied. On the basis of studying above practical software, the operation of power system and intelligent control level are improved.

Distribution Network

Distribution network of automation technology adopts the public information model, which is the minimum standard of internationally recognized. Transmission network adopts the theoretical algorithm, which has combined the practical operation with advanced application software.

Adding the coordination of gray neuron algorithms for artificial intelligence in the process of load forecasting, and using the recursive virtual-flow algorithms of distribution network, the calculation is completed in the final process of power flow calculation.

The major breakthroughs of distribution network automation technology in distribution system are shown in advanced application software, information distribution network integration, medium and low voltage network digital and distribution network model, which have effectively solved the carrier in the distribution network of the decline, routing and other technical problems. It is mainly because of the application of digital signal processing technology that the sensitivity of carrier receiving and receiving is improved [3] .

APPLICATION ADVANTAGES

High Controllability

With the continuous development and progress of social economy, science and technology in China, breakthrough has been made in power industry and power automation technology. It is also due to the acceleration of the development of modern cities, the dependence of automation technology on the development of many industries is becoming stronger and stronger. Therefore, electrical automation technology occupies an increasingly high position. With the increasing expansion of electricity market, the controllability of electrical automation technology also faces higher requirements from people. Because electrical automation technology is an indispensable key link in the power system, a lot of peripheral information can be obtained when using electrical automation technology. These information needs to be processed to form an information control management system with strong controllability, which improves the controllability of electrical automation technology and makes the system more stable. Improving the controllability of electrical automation can not only promote the safety and stability of power system operation, but also greatly promote the development of China's power industry.

Integration

With widely use in many industries and fields, automation technology achieves quite ideal effect [4] . However, because the process flow of automation technology is complex, technical researchers optimize and improve it to promote the further use of automation technology in various industries. The most obvious change is to simplify the links of electrical control, measurement, power protection and so on. The optimization and improvement of electrical automation technology not only promotes the improvement of production quality and efficiency of electrical enterprises, but also effectively reduces the workload of manual production, thus greatly reducing the error rate of technical operation in the manual production process, and to a certain extent, reducing the probability of safety accidents in the production and operation process of electric enterprises.

The Support of Advanced Information Technology

Information technology is a kind of protective operation of power information system and its development and progress benefits from the application of electrical automation technology in power system to a certain extent. Therefore, the development of information technology in power system can not only make the power system more convenient, but also promote it to deal with the system fault more effectively. Furthermore, as far as enterprise is concerned, it can improve the role in management system of power information and promote the healthy, sustainable and stable development of enterprises.

PRACTICAL APPLICATION

Application of Computer Technology

Technology of Smart Grid

The application of information management system is one of the most widely-used technologies in computer technology. The combination of computer technology and electrical automation technology will form an intelligent control technology for the whole called smart grid technology. Smart grid technology is a typical management technology, which covers power generation, transmission and transformation, dispatching, distribution, user and other links.

Among the technology, the stability control system and automation system of substation are widely used in computer technology system. To a certain extent, it can be regarded as the prototype of the current smart grid, which also lays a certain foundation for the construction of smart grid in China.

The communication technology, as the representative of smart grid, needs to rely on a great number of computer technologies in the process of construction. Therefore, it is necessary to possess the characteristics of reliability, bi-directionality and real-time. This system needs the modern advanced network communication technology, and its existence completely depends on the computer technology, so it has the information management system at the same time. Table 1 is the introduction about the process management of real-time data.

Table 1. The process management of real-time data.

Storage Introduction	Real-Time Library
Users Management	Job Processing
Report	Graphic Browsing
Event Querying	Process Management
Collocation Tool	Calculating Point
Graphic Edition	Preposition

Substation

The realization of substation automation depends on the development of computer technology to some extent. To realize the modernization of power production, it is necessary to realize substation automation, which is the most important and indispensable key link. The realization of substation automation relies on computer technology. In this process, the computer is fully utilized to realize the digitalization, networking and integration of secondary equipment. It is the signal cable that completely replaces the optical fiber or cable of the computer with the power.

The realization of automation on substation achieves the computer screen as well as the automation record statistics and the movement management. The operation and monitoring are also involved in it. The reason why the automation of substation could be achieved is that the improvement of so many components. In order to facilitate the connection between power users and power plants, transformer substations and related lines for transmission and distribution are essential. The automatic management of substation is realized not only to form the automation of dispatching, but also to meet the needs of different operations in substation.

PLC Technology

Information Processing

PLC (Programmable Logic Controller) technology can complete the collecting, analyzing and processing of information and possesses the function of table checking, sorting, data converting, computing, bit manipulation and data delivering. The collected and processed data can be transmitted to different intelligent devices with communication functions to complete some control operations, compared with the existing reference values in the packet, and tabulated for printing. These data can also be applied in the process control system to process the data of some large control systems, such as the unmanned flexible manufacturing system [5] . Figure 1 is the typical circuit structure diagram of PLC.

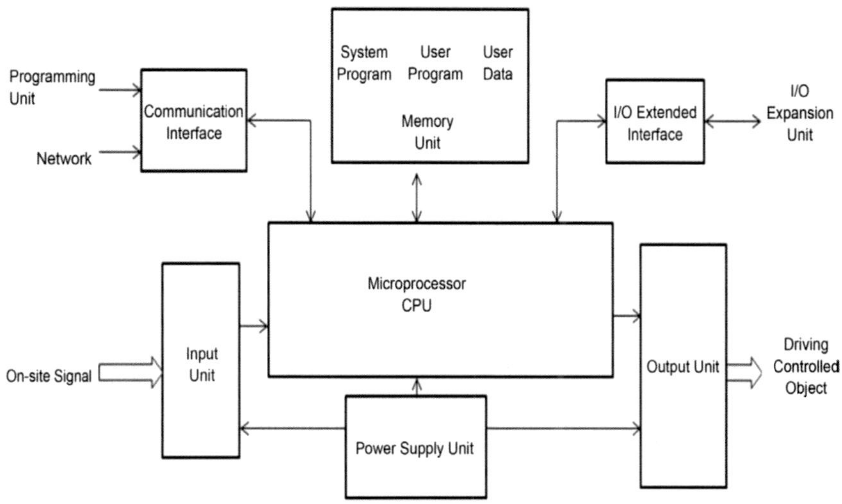

Figure 1. A typical circuit structure diagram of PLC.

Close-Loop Process Control

Close-loop process control refers to the close-loop control of continuous change analog quantity such as flow rate, temperate and pressure. PLC technology carries out closed-loop PID control of analog quantity through analog quantity I/O module, and realizes D/A and A/D conversion between analog quantity and digital quantity. The close-loop process control can be achieved by specialized PID module or transformed by the subroutine of PID (Packet Identifier).

On-Off Control

The on-off control order is the most frequently used measures of internal auxiliary process of thermal power system. At present, many domestic industries apply PLC to realize the on-off control such as motor control, automobile assembly line and machine tool electrical control, beer filling production line and elevator operation control and so on.

ON/OFF is the switching signal of input and output for PLC technology. The logic control of on-off is quite frequently used in the industrial practices. The number of input and output points can be realized by expansion from a dozen to a hundreds of thousands that can do the automatic control without limits, which saves the human resources and improves the efficiency greatly.

CONCLUSION

In a word, with the development of science and technology, people's demand on electricity has increased day by day. Therefore, power enterprises need to improve the electrical automation technology and operation efficiency of power system, so as to improve the practical application efficiency of electrical automation technology.

REFERENCES

1. Luo, Y. (2017) Brief Introduction to the Application of Electrical Automation Technology in Power System. China New Communication, No. 1, 105.
2. Wu, Z. and Yu, G. (2017) Application of Electrical Automation Technology in Power System. Internal Combustion Engine and Accessories, No. 8, 66-67.
3. Dong, G.S. (2018) Application of Electrical Automation Technology in Power System and Thermal Power Generation. China Equipment Engineering, 405, 230-231.
4. Wang, B.L. (2018) Application of Electrical Automation Technology in Power System. Electronic Technology and Software Engineering, No. 14, 132.
5. Yi, Z.C. (2017) Application Strategy of Electrical Automation Technology in Power System Operation. Time Agricultural Machinery, 44, 39.

INDEX